Biology, Ecology and Management of Aquatic Macrophytes and Algae

Biology, Ecology and Management of Aquatic Macrophytes and Algae

Guest Editors

Jinlin Liu
Shuang Zhao
Wei Liu

MDPI

Basel • Beijing • Wuhan • Barcelona • Belgrade • Novi Sad • Cluj • Manchester

Guest Editors

Jinlin Liu
State Key Laboratory of
Marine Geology
Tongji University
Shanghai
China

Shuang Zhao
Ocean College
Fujian Polytechnic
Normal University
Fuqing
China

Wei Liu
School of Environmental and
Chemical Engineering
Shanghai University
Shanghai
China

Editorial Office
MDPI AG
Grosspeteranlage 5
4052 Basel, Switzerland

This is a reprint of the Special Issue, published open access by the journal *Biology* (ISSN 2079-7737), freely accessible at: https://www.mdpi.com/journal/biology/special_issues/1096EF6XUW.

For citation purposes, cite each article independently as indicated on the article page online and as indicated below:

Lastname, A.A.; Lastname, B.B. Article Title. *Journal Name* **Year**, *Volume Number*, Page Range.

ISBN 978-3-7258-5977-1 (Hbk)
ISBN 978-3-7258-5978-8 (PDF)
https://doi.org/10.3390/books978-3-7258-5978-8

Cover image courtesy of Shuang Zhao

Contents

About the Editors

Jinlin Liu

Jinlin Liu is a scientist in the State Key Laboratory of Marine Geology at Tongji University in Shanghai, China. As a biologist, his research interests are extensive, driven by his passion for life and pursuit of understanding its essence. His work focuses on key areas such as *Ulva* and *Sargassum*, algal bloom dynamics, ecological restoration, marine development and management, resource utilization, hydrophytes, intertidal zones, biodiversity, water quality, nitrogen and phosphorus cycles, and epizoans and epiphytes. His distinguished accolades include the Second Prize of the Shanghai Ocean Science and Technology Award and the "Outstanding Reviewer Award 2024 (Resources)". To date, Jinlin Liu has contributed to over 80 academic publications, patents, and scholarly monographs, including 3 ESI Highly Cited Papers ranking in the global top 1%. Among these, he has authored or co-authored more than 40 publications as first author, corresponding author, or co-first author in SCI, ESCI, EI, CSCD-indexed journals and specialized monographs. His SCI publications have garnered approximately 1000 citations, with an H-index of 18. Additionally, he has reviewed over 130 manuscripts.

Shuang Zhao

Shuang Zhao is a Young Scientist and a Class C High-Level Talent in Fujian Province, affiliated with the Ocean College of Fujian Polytechnic Normal University. His research focuses on marine algal ecology and sustainable coastal solutions, with core areas including macroalgal physiological ecology, seaweed genetic breeding, and the dynamics and management of harmful algal blooms. Dr. Zhao obtained his Bachelor of Engineering in Bioengineering from Qingdao University of Science and Technology in 2015, followed by a Master's degree (2018) and a doctoral degree (2023) in Marine Science from Shanghai Ocean University. Currently, he teaches courses such as Marine Biology, Marine Resources & Environment, Aquatic Biology, Phycology, and Marine Survey Techniques. He has participated in over eight national-level projects, including those under the National Key R&D Program and the green tide research initiatives of the Ministry of Natural Resources. Dr. Zhao has authored more than 40 SCI-indexed papers, over 10 of which are as first author or corresponding author. He also holds two patents related to *Ulva prolifera* clearance technology. Additionally, he serves as a reviewer for peer-reviewed journals such as *Marine Pollution Bulletin, Marine Environmental Research, Ocean and Coastal Management, Science of the Total Environment, Science of Remote Sensing, Plants*, and *Diversity*.

Wei Liu

Wei Liu, a young scientist at the School of Environmental and Chemical Engineering, Shanghai University, formerly served as a researcher at Wenzhou Academy of Agricultural Sciences. As an ecologist, he pursues broad research interests centered on the impacts of environmental pollutants on organisms and the adaptive responses of biota to environmental stressors. His primary research foci encompass biodiversity dynamics, plant physiology and photosynthesis, multi-omics integration, and systematic ecological restoration engineering. He lectures on courses including Environmental Ecology, Agricultural Ecology and Environment, and Organic Chemistry. To date, he has published multiple SCI-indexed papers, including 10 articles as first or corresponding author. He has participated in national and provincial-level research and demonstration projects, while serving as a reviewer for international journals such as *Marine Life Science & Technology, Frontiers in Marine Science, Ecological Informatics, Ecology and Evolution*, and *Biology*.

Preface

Aquatic macrophytes and algae are integral to the functioning and health of aquatic ecosystems worldwide. They significantly improve water quality by assimilating excess nutrients, thus mitigating eutrophication and associated issues such as hypoxia and the formation of dead zones. Additionally, they inhibit the proliferation of harmful microalgal blooms, which can have severe impacts on aquatic biodiversity and human health. Beyond these critical functions, aquatic macrophytes and algae provide essential habitats and foraging grounds for a wide variety of aquatic animal species, thereby maintaining ecosystem stability and biodiversity levels. These multifaceted roles make them valuable tools for the restoration of intertidal zones, lakes, and river ecosystems.

However, the ecological roles of aquatic macrophytes and algae are not solely beneficial. In some cases, certain species can become dominant, leading to large-scale ecological disasters such as green tides (macroalgal blooms dominated by *Ulva* spp.) and golden tides (caused by *Sargassum* spp.). These events can severely impact the management and sustainable development of regional ecosystems by causing oxygen depletion, blocking sunlight, and adversely affecting other marine organisms. The dual nature of aquatic macrophytes and algae—both as beneficial components and potential sources of ecological problems—highlights the complexity of their roles and the need for a comprehensive understanding of their interactions within aquatic environments.

This Special Issue, "Biology, Ecology and Management of Aquatic Macrophytes and Algae", aims to provide a platform for interdisciplinary research and discussion on the ecology of macrophytes and algae. It seeks to explore their various living environments and interactions with diverse habitats, including marine, coastal zones, and climate-vulnerable ecosystems. By integrating perspectives from biology, systematics, environmental science, and aquaculture management, this Reprint facilitates a holistic understanding of the roles of aquatic macrophytes and algae in different aquatic environments.

The motivation for this Reprint stems from the urgent need to address the dual nature of aquatic macrophytes and algae. While they offer substantial benefits for water quality and biodiversity, their potential to cause ecological disasters necessitates careful management and in-depth research. This Reprint is intended for researchers, policymakers, and practitioners involved in aquatic ecosystem management, ecological restoration, and environmental science. It aims to provide valuable insights and evidence-based practical guidance for those working to balance the benefits and risks associated with aquatic macrophytes and algae.

We are pleased to present this collection of articles, which comprises a range of studies that reflect the complexity and diversity of aquatic macrophytes and algae. These contributions address various aspects of their biology, ecology, and management, offering innovative perspectives and solutions for the challenges they present.

We extend our sincere gratitude to all the authors who have contributed to this Special Issue, as well as to the reviewers whose expertise has ensured the high quality of the published articles. We are also grateful to the editorial team at *Biology* for their support and assistance throughout the publication process.

We hope that this Special Issue will serve as a valuable resource for the scientific community and beyond, stimulating further research and collaboration in the field of aquatic macrophytes and algae.

Jinlin Liu, Shuang Zhao, and Wei Liu
Guest Editors

biology

MDPI

Editorial

Biology, Ecology and Management of Aquatic Macrophytes and Algae (Volume I)

Jinlin Liu [1,2,3], Wei Liu [4,*] and Shuang Zhao [1,*]

1 Fujian Provincial Key Laboratory of Coastal Basin Environment, Fujian Polytechnic Normal University, Fuqing 350300, China; jlliu@tongji.edu.cn
2 Project Management Office of China National Scientific Seafloor Observatory, Tongji University, Shanghai 200092, China
3 State Key Laboratory of Marine Geology, Tongji University, Shanghai 200092, China
4 College of Environmental and Chemical Engineering, Shanghai University, Shanghai 200444, China
* Correspondence: hsliuwei@shu.edu.cn (W.L.); zhaos@fpnu.edu.cn (S.Z.)

1. Introduction

Aquatic macrophytes and algae constitute essential components of aquatic ecosystems, fulfilling diverse and critical roles in sustaining ecological integrity and equilibrium [1–3]. Among their most significant ecological functions is their capacity to improve water quality [4]. Through the process of photosynthesis, these organisms assimilate excess nutrients, particularly nitrogen and phosphorus, which are recognized as primary drivers of water pollution. By effectively reducing nutrient concentrations, they play a pivotal role in mitigating eutrophication—a phenomenon that can lead to hypoxic conditions and the formation of dead zones in aquatic systems [5,6]. Moreover, as demonstrated by macrophytes and macroalgae, these organisms exhibit inhibitory effects on the proliferation of microalgal blooms [7–9], which are known to pose substantial threats to aquatic biodiversity and human health [10].

In addition to their critical function in water quality regulation, aquatic macrophytes and algae play a vital role as habitats and foraging resources for a diverse array of aquatic fauna [11–13]. Their intricate underwater structures provide refuge for fish, invertebrates, and other aquatic organisms, offering protection from predation and establishing optimal conditions for reproduction [14–16]. This ecological function, in turn, sustains complex trophic networks, ranging from microscopic plankton to higher trophic-level fish species [4]. Furthermore, the presence of these plants and algae contributes significantly to ecosystem stability by fostering biodiversity.

Given these multifunctional benefits, aquatic macrophytes and algae have become valuable tools in ecological restoration projects. They are extensively employed in the rehabilitation of intertidal zones, lacustrine systems, and riverine ecosystems, where they play a fundamental role in restoring natural habitats and enhancing water quality [17–19]. For instance, in intertidal zones, macrophytes contribute to sediment stabilization, mitigate coastal erosion, and promote the recovery of marine biodiversity [20]. Similarly, in freshwater systems, macrophytes improve water clarity and support the proliferation of native species, thereby maintaining ecological balance [21,22].

However, it is crucial to emphasize that the introduction of these macrophytes and algae must be carefully managed. Non-native species, if not properly controlled, can become invasive and disrupt local ecosystems [23,24]. Therefore, ecological restoration projects involving aquatic macrophytes and algae require meticulous planning, continuous monitoring, and

 https://doi.org/10.3390/biology14030246

adaptive management strategies to ensure their success. When implemented appropriately, these organisms serve as powerful tools in combating environmental degradation.

Meanwhile, in natural ecosystems, certain aquatic macrophytes and algae can become dominant species, often leading to large-scale ecological disasters. For instance, green tides and red tides are two prominent examples [25,26]. These phenomena not only disrupt the ecological balance, but also pose significant challenges to the management and sustainable development of regional ecosystems [27]. The excessive proliferation of these dominant species can result in oxygen depletion, block sunlight penetration, and adversely impact other marine organisms [28]. Consequently, the overall health and biodiversity of the affected ecosystems are severely compromised. These specific ecological disasters primarily include the following:

Firstly, the outbreak of green tides in coastal waters is becoming increasingly severe [29,30], with the genus *Ulva* spp. being the main disaster-causing species. The Southern Yellow Sea of China has become the largest green tide disaster area in the world due to the sharp increase in land-derived nutrient inputs and the suitable hydrodynamic conditions. Meanwhile, areas with severe eutrophication, such as harbors [31], aquaculture zones [32], and intertidal zones with relatively simple habitats [33], are also prone to small-scale green tide disasters.

Secondly, in recent years, the occurrence of golden tides in global oceans has shown a significant increasing trend, with the large floating brown algae *Sargassum* spp. being the main disaster-causing species [34]. Currently, golden tides are primarily observed in the Atlantic and Pacific Oceans (e.g., the Gulf of Mexico, Caribbean Sea, Yellow Sea, East China Sea, and other localized areas), affecting the ecological security of coastal waters in countries such as the United States, Mexico, China, and South Korea [35,36].

Thirdly, red tides, as a global ecological disaster frequently occurring in eutrophic coastal waters, have attracted widespread attention from the academic community due to their increasing frequency, spatial scale, and ecological hazards. Red tides frequently occur in coastal areas near industrial or densely populated zones, as well as in estuaries, in countries such as the United States, Japan, Canada, and China [37–40]. Red tides not only disrupt the marine ecological balance, causing the death of organisms, but also pose potential threats to human health through the food chain, becoming an urgent global environmental issue that requires transregional collaborative management.

Fourthly, the frequent outbreak of algal blooms (such as cyanobacterial blooms) in lakes, rivers, and brackish water mixing zones worldwide has become a significant challenge for global aquatic ecosystem management [41–43]. For example, nearly 50% of large lakes globally have experienced algal bloom events, with European lakes having the highest frequency and severity of algal blooms, followed by North America. In addition, lakes in Central Asia, East Asia, southeastern South America, and eastern and southern Africa also commonly experience algal bloom outbreaks [44,45].

Fifthly, due to poor management, aquatic macrophytes in inland water bodies may also experience small-scale ecological outbreaks. Invasive species with strong reproductive capabilities, such as *Eichhornia crassipes* [46], can rapidly cover the water surface in eutrophic waters, often leading to secondary ecological risks such as river blockages and water quality deterioration, and significantly reducing the flood discharge efficiency of flood control projects. It is worth noting that even *Hydrilla verticillata*, which is commonly used in ecological restoration projects, can accumulate excessive biomass and form "underwater meadows" in eutrophic waters, trapping fishing boats and obstructing water flow, making cleanup difficult.

As demonstrated above, aquatic macrophytes and algae play a dual role in ecosystems. On the one hand, they play an important role in maintaining ecosystem stability and biodiversity. On the other hand, some species may trigger ecological disasters, posing threats to the healthy development of regional ecosystems. Therefore, it is highly urgent for researchers to conduct timely studies on aquatic macrophytes and algae. As three researchers who have long been engaged in the field of ecology, we are well aware that a single article often represents only a very limited range of knowledge points. Only through the format of a Special Issue can we further promote integration and interaction among different sub-disciplines. From a long-term perspective, the establishment of a Special Issue is conducive to the development of the discipline, and thus we have always had the vision of creating a Special Issue.

Fortunately, upon invitation from the editorial department of Biology, we explored the possibility of creating this Special Issue and ultimately determined the relevant details of its establishment, with the aim of providing an academically inclusive platform for scholars in the relevant research fields to publish their academic achievements and promote open discussions. Initially, this Special Issue was named "Biology, Ecology, and Management of Aquatic Macrophytes". With the development of modern taxonomy, scientists have come to recognize that some algae (e.g., *Saccharina japonica*) do not belong to the plant kingdom [47], but rather to the protist kingdom, and thus the term "algae" is more accurate than "macrophytes". Therefore, the name of this Special Issue was ultimately changed to "Biology, Ecology, and Management of Aquatic Macrophytes and Algae". Although the modern taxonomic system is in a continuous state of dynamic change with the development of molecular biological classification techniques [48,49], the name of this Special Issue, determined based on the current taxonomic system, is accurate.

Meanwhile, it is also crucial to determine the research topics for this Special Issue. There are hundreds of research directions related to the theme of "Biology, Ecology, and Management of Aquatic Macrophytes and Algae". Combining our respective strengths in our research fields, the three Guest Editors have decided to focus on the following five main research topics: (1) the key role of aquatic macrophytes and algae in ecosystems; (2) the application of aquatic macrophytes and algae in ecological restoration engineering; (3) the allelopathy of aquatic macrophytes and algae toward algal bloom species (red tide and cyanobacterial bloom); (4) ecological risks and management measures associated with macroalgal blooms or invasive species; and (5) biodiversity conservation and ecological management issues to be considered in the process of ecological restoration. Whether our Special Issue will receive submissions on these topics depends partly on serendipity and a bit of luck.

Luckily, within half a year, we received approximately 20 manuscript submissions or pre-reviews, which far exceeded our expectations and by which we are truly honored. After undergoing a rigorous multi-round anonymous peer-review process conducted by the editorial department, 10 articles were ultimately accepted for publication. We extend our congratulations to all the authors of these 10 papers. Meanwhile, for those whose manuscripts were unfortunately rejected, we also believe that your submissions are academically significant. Perhaps you lacked some luck during the peer-review stage. We look forward to your continued support of our Special Issue in the future.

2. An Overview of the Published Articles

The 10 articles ultimately accepted and published in this Special Issue were authored primarily by researchers from three countries: China, Peru, and the United States. A total of 14 research institutions were involved, with 12 from China, 1 from Peru, and 1 from the United States. These institutions are all recognized as famous scientific research organizations globally.

This Special Issue mainly focuses on the ecology of macrophytes and algae, including various living environments of macrophytes and algae, systematically exploring their interactions with diverse habitats (marine, coastal soils, and climate-vulnerable ecosystems) through interdisciplinary lenses encompassing biology, systematics, environmental science, and aquaculture management. Figure 1 highlights the foci of this Special Issue: macroalgae and the marine environment, particularly *Ulva prolifera*. For the sake of argument, we have divided the 10 articles into 5 parts: macrophytes, macroalgae, microalgae, ecological management of hazardous macroalgae, and ecological disasters and disaster prevention/mitigation.

Figure 1. Word cloud of the 10 articles in the Special Issue.

2.1. Macrophytes

Macrophytes play indispensable roles in global ecosystems by sustaining biodiversity, maintaining ecological equilibrium, and promoting human well-being. They deliver critical ecosystem services, including water purification, carbon sequestration, habitat provision for aquatic organisms, and flood regulation through the root stabilization of sediments. These services are vital for both environmental health and sustainable human development.

Macrophytes exhibit sophisticated stress-resistance mechanisms when confronting environmental adversities such as salinity, heat stress, drought, and low-temperature stress, endowing them with remarkable adaptive plasticity [50]. However, this formidable adaptive capacity, compounded by globalization, has facilitated the global proliferation of invasive macrophyte species. Excessive colonization by invasive species can obstruct waterways, block sunlight, cause oxygen deprivation in the water, and directly threaten native biodiversity. Counterintuitively, Zamora-Aranda and Aponte [51] reevaluated the invasive macrophyte *Pistia stratiotes* in Peru's Santa Rosa Wetland through an ecosystem service lens. Employing field measurements and spatial modeling, they quantified the wetland's *P. stratiotes* biomass at 37,809.99 metric tons, equivalent to 1075.25 tons of sequestered carbon, 3942.57 tons of CO_2 mitigation, 2134.41 tons of forage potential, and 339.05 tons of protein storage. This finding underscores *P. stratiotes*' dual service potential in carbon capture and nutritional provisioning, demonstrating how strategically managed macrophytes can function as "low-cost ecological engineers" even in arid coastal zones with water scarcity.

However, we must also recognize that the management of macrophytes requires a cautious approach. Some macrophytes have strong reproductive capabilities, which can lead to the formation of dominant populations and even result in them becoming invasive species that threaten local ecosystems. Therefore, when utilizing the ecosystem services provided by macrophytes, it is essential to fully consider their potential risks and implement scientifically sound management measures to achieve a balance between ecological and economic benefits.

2.2. *Multidimensional Research on Macroalgae Cultivation and Marine Environmental Changes*

In recent years, the intensification of global marine environmental changes, coupled with the rapid expansion of the macroalgae cultivation industry, has underscored the need to optimize cultivation practices, enhance ecological and economic value, and address the challenges posed by shifting environmental conditions. These priorities have emerged as critical focal points in marine scientific research. In this Special Issue of Biology, four studies explore the key dimensions of macroalgae cultivation, including rotational cultivation models, the influence of environmental factors on growth dynamics, and the potential applications of macroalgae-derived polysaccharides in biomedicine. Collectively, these investigations provide robust theoretical foundations and practical insights to advance the sustainable development of macroalgae cultivation.

2.2.1. Research on Rotational Cultivation Models of Macroalgae Based on Photosynthetic Physiological Characteristics

Currently, global macroalgae cultivation is predominantly characterized by monoculture practices, which face challenges such as a lack of industrial diversity and limited resilience to risks. To address these issues, Chen et al. [52] conducted a study focusing on the rotational cultivation of three macroalgae species—*Hizikia fusiformis*, *S. japonica*, and *Gracilariopsis lemaneiformis*—in Xiangshan Bay, Zhejiang Province. The study involved in situ experiments to monitor these species' photosynthetic activity responses under varying temperature and light conditions. The results revealed significant differences in the tolerance of the three species to light and temperature: the effective quantum yield of photosynthesis in *H. fusiformis* and *G. lemaneiformis* remained stable under fluctuating light conditions, while that of *S. japonica* initially decreased before recovering. In terms of temperature, *H. fusiformis* and *S. japonica* exhibited their highest relative electron transport rates in May (20.3 °C), whereas *G. lemaneiformis* showed optimal growth in September (27.5 °C).

Based on these findings, Chen et al. [52] developed a rotational cultivation model for macroalgae, tailored to the seasonal temperature variations in the coastal waters of the Shandong, Zhejiang, and Fujian provinces in China. This model selects suitable macroalgae species according to specific temperature conditions, aiming to achieve stable year-round production and provide theoretical support for the construction of artificial macroalgae fields in marine ranching systems. Such an approach enhances both ecological and economic benefits. Chen et al. [52] offer innovative insights into overcoming the limitations of monoculture practices, highlighting the importance of rotational cultivation models based on ecological and physiological characteristics for improving the sustainability of macroalgae cultivation. However, the research primarily focused on the effects of temperature and light on photosynthetic physiology. Future studies could further investigate the integrated impacts of additional environmental factors, such as nutrient availability and water flow, on macroalgae growth.

2.2.2. Impact of Environmental Changes on *Saccharina japonica* Cultivation

In practical applications of macroalgae cultivation, changes in environmental factors significantly influence cultivation outcomes. Bao and Xu [53] conducted an in-depth study on the effects of environmental changes on *S. japonica* cultivation in Xiangshan Bay. The research, conducted from November 2020 to May 2021, involved five observational cruises to systematically analyze the variations in physicochemical environmental factors, including temperature, salinity, light, nutrient concentrations, and ocean currents, and their impacts on the growth suitability of *S. japonica*.

The results revealed that temperature is the decisive factor influencing *S. japonica* cultivation in Xiangshan Bay. The suitability index for temperature ranged from 0.02 to 0.94 during the observation period, indicating that temperature fluctuations had the most significant impact on growth. In contrast, the suitability indices for salinity and light conditions ranged from 0.96 to 0.99 and 0.97 to 1.00, respectively, demonstrating that the salinity and light conditions in Xiangshan Bay are highly favorable for *S. japonica* cultivation, and do not limit the expansion of the cultivation time window. Additionally, nutrient concentrations remained at high suitability levels during the later stages of cultivation (DIN: 0.96–0.97; DIP: 0.92–0.95), suggesting that the current scale of *S. japonica* cultivation in Xiangshan Bay has not yet reached the carrying capacity of the area, leaving room for further development. Although *S. japonica* cultivation led to a noticeable reduction in surface current intensity, it did not significantly affect nutrient replenishment, indicating that cultivation has not yet exhibited a pronounced self-limiting effect in Xiangshan Bay.

In summary, temperature is the key limiting factor for *S. japonica* cultivation in Xiangshan Bay, while salinity and light conditions are highly conducive to its growth. Future research should focus on the risks posed by temperature fluctuations to *S. japonica* cultivation and explore the breeding of heat-resistant varieties to further enhance the efficiency and sustainability of cultivation in Xiangshan Bay.

2.2.3. Coupled Effects of Ocean Acidification and Temperature on Macroalgae

In the context of global climate change, ocean acidification and rising temperatures pose dual threats to marine ecosystems. Wang et al. [54] focused on this issue, using *Ulva fasciata* and *Sargassum horneri* collected from Ma'an Archipelago as experimental subjects. Through a series of cultivation experiments under different temperatures (15 °C and 20 °C) and carbon dioxide concentration (400 μL/L and 1000 μL/L) conditions, they investigated the interactive effects of ocean acidification and temperature changes on the photosynthetic physiology of these two macroalgae. Key indicators such as rapid light curves, fluorescence induction parameters, relative growth rates, dissolved organic carbon release, and photosynthetic pigment content were measured to comprehensively assess the responses of the macroalgae.

The study found that at 20 °C, ocean acidification reduced the tolerance of *U. fasciata* to high light intensity, exacerbating photoinhibition. However, at 15 °C, while temperature significantly inhibited the growth and photosynthetic activity of *U. fasciata*, ocean acidification alleviated this inhibition and promoted its growth. For *S. horneri*, ocean acidification similarly reduced its tolerance to high light intensity at 20 °C, but this inhibitory effect was less pronounced at 15 °C, where ocean acidification further enhanced its growth. Additionally, the relative growth rate, photosynthetic pigment content, and dissolved organic carbon release of *U. fasciata* were primarily influenced by temperature, with 20 °C being more conducive to its growth and metabolite synthesis. In contrast, *S. horneri* exhibited faster growth and higher metabolic activity at 15 °C.

Overall, ocean acidification reduces the tolerance of macroalgae to high light intensity under optimal growth temperatures, while under suboptimal temperatures, it mitigates the inhibitory effects of temperature and promotes growth. This study provides valuable insights into how macroalgae may respond to future oceanic climate changes in natural environments. Future research should focus on the long-term adaptive changes in macroalgae under the coupled effects of ocean acidification and temperature, and explore the cultivation of macroalgae varieties adapted to changing marine environments, thereby enhancing the sustainability of marine ecosystems.

2.2.4. Antioxidant Activity and Application Potential of *Sargassum horneri* Polysaccharides

In addition to the ecological and environmental effects of seaweed cultivation, the bioactive substances derived from seaweeds also hold significant developmental value. The study by Wei et al. [55] focused on the antioxidant activity of *S. horneri* polysaccharides (SHP). With increasing environmental complexity, excessive production of reactive oxygen species (ROS) in the human body can lead to oxidative stress, thereby harming health. Given that polysaccharides extracted from seaweeds possess various biological activities and the proliferation of *S. horneri* poses environmental challenges, the development of SHP resources is of great importance.

The study evaluated the antioxidant activity of SHP by inducing oxidative stress in African green monkey kidney cells (Vero cells) and zebrafish using hydrogen peroxide (H_2O_2). The experimental results demonstrated that SHP exhibited no significant toxicity to Vero cells and showed a protective effect; it significantly increased the viability of treated Vero cells, effectively reduced intracellular ROS levels, enhanced superoxide dismutase (SOD) activity, and decreased malondialdehyde (MDA) content. In zebrafish experiments, SHP also demonstrated protective effects, improving the survival rate of H_2O_2-induced zebrafish embryos, reducing embryonic malformations, lowering heart rate, and enhancing the activities of SOD, catalase (CAT), and glutathione peroxidase (GSH-PX), while reducing MDA and ROS levels.

In summary, SHP effectively protected Vero cells and zebrafish from H_2O_2-induced oxidative damage, demonstrating its potential as a natural antioxidant. This study provides new insights into the mechanisms of antioxidant activity in biological systems and offers a novel approach to mitigating the environmental pressure caused by *S. horneri* and realizing its resource utilization. Future research should focus on the long-term antioxidant effects of SHP in different biological systems and complex environments, and actively explore its potential for application in industries such as biomedicine, cosmetics, and food, thereby further enhancing its positive impacts on ecosystems and human health.

2.3. *Microalgae*

Microalgae, as the most abundant biological group in aquatic ecosystems, serve as primary producers and play crucial roles in material cycling and transformation, energy flow, and ecological regulation [56]. Their high sensitivity to environmental changes makes them effective bioindicators. Intensive human activities, particularly rapid urbanization and population aggregation in coastal bay areas, frequently lead to environmental challenges such as water eutrophication and red tide outbreaks. Therefore, a comprehensive understanding of the diversity distribution, spatial heterogeneity patterns, and environmental gradient responses of freshwater, estuarine, and marine microalgae communities holds immeasurable value for pollution source tracking, water quality assessment, and ecosystem health evaluation.

In China's highly urbanized Pearl River Estuary, Xia et al. [57] employed in situ fluorescence parameter measurements coupled with 18S rRNA sequencing to reveal that Chlorophyta and Cryptophyta dominate freshwater areas, while Dinophyta and Haptophyta prevail in marine zones. The algal distribution demonstrates significant correlations with temperature and salinity gradients. Notably, nutrient levels (NO_3^--N, PO_4^{3-}-P, and SiO_3^{2-}-Si) show marked associations with the Simpson index, suggesting a linear relationship between eutrophication intensity and microalgae community structure. These findings provide valuable references for water management strategies in this region.

In conclusion, complex interactions exist between microalgae community structure and environmental factors in the Pearl River Estuary. Future research should focus on three key directions: (1) elucidating the response mechanisms of specific algal taxa to particular pollutants to establish refined bioindicator systems; (2) integrating long-term monitoring data to predict climate change and the anthropogenic impacts on microalgae communities, informing adaptive management strategies; and (3) applying metagenomics and other advanced techniques to decode the functional gene composition of microalgae communities, thereby comprehensively assessing their roles in material cycling and energy flow. Such multidisciplinary approaches will enhance our understanding of estuarine ecosystem dynamics.

2.4. Ecological Management of Hazardous Macroalgae

2.4.1. Biological Characteristics of the Harmful Macroalga *Ulva prolifera*

A prominent focus of this Special Issue is the recurrent discussion of *U. prolifera*, a disaster-causing macroalga. Under favorable environmental conditions, *U. prolifera* can rapidly proliferate and form massive floating aggregates, a phenomenon termed "green tides". While water eutrophication is recognized as a key trigger for green tide outbreaks, the genetic traits of *U. prolifera* itself have emerged as a critical research frontier in green tide biology. As a euryhaline species, *U. prolifera* thrives in environments with drastic salinity fluctuations, such as intertidal zones and estuaries. Its salinity tolerance mechanisms have long attracted scientific attention. He et al. [58] recently elucidated the molecular basis of its adaptation to high salt stress. Their study cloned and analyzed the zeaxanthin epoxidase (ZEP) gene (UpZEP) in *U. prolifera*, revealing that high salinity significantly enhances xanthophyll accumulation and induces UpZEP expression. This gene plays a central role in protecting the photosynthetic apparatus from photodamage under abiotic stress by regulating xanthophylls synthesis, thereby conferring a competitive advantage over other algae in high-salinity environments. This discovery offers novel insights into the mechanisms driving green tide formation.

While He et al. [58] uncovered the genetic foundations of *U. prolifera*'s salt adaptation, Zeng et al. [59] further explored its ecological competitiveness through allelopathy. In their review, Zeng et al. [59] systematically cataloged allelochemicals produced by *U. prolifera*, including fatty acids, aldehydes, phenols, and terpenoids. These compounds exert broad environmental impacts, not only inhibiting the growth of competing algae, but also influencing marine microbial communities, mycorrhizal fungi, and invertebrates, thereby reshaping ecosystems. The development of rapid detection technologies for these allelochemicals could enhance the real-time monitoring of green tide dynamics, unravel complex interactions between algal blooms and their environments, and advance early warning systems for coastal ecological disasters.

Notably, some allelochemicals highlighted in this review demonstrate dual regulatory effects on green tides, red tides, and the growth of invasive species like *Spartina alterniflora*. This underscores the inseparable interplay between marine and terrestrial ecosystems in coastal zones. Consequently, mitigating marine ecological disasters requires integrated land–sea management strategies. Reducing pollutant and nutrient discharges into coastal waters, coupled with enhancing carbon, nitrogen, and phosphorus sequestration in terrestrial soils, could alleviate eutrophication pressures and curb the frequency and intensity of algal blooms. Such interdisciplinary approaches are essential for balancing ecological health with sustainable coastal development.

2.4.2. Disaster Prevention and Control from a Land–Sea Integrated Perspective

The frequent occurrence of green tides poses severe threats to coastal economies and ecological security. These events are not isolated marine environmental issues, but rather manifestations of systemic imbalances between terrestrial and marine ecosystems. Current mitigation strategies predominantly focus on offshore interception (e.g., the removal of nori aquaculture zones in Jiangsu Province) or end-of-pipe pollution control. However, Niu et al. [60] propose an innovative approach: enhancing the carbon sequestration capacity of urban green spaces to improve terrestrial nutrient retention and purification, thereby reducing the flux of nitrogen, phosphorus, and other pollutants into marine environments and indirectly alleviating coastal eutrophication.

The study explores the effects of Fe-modified biochar combined with plant growth-promoting bacteria (PGPB) on the fertility and carbon retention of urban green space soils. The results indicate that, compared to the application of biochar or PGPB alone, the combination of iron-modified biochar and PGPB significantly reduces soil pH, mitigates soil alkalization, increases the content of alkaline nitrogen in the soil, and improves soil aggregate stability. These changes enhance soil fertility and ecological function. Furthermore, this treatment significantly increases the levels of soil organic carbon (SOC), particulate organic carbon (POC), and soil inorganic carbon (SIC), thereby boosting the soil's carbon retention capacity. This approach holds promise for reducing the entry of nutrients such as carbon, nitrogen, and phosphorus into the ocean.

In summary, adopting a land–sea integrated perspective for disaster prevention and control is essential. By enhancing the ecological functions of terrestrial systems, we can create a more balanced land–sea interaction, ultimately contributing to the mitigation of marine ecological disasters such as green tides. This holistic approach not only addresses the symptoms of the problem, but also targets the underlying causes, paving the way for sustainable coastal management and ecological resilience.

2.5. Ecological Disasters and Disaster Prevention/Mitigation

Yao et al. [61] conducted a comprehensive review of the current status of marine ecological disasters in China, with a focused analysis on disasters triggered by marine organisms. Their study systematically examined the impacts of these disasters on marine ecosystems, fisheries, tourism, and coastal industries. Yao et al. [61] emphasized that global climate change and intensified human activities have disrupted the marine ecosystem equilibrium, leading to frequent occurrences of marine biological disasters. Beyond commonly observed green tides, golden tides, and red tide outbreaks, proliferations of marine organisms such as Cnidaria (jellyfish), Annelida (*Urechis unicinctus*), Mollusca (*Philine kinglipini*), Arthropoda (*Acetes chinensis*), and Echinodermata (*Acaudina molpadioides*, Asteroidea, and Ophiuroidea) have significantly impacted the sustainability of marine ecological resources and human socio-economic activities. These biological outbreaks not only destabilize marine ecosystem integrity and biodiversity, but also pose substantial threats to the safety of coastal nuclear power plants, recreational zones, and nearshore infrastructure.

Yao et al. [61] provided a detailed analysis of the environmental drivers, causative mechanisms, and ecological consequences of these marine organism outbreaks. To address these challenges, Yao et al. [61] proposed strategic recommendations, including the following: enhanced monitoring and early warning systems for marine ecological disasters, with expanded coverage of biological threats; strengthened resource utilization strategies; and foundational research initiatives. These measures aim to improve preparedness for future marine biological disasters while safeguarding the health and sustainable development of marine ecosystems.

3. Summary and Outlook

We are pleased to report that the Special Issue entitled "Biology, Ecology, and Management of Aquatic Macrophytes and Algae" has achieved significant progress. A total of 10 articles were accepted and published, garnering widespread recognition from the research community. Notably, this Special Issue has begun to establish its academic influence within the field, which is particularly encouraging for the editorial team.

While the Special Issue encompasses both aquatic macrophytes and algae, the submitted content did not fully align with our initial expectations. Notably, manuscripts focusing on aquatic macrophytes were underrepresented, with algae-related studies dominating the issue. Furthermore, research on marine environments prevailed, while studies on freshwater, estuarine, and other habitat types were limited. We attribute this imbalance to several factors: (1) the disciplinary backgrounds of the Guest Editors, which are centered on marine ecosystems and algae, likely influenced the thematic focus; (2) insufficient promotion of the Special Issue within the aquatic macrophyte research community; and (3) lingering reservations among some scholars about publishing in OA journals. Despite these challenges, we persevered and successfully completed this scholarly work. Looking ahead, we are excited about the direction of submissions for the next Special Issue (Volume II). We believe that the next issue can focus more on the following four research themes:

① Advancing Research on Macrophytes in Freshwater and Coastal Environments

Future studies should emphasize the ecological roles of macrophytes in freshwater and coastal ecosystems, particularly their responses to environmental changes and applications in ecological restoration. Examples include submerged macrophytes' remediation potential in eutrophic waters or emergent macrophytes' habitat functions in wetland ecosystems. Submissions may address ecological functions, community dynamics, climate change responses, or optimization of restoration strategies, though other relevant topics are also welcome.

② Expanding Research on Allelopathy and Biodiversity Conservation

We encourage studies on allelopathic interactions between aquatic macrophytes/algae and harmful algal bloom species, with emphasis on the mechanisms and applications for bloom suppression. Concurrently, research on balancing biodiversity conservation with ecosystem functionality during restoration processes should be prioritized. Potential topics include the identification of allelochemicals, biodiversity protection strategies, and sustainable management in restoration contexts.

③ Promoting Integrated Land–Sea Management Approaches

Given the interconnectedness of terrestrial and marine ecosystems, future research should adopt a holistic perspective to improve aquatic ecosystem health through the systemic management of agricultural and urban runoff. Submissions may explore the comprehensive impacts of land-based pollution, integrated management frameworks, or cross-regional restoration practices.

④ Enhancing Long-Term Monitoring and Adaptive Management

The long-term monitoring of aquatic macrophytes and algae is critical for understanding their responses to environmental shifts and developing adaptive management strategies. The incorporation of advanced technologies (e.g., remote sensing, molecular biology, or metagenomics) to track community structure and functional changes is strongly encouraged. Topics may include the establishment of ecological monitoring networks, adaptive management frameworks, or assessments of ecosystem services.

Finally, we extend our deepest gratitude to all authors, anonymous reviewers, assistant editors (especially Joey Li for his exceptional support), and academic editors involved in this publication. We regret that some scholars could not submit manuscripts within the tight timeframe and look forward to their contributions in the next Special Issue (Volume II). See you then.

Funding: This work was supported by the Fujian Province Science and Technology Plan Guiding Project (2024Y0031) and the Youth Development Fund Project of the State Key Laboratory of Marine Geology.

Acknowledgments: Special thanks are extended to Peimin He, Rui Jia, Jianheng Zhang, Kejun Li, Liu Shao, and Meiqin Wu from Shanghai Ocean University for their academic mentorship and encouragement throughout the research journey. Wei Liu expressed gratitude to Chiquan He's laboratory at Shanghai University, and also acknowledged the funding and support from the Wenzhou Federation of Philosophy and Social Sciences and the Wenzhou Academy of Agricultural Sciences. Meanwhile, Jinlin Liu thanked Qunhui Yang (Project Management Office of China National Scientific Seafloor Observatory) for providing a suitable working environment.

Conflicts of Interest: The authors declare no conflicts of interest.

References

1. Abell, J.; Özkundakci, D.; Hamilton, D.; Reeves, P. Restoring shallow lakes impaired by eutrophication: Approaches, outcomes, and challenges. *Crit. Rev. Environ. Sci. Technol.* **2020**, *52*, 1199–1246. [CrossRef]
2. Scheffer, M.; Hosper, S.; Meijer, M.; Moss, B.; Jeppesen, E. Alternative equilibria in shallow lakes. *Trends Ecol. Evol.* **1993**, *8*, 275–279. [CrossRef]
3. Steneck, R.; Graham, M.; Bourque, B.; Corbett, D.; Erlandson, J.; Estes, J.; Tegner, M. Kelp forest ecosystems: Biodiversity, stability, resilience and future. *Environ. Conserv.* **2002**, *29*, 436–459. [CrossRef]
4. Liu, J.; Yuan, H.; Xia, Z.; He, P. Paying attention to the safety of global edible seaweeds after the discharge of nuclear-contaminated water from Japan. *Algal Res. Biomass Biofuels Bioprod.* **2024**, *84*, 103811. [CrossRef]
5. Wurtsbaugh, W.; Paerl, H.; Dodds, W. Nutrients, eutrophication and harmful algal blooms along the freshwater to marine continuum. *Wiley Interdiscip. Rev. Water* **2019**, *6*, e1373. [CrossRef]
6. Altieri, A.; Gedan, K. Climate change and dead zones. *Glob. Change Biol.* **2015**, *21*, 1395–1406. [CrossRef] [PubMed]
7. Qin, B.; Zhang, Y.; Zhu, G.; Gao, G. Eutrophication control of large shallow lakes in China. *Sci. Total Environ.* **2023**, *881*, 163494. [CrossRef] [PubMed]
8. Kibuye, F.; Zamyadi, A.; Wert, E. A critical review on operation and performance of source water control strategies for cyanobacterial blooms: Part II-mechanical and biological control methods. *Harmful Algae* **2021**, *109*, 102119. [CrossRef]
9. Patil, V.; Huang, L.; Liang, J.; Sun, L.; Wang, D.; Gao, Y.; Chen, C. The allelopathic potential of red macroalga *Pyropia haitanensis* solvent extracts on controlling bloom-forming microalgae: Insights into the inhibitory compounds. *Ecotoxicol. Environ. Saf.* **2024**, *272*, 116083. [CrossRef] [PubMed]
10. Shen, X.; Jiang, R.; Liu, J.; Zhao, D.; Wang, L.; Liu, Y.; Yin, Y.; Zhang, J.; Shao, L.; He, W.; et al. Algae extermination by a novel algicide (DMPAI) with low-dose and field experiment. *Algal Res. Biomass Biofuels Bioprod.* **2023**, *75*, 103264. [CrossRef]
11. Liu, J.; Zhuang, M.; Zhao, L.; Liu, Y.; Wen, Q.; Fu, M.; Yu, K.; Zhang, J.; He, P. Taxonomy and genetic diversity of amphipods living on *Ulva lactuca* L. from Gouqi Coast, China. *Pac. Sci.* **2020**, *74*, 137–146. [CrossRef]
12. Stachowicz, J. Facultative mutualism between an herbivorous crab and a coralline alba: Advantages of eating noxious seaweeds. *Oecologia* **1996**, *105*, 377–387. [CrossRef]
13. Villamagna, A.; Murphy, B. Ecological and socio-economic impacts of invasive water hyacinth (*Eichhornia crassipes*): A review. *Freshw. Biol.* **2010**, *55*, 282–298. [CrossRef]
14. Whitfield, A. The role of seagrass meadows, mangrove forests, salt marshes and reed beds as nursery areas and food sources for fishes in estuaries. *Rev. Fish Biol. Fish.* **2017**, *27*, 75–110. [CrossRef]
15. Xia, J.; Li, C.; Tang, Y.; Li, J.; Wu, T.; Liu, J.; Zhang, J. Epizoans on floating golden tide macroalgae in the Southern Yellow Sea. *J. Mar. Sci. Eng.* **2023**, *11*, 479. [CrossRef]
16. Behera, D.; Ingle, K.; Mathew, D.; Dhimmar, A.; Sahastrabudhe, H.; Sahu, S.; Krishnan, M.; Shinde, P.; Ganesan, M.; Mantri, V. Epiphytism, diseases and grazing in seaweed aquaculture: A comprehensive review. *Rev. Aquac.* **2022**, *14*, 1345–1370. [CrossRef]
17. Bai, G.; Liu, Y.; Liu, Z.; Kong, L.; Tang, Y.; Ding, Z.; Zou, Y.; Wang, C.; Zhang, C.; Chen, D.; et al. Effects of lake geo-engineering on plankton in a typical shallow urban lake: Evidence based on 10-year data. *ACS EST Eng.* **2022**, *3*, 105–120. [CrossRef]
18. Shan, H.; Chou, Q.; Lv, C.; Tian, Y.; Wang, H.; Shi, L.; Wen, Z.; Wang, W.; Zhang, X.; Li, K.; et al. How do the growth forms of macrophytes affect the homogeneity of nearshore and open water areas? *Sci. Total Environ.* **2024**, *908*, 168165. [CrossRef] [PubMed]
19. Liu, Z.; Bai, G.; Liu, Y.; Zou, Y.; Ding, Z.; Wang, R.; Chen, D.; Kong, L.; Wang, C.; Liu, L.; et al. Long-term study of ecological restoration in a typical shallow urban lake. *Sci. Total Environ.* **2022**, *846*, 157505. [CrossRef]
20. Milbrandt, E.; Thompson, M.; Coen, L.; Grizzle, R.; Ward, K. A multiple habitat restoration strategy in a semi-enclosed Florida embayment, combining hydrologic restoration, mangrove propagule plantings and oyster substrate additions. *Ecol. Eng.* **2015**, *83*, 394–404. [CrossRef]
21. Larned, S.; Suren, A.; Flanagan, M.; Biggs, B.; Riis, T. Macrophytes in urban stream rehabilitation: Establishment, ecological effects, and public perception. *Restor. Ecol.* **2006**, *14*, 429–440. [CrossRef]

22. Chen, K.; Bao, C.; Zhou, W. Ecological restoration in eutrophic Lake Wuli: A large enclosure experiment. *Ecol. Eng.* **2009**, *35*, 1646–1655. [CrossRef]
23. Londe, V.; Prince, C.; Flory, S. Coexistence in long-term managed lakes: Limited evidence of negative impacts of invasive macrophytes on fish communities. *Ecol. Indic.* **2024**, *169*, 112931. [CrossRef]
24. Rodrigues, T.; Kratina, P.; Setubal, R.; Ferro, J.; Abe, D.; Costa, L.; Nova, C.; Farjalla, V.; Pires, A. Interaction between climate change scenarios and biological invasion reveals complex cascading effects in freshwater ecosystems. *Glob. Change Biol.* **2024**, *30*, e17540. [CrossRef]
25. Liu, J.; Xia, J.; Zhuang, M.; Zhang, J.; Yu, K.; Zhao, S.; Sun, Y.; Tong, Y.; Xia, L.; Qin, Y.; et al. Controlling the source of green tides in the Yellow Sea: NaClO treatment of *Ulva* attached on *Pyropia* aquaculture rafts. *Aquaculture* **2021**, *535*, 736378. [CrossRef]
26. Shang, X.; Yang, S.; Sun, J. Succession of phytoplankton communities from macro-scale to micro-scale in coastal waters of Qinhuangdao, China. *Front. Mar. Sci.* **2024**, *11*, 1371196. [CrossRef]
27. Liu, J.; Liu, W.; Xia, J. Perplexity and choice: Challenges and future development of laver cultivation in Jiangsu Province, China. *Front. Mar. Sci.* **2025**, *12*, 1526933. [CrossRef]
28. Xia, Z.; Liu, J.; Zhao, S.; Sun, Y.; Cui, Q.; Wu, L.; Gao, S.; Zhang, J.; He, P. Review of the development of the green tide and the process of control in the southern Yellow Sea in 2022. *Estuar. Coast. Shelf Sci.* **2024**, *302*, 108772. [CrossRef]
29. Zhang, Y.; He, P.; Li, H.; Li, G.; Liu, J.; Jiao, F.; Zhang, J.; Huo, Y.; Shi, X.; Su, R.; et al. *Ulva prolifera* green-tide outbreaks and their environmental impact in the Yellow Sea, China. *Natl. Sci. Rev.* **2019**, *6*, 825–838. [CrossRef]
30. Zhao, S.; Xia, Z.; Liu, J.; Sun, J.; Zhang, J.; He, P. Morphology, growth, and photosynthesis of *Ulva prolifera* O.F. Müller (Chlorophyta, Ulvophyceae) gametophytes, the dominant green tide species in the Southern Yellow Sea. *J. Sea Res.* **2023**, *193*, 102375. [CrossRef]
31. Xia, Z.; Yuan, H.; Liu, J.; Zhao, S.; Tong, Y.; Sun, Y.; Li, S.; Li, A.; Cao, J.; Xia, J.; et al. Biomass and species composition of green macroalgae in the Binhai Harbor intertidal zone of the Southern Yellow Sea. *Mar. Pollut. Bull.* **2023**, *186*, 114407. [CrossRef] [PubMed]
32. Silva, P.; McBride, S.; de Nys, R.; Paul, N. Integrating filamentous 'green tide' algae into tropical pond-based aquaculture. *Aquaculture* **2008**, *284*, 74–80. [CrossRef]
33. Liu, J.; Xia, Z.; Zeng, Y.; Xia, J.; He, P. Exploration and implication of green macroalgal proliferation in the Nanhui-east-tidal-flat: An investigation of post-reclamation mudflat wetlands. *Front. Mar. Sci.* **2025**, *12*, 1505586. [CrossRef]
34. Wu, H.; Zhang, J.; Li, H.; Li, S.; Pan, C.; Yi, L.; Xu, J.; He, P. Ocean warming enhances the competitive advantage of *Ulva prolifera* over a golden tide alga, *Sargassum horneri* under eutrophication. *Front. Mar. Sci.* **2024**, *11*, 1464511. [CrossRef]
35. Xiao, J.; Wang, Z.; Liu, D.; Fu, M.; Yuan, C.; Yan, T. Harmful macroalgal blooms (HMBs) in China's coastal water: Green and golden tides. *Harmful Algae* **2021**, *107*, 102061. [CrossRef] [PubMed]
36. Luo, H.; Yang, Y.; Xie, S. The ecological effect of large-scale coastal natural and cultivated seaweed litter decay processes: An overview and perspective. *J. Environ. Manag.* **2023**, *341*, 118091. [CrossRef] [PubMed]
37. Xu, Y.; Chen, J.; Yang, Q.; Jiang, X.; Fu, Y.; Pan, D. Trend of harmful algal bloom dynamics from GOCI observed diurnal variation of chlorophyll *a* off Southeast coast of China. *Front. Mar. Sci.* **2024**, *11*, 1357669. [CrossRef]
38. Jiang, W.; Liu, C.; Yang, D.; Xu, L.; Yin, B. Causes analysis of red tide event in the offshore sea of Rongcheng, Shandong Province, China based on a coupled physical-biological model. *Front. Mar. Sci.* **2024**, *11*, 1417667. [CrossRef]
39. Zahir, M.; Su, Y.; Shahzad, M.; Ayub, G.; Rahman, S.; Ijaz, J. A review on monitoring, forecasting, and early warning of harmful algal bloom. *Aquaculture* **2024**, *593*, 741351. [CrossRef]
40. Anderson, D.; Fensin, E.; Gobler, C.; Hoeglund, A.; Hubbard, K.; Kulis, D.; Landsberg, J.; Lefebvre, K.; Provoost, P.; Richlen, M.; et al. Marine harmful algal blooms (HABs) in the United States: History, current status and future trends. *Harmful Algae* **2021**, *102*, 101975. [CrossRef]
41. Harris, T.; Reinl, K.; Azarderakhsh, M.; Berger, S.; Berman, M.; Bizic, M.; Bhattacharya, R.; Burnet, S.; Cianci-Gaskill, J.; Domis, L.; et al. What makes a cyanobacterial bloom disappear? A review of the abiotic and biotic cyanobacterial bloom loss factors. *Harmful Algae* **2024**, *133*, 102599. [CrossRef] [PubMed]
42. Hao, A.; Iseri, Y.; Yan, Z.; Liu, X.; Zhao, M. Germination of phytoplankton resting cells in surface sediments collected from the tributary and littoral zones of Lake Taihu, China. *Harmful Algae* **2025**, *142*, 102780. [CrossRef] [PubMed]
43. Huisman, J.; Codd, G.; Paerl, H.; Ibelings, B.; Verspagen, J.; Visser, P. Cyanobacterial blooms. *Nat. Rev. Microbiol.* **2018**, *16*, 471–483. [CrossRef]
44. Wang, Y.; Zhao, D.; Woolway, R.; Yan, H.; Paerl, H.; Zheng, Y.; Zheng, C.; Feng, L. Algal blooms intensify in global large lakes over the past two decades. *Natl. Sci. Rev.* **2025**, *12*, nwaf011. [CrossRef]
45. Ho, J.; Michalak, A.; Pahlevan, N. Widespread global increase in intense lake phytoplankton blooms since the 1980s. *Nature* **2019**, *574*, 667–670. [CrossRef]
46. Feng, W.; Xiao, K.; Zhou, W.; Zhu, D.; Zhou, Y.; Yuan, Y.; Xiao, N.; Wan, X.; Hua, Y.; Zhao, J. Analysis of utilization technologies for *Eichhornia crassipes* biomass harvested after restoration of wastewater. *Bioresour. Technol.* **2017**, *223*, 287–295. [CrossRef] [PubMed]
47. Walker, T. *Plants: A Very Short Introduction*; Oxford University Press: Oxford, UK, 2010.

48. LaJeunesse, T.; Parkinson, J.; Gabrielson, P.; Jeong, H.; Reimer, J.; Voolstra, C.; Santos, S. Systematic revision of Symbiodiniaceae highlights the antiquity and diversity of coral endosymbionts. *Curr. Biol.* **2018**, *28*, 2570–2580. [CrossRef]
49. Vogler, A.; Monaghan, M. Recent advances in DNA taxonomy. *J. Zool. Syst. Evol. Res.* **2007**, *45*, 1–10. [CrossRef]
50. Liu, W.; Liu, J.; Zhang, M.; Zhang, J.; Sun, B.; He, C.; He, P.; Zhang, W. 1+ 1< 2: Combined effect of low temperature stress and salt stress on *Sesuvium portulacastrum* L. *Plant Physiol. Biochem.* **2025**, *219*, 109404.
51. Zamora-Aranda, A.; Aponte, H. Evaluation of two ecosystem services provided by a *Pistia stratiotes* population on the Pacific Coast of South America. *Biology* **2024**, *13*, 573. [CrossRef]
52. Cheng, X.; Zhao, X.; Lin, J.; Zhang, S.; Wang, Z.; Huang, H.; Wang, K.; Chen, J. Rotation culture of macroalgae based on photosynthetic physiological characteristics of algae. *Biology* **2024**, *13*, 459. [CrossRef]
53. Bao, Y.; Xu, P. Implications of environmental variations on *Saccharina japonica* cultivation in Xiangshan Bay, China. *Biology* **2025**, *14*, 175. [CrossRef]
54. Wang, K.; Tao, X.; Zhang, S.; Zhao, X. Effects of ocean acidification and temperature coupling on photosynthetic activity and physiological properties of *Ulva fasciata* and *Sargassum horneri*. *Biology* **2024**, *13*, 640. [CrossRef] [PubMed]
55. Wei, S.; Wang, L.; Yang, J.; Xu, R.; Jia, R.; He, P. Protective effect of polysaccharides isolated from *Sargassum horneri* against H_2O_2-Induced oxidative stress both in vitro, in Vero cells, and in vivo in zebrafish. *Biology* **2024**, *13*, 651. [CrossRef]
56. Liu, W.; Bao, Y.; Li, K.; Yang, N.; He, P.; He, C.; Liu, J. The diversity of planktonic bacteria driven by environmental factors in different mariculture areas in the East China Sea. *Mar. Pollut. Bull.* **2024**, *201*, 116136. [CrossRef] [PubMed]
57. Xia, J.; Hu, H.; Gao, X.; Kan, J.; Gao, Y.; Li, J. Phytoplankton diversity, spatial patterns, and photosynthetic characteristics under environmental gradients and anthropogenic influence in the Pearl River Estuary. *Biology* **2024**, *13*, 550. [CrossRef] [PubMed]
58. He, H.; Yang, X.; Zeb, A.; Liu, J.; Gu, H.; Yang, J.; Xiang, W.; Shen, S. Cloning and functional analysis of a zeaxanthin epoxidase gene in *Ulva prolifera*. *Biology* **2024**, *13*, 695. [CrossRef]
59. Zeng, Y.; Yang, X.; Xia, Z.; Chen, R.; He, F.; Zhang, J.; He, P. Review of allelopathy in green tides: The case of *Ulva prolifera* in the South Yellow Sea. *Biology* **2024**, *13*, 456. [CrossRef] [PubMed]
60. Niu, G.; He, C.; Mao, S.; Chen, Z.; Ma, Y.; Zhu, Y. Enhanced soil fertility and carbon sequestration in urban green spaces through the application of Fe-Modified biochar combined with plant growth-promoting bacteria. *Biology* **2024**, *13*, 611. [CrossRef]
61. Yao, L.; He, P.; Xia, Z.; Li, J.; Liu, J. Typical marine ecological disasters in China attributed to marine organisms and their significant insights. *Biology* **2024**, *13*, 678. [CrossRef]

biology

MDPI

Article

Evaluation of Two Ecosystem Services Provided by a *Pistia stratiotes* Population on the Pacific Coast of South America

Adela Zamora-Aranda [1] and Héctor Aponte [2,*]

1 Carrera de Ingeniería Ambiental, Universidad Científica del Sur, Lima 15842, Peru; adelabzamora@gmail.com
2 Carrera de Biología Marina, Universidad Científica del Sur, Lima 15842, Peru
* Correspondence: haponte@cientifica.edu.pe; Tel.: +51-994-990-121

Simple Summary: Ecosystems store large amounts of carbon, contributing to the control of the gases that lead to climate change. At the same time, certain species provide beneficial materials for the production of fodder or organic fertilizers. This research measured the amount of carbon stored by, and the biomass provision potential of, a water lettuce (*Pistia stratiotes*) population in a disturbed and unprotected Peruvian coastal wetland ecosystem. The results indicated that this population stored 3942.57 tCO_2 and that a potential 2132.41 tons of biomass could be obtained for fodder. This pioneering research in Peru measured these ecosystem services, demonstrating the potential of this population of floating aquatic plants to provide both services.

Abstract: One of the most fascinating wetlands on Peru's central coast is the Santa Rosa wetland (Chancay, Lima), an ecosystem threatened by anthropogenic activities. Some of these impacts have led to the uncontrolled growth of *Pistia stratiotes*, an invasive aquatic plant. This study sought to quantify the regulation and provisioning of ecosystem services provided by *P. stratiotes* using carbon storage and the provision of biomass as indicators. To this end, the biomasses of 50 plots measuring 0.0625 m^2 were weighed and georeferenced and the percentages of dry biomass (%DB) and total organic carbon in the biomass (%C) were quantified. The biomass and its coordinates were entered into ArcGIS and a Kriging interpolation technique was applied to determine the total amount of biomass (B). It was found that *P. stratiotes* stored 3942.57 tCO_2 and that 2132.41 tons of biomass could be obtained for fodder. The total carbon stored by this aquatic plant represented 28.46% of the total carbon sequestered in the wetland ecosystem by vascular plants, suggesting that its contribution to the carbon cycle is significant. This is the first study to estimate the biomass of a floating aquatic plant population in a coastal Peruvian wetland and is a pioneering study addressing the in situ carbon estimation of Peruvian floating aquatic plants. The results and methods proposed in this research will serve in the evaluation of the potential of ecosystem services among similar populations of floating aquatic species. In addition, the data presented can be used to establish plans for the management and use of this biomass in the production of soil fertilizers and cattle forage.

Keywords: carbon storage; coastal wetlands; water lettuce; aquatic plants; forage provision

1. Introduction

Ecosystem services are the benefits obtained by people from ecosystems through interactions with nature and are further subdivided into provisioning, regulating, supporting and cultural services [1]. These services are capable of generating food, medicines and energy, contributing to the regulation of all the biogeochemical cycles involved [2]. In this context, the endemic biodiversity of wetlands plays an essential role by offering services with a significant social, economic and environmental value such as tidal control, fiber provision, the maintenance of cultural goods and the cyclical control of nutrients [3]. However, such ecosystems are increasingly being threatened by urban growth, building projects, increases in invasive species and the development of port industries [4]. All of these factors combine to generate the loss and degradation of wetlands, thereby limiting the potential benefits of their ecosystem services among local populations [5]. In Peru, several wetland areas have undergone considerable reductions in area as a result of such threats [6–8].

The Santa Rosa wetland (SRW) is one of the most fascinating ecosystems on Peru's central coast. The diversity of organisms found in this wetland (it has one of the highest concentrations of bird and plant species per unit area on the entire Lima coast) is of major importance [9–11]. However, the natural cycles of this wetland have been altered by anthropogenic activities in the area; these include agriculture, solid-waste dumping, land clearance, pig farming and cattle grazing [12]. These activities have resulted in the growth of invasive species, capable of rapidly colonizing and threatening native species, thereby leading to environmental damage [13]. Among the 123 species of vascular plants in the SRW, 2 are introduced species, 31 are invasive and 12 are potential invaders [12]. One of these invasive species is *Pistia stratiotes*.

P. stratiotes is a floating aquatic plant species from the Araceae family, which is native to tropical and subtropical regions. Distributed in ponds, freshwater rivers and lakes, *P. stratiotes* forms rosettes up to 30 cm in size composed of thick and fleshy light-green leaves with tiny hairs that protect the plant from dehydration [14]. It has long, thin and fibrous roots and these hang in the water and can branch out and spread in several directions. Its reproduction can be sexual or vegetative through runners or stolons that enable the plant to form large colonies, thereby ensuring rapid propagation [15,16]. Due to the characteristics of its leaves and its capacity for rapid growth, this species is able to adapt to water-stress conditions, giving it an advantage over other aquatic species in terms of survival [17–20].

In the SRW, *P. stratiotes* is the predominant floating aquatic species [21]. Its uncontrolled proliferation has led to it covering around 80% of the lagoon and it seems likely that excessive nutrient input from drainage and agriculture is responsible for this condition [22]. Alternatives for the control and use of this species have been proposed; however, despite the existing legal framework for its conservation, no large-scale use has been developed to date [23].

It has been demonstrated that aquatic species such as *Eichhornia crassipes, Salvinia molesta, Azolla* [24,25] and *Elodea potamogeton* [26] have the potential for use as forage, while species such as *Typha domingensis* and *Schoenoplectus americanus* [27], among others, could potentially be used for carbon storage [28]. The carbon capturing of some floating aquatic species in wild conditions is also known; for example, studies of *Eichhornia azurea, Nymphaea rudgeana* [29,30], *Pontederia cordata* [31] and *Nymphoides indica* [32] have indicated values of between 3.5 T/ha and 17.5T/ha of carbon. *P. stratiotes* also has the potential to provide ecosystem services such as regulation through carbon storage [33] and provisioning through the production of forage [25]. Clearly, these two services (carbon capture and forage provision) could also be assessed in the population within the SRW. Previous studies have described significant carbon sequestration present in the SRW [27]; however, aquatic populations have not been evaluated, given that the instruments and methodology required are very different from those used for floating aquatic plant populations.

In this context, the aim of this pioneering study—addressing the in situ carbon estimation of Peruvian floating aquatic plants using spatial modeling combined with ecological techniques—was to (a) quantify the biomass of *P. stratiotes* in the SRW, (b) estimate the carbon stored in this habitat and (c) calculate the forage provisioning potential of the biomass.

2. Materials and Methods

2.1. Study Area

The study was conducted in the SRW, located in Chancay in the province of Huaral in the department of Lima (Figure 1a,b). This ecosystem receives average annual rainfall of 18 mm, has an average temperature of between 18 and 19 °C and is classified as a subtropical desert [34]. In terms of fauna, 89 species of resident and migratory birds have been registered in the area [10] along with 2 species of reptiles [34]. There are 57 species of vascular plants, of which 98% are herbaceous and 2% are shrubs. Among the aquatic herbs, there are two emergent species and six floating aquatic species [21].

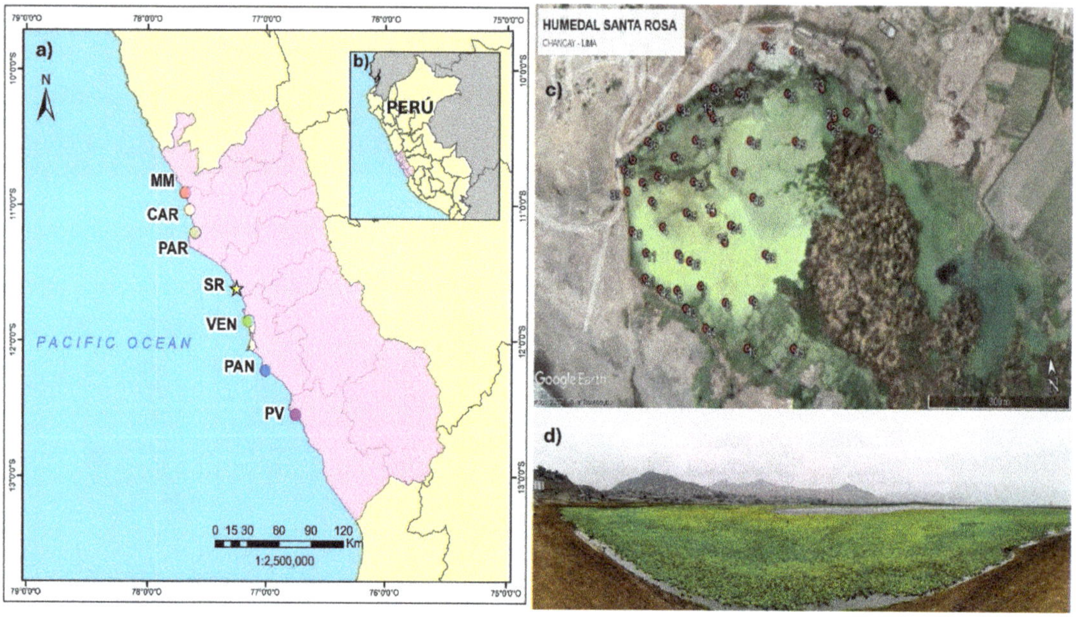

Figure 1. (**a**) Location of Santa Rosa wetland (★) on Peru's central coast; also shown are other important wetlands, which together constitute an ecological corridor on Lima's coast. MM: Albufera de Medio Mundo; CAR: Humedal de Carquín; PAR: Laguna El Paraíso; SR: Humedal de Santa Rosa; VEN: Humedales de Ventanilla; PAN: Los Pantanos de Villa; PV: Humedales de Puerto Viejo. In (**b**), the location of SRW in Peru is shown; the purple area is the department of Lima. (**c**) Distribution of plots in the lagoon. Satellite image taken from Google Earth (date of capture: 04 March 2021). (**d**) View of the lagoon, where the population of *Pistia stratiotes* can be seen (date of capture: 24 July 2021).

The SRW covers an area of 61 hectares [9], bordered to the north and partially to the northwest by the hill known as Cerro El Cascajo, to the west and partially to the south by the Pacific Ocean, to the south and southeast by agricultural areas (Peralvillo and Salinas Altas) and to the east by Cerro Salinas. Elevated areas are used for grazing. It is fed by water from the Chancay River via a channel and groundwater infiltration [21].

The evaluation was conducted in the main lagoon, which has a surface area of 17.5 hectares and is the primary habitat of *P. stratiotes* in this ecosystem (Figure 1).

2.2. *In Situ Biomass Assessment*

The evaluation was carried out between August and September 2021. A total of 50 plots of 0.0625 m^2 (0.25 m $\times$ 0.25 m) were located at a distance of $\approx$50 m from each other (Figure 1c). In each plot, the wet biomass of *P. stratiotes* was weighed using a JPSYSTEMS electronic digital hand scale (accurate to 0.01 kg) by removing water from the samples (without mistreating the plants) and excluding any other plant species that were present. Using the wet biomass, a performance curve was used to determine if the number of sampling points was representative of the population. The performance curve stabilized after the 30th sample; however, the measurement was extended to N°50 in order to cover the entire study area (Supplementary Material, Figure S1).

2.3. Analysis of Plant Tissue

Six samples of 600 g each of wet biomass were collected and taken to the Universidad Científica del Sur laboratories, where the gravimetric method was applied in order to determine the percentage of dry biomass (%DB) [35].

Four samples of 200 g each were taken to the Soil, Plant, Water and Fertilizer Analysis Laboratory of the La Molina National Agrarian University (LASPAF-UNALM). In order to estimate the percentage of carbon in the biomass (%C), these samples were analyzed using the Walkley and Black method, which measures the organic carbon in samples through oxidation methods, using a sample containing sulfur and chromium [36]. The result was multiplied by a correction factor of 1.32 in order to obtain the oxidizable carbon in the biomass (%CFO) [37].

2.4. Statistical Analysis and Spatial Modeling

Once the average wet biomass per plot had been calculated, ARC GIS software (Version 10.8) and Arc Map 10.8 were used to determine the amount of total biomass (B) in the lagoon through interpolation methods. The coordinates and biomass were entered into an information table and a point shapefile was created, after which the limits of the lagoon were established (creating a polygon shapefile) through an image basemap.

The spherical, exponential and circular models were tested using the Kriging interpolation tool (as there were many points with a certain average distance, the ordinary method was chosen) [38,39]. The model with the lowest mean error was selected. The correlation between the measured values and the predictions was also calculated using the selected model. This process has been used in previous studies to calculate plant biomass and carbon stock [40,41]. The modeling values (see the Results section) meant that the ordinary method was used. Using this method, the statistical relationships between the measured points were determined, thereby quantifying the spatial relationship between them. The figure established using the Kriging process was exported to a raster with a pixel size of 0.0625 (representing the plot size). The total biomass of the lagoon (B) was obtained by adding the biomass values of the total pixels.

2.5. Carbon Stock and Forage Supply Potential

The carbon stock (CS) was determined by the product of B, %DB and %C using the following equation:

$$CS = B \times \%DB \times \%C \quad (1)$$

where B is the total biomass (obtained from spatial modeling), %DB is the percentage of dry biomass and %C is the percentage of carbon in the biomass.

Available forage (F) was determined from the dry biomass in the entire lagoon by multiplying B and %DB using the following equation:

$$F = B \times \%DB \quad (2)$$

The amount of protein provided by *P. stratiotes* was obtained by multiplying F and the percent of protein on a dry basis (15.9% [42]).

3. Results

Table 1 details the biomass per unit area of each plot. The variability of the biomass in the lagoon was apparent (min = 0 kg; max = 1.4 kg/plot; coeff. var. = 36.90), with a non-homogeneous distribution across the lagoon. The average biomass per plot was 0.82 kg/plot, equal to 13.20 kg/m^2.

Table 1. UTM coordinates, biomass weight and carbon stored per plot.

N°	X	Y	Weight (kg/plot)	Stored Carbon (kg)
1	0252340	8717547	0.80	0.02
2	0252357	8717508	0.00	0.00
3	0252375	8717475	0.85	0.02
4	0252395	8717431	1.05	0.03
5	0252424	8717384	1.15	0.03
6	0252416	8717327	1.05	0.03
7	0262467	8717328	0.50	0.01
8	0252518	8717296	0.80	0.02
9	0252575	8717293	0.60	0.02
10	0252556	8717219	0.80	0.02
11	0252355	8717394	1.40	0.04
12	0252392	8717514	0.70	0.02
13	0252468	8717496	0.65	0.02
14	0252506	8717442	0.65	0.02
15	0252527	8717391	1.30	0.04
16	0252447	8717370	0.85	0.02
17	0252431	8717542	0.60	0.02
18	0252479	8717564	0.80	0.02
19	0252517	8717607	0.35	0.01
20	0252582	8717632	0.40	0.01
21	0252522	8717596	0.70	0.02
22	0252688	8717617	0.70	0.02
23	0252764	8717617	0.90	0.03
24	0252542	8717417	0.45	0.01
25	0252448	8717448	0.00	0.00
26	0252864	8717538	0.55	0.02
27	0252812	8717572	1.20	0.03
28	0252785	8717565	1.20	0.03
29	0252775	8717553	1.05	0.03
30	0252710	8717687	0.95	0.03
31	0252649	8717702	0.80	0.02
32	0252614	8717670	1.15	0.03
33	0252529	8717646	1.15	0.03
34	0252451	8717620	0.80	0.02
35	0252405	8717596	0.70	0.02
36	0252375	8717574	0.50	0.01
37	0252338	8717548	1.20	0.03
38	0252320	8717533	0.65	0.02
39	0252323	8717497	0.75	0.02
40	0252328	8717432	1.00	0.03
41	0252350	8717354	0.65	0.02
42	0252381	8717333	0.60	0.02
43	0252432	8717296	0.90	0.03
44	0252472	8717260	1.05	0.03
45	0252655	8717207	0.80	0.02
46	0252598	8717549	1.20	0.03
47	0252559	8717502	1.05	0.03
48	0252594	8717456	0.97	0.03
49	0252697	8717537	1.18	0.03
50	0252611	8717362	1.15	0.03

The mean %DB was 5.64% (min = 4.71%, max = 7.32% and coeff. var. = 20.25; for details, see Table 2). The mean %CFO was 38.2% (min = 36.57%, max = 39.63% and coeff. var. = 4.11; details in Table 3). This value showed less variability than the %DB. Using %CFO and the correction factor, the %C was calculated to be 50.42%.

Table 2. Percentage of dry biomass (%DB) in the samples.

Sample	%DB
1	6.87
2	5.07
3	4.99
4	7.32
5	4.88
6	4.71
Mean	5.64

Table 3. Percentage of readily oxidizable carbon in the biomass (%CFO) of the samples.

Sample	%CFO
1	39.46
2	36.57
3	37.14
4	39.63
Mean	38.2

B was estimated using the spherical model (error = 0.004; correlation = 0.84; 37,809.99 tons; Figure 2). Other model details are described in Tables S1–S5 and Equation (S1) in the Supplementary Material. Using %DB, %C and B, a total of 1075.25 tons of carbon, equivalent to 3942.57 tons of CO_2, was calculated. Using %DB and B, forage (F) was found to be 2132.41 tons, equivalent to 339.05 tons of stored protein, resulting in 19.38 Tn/ha.

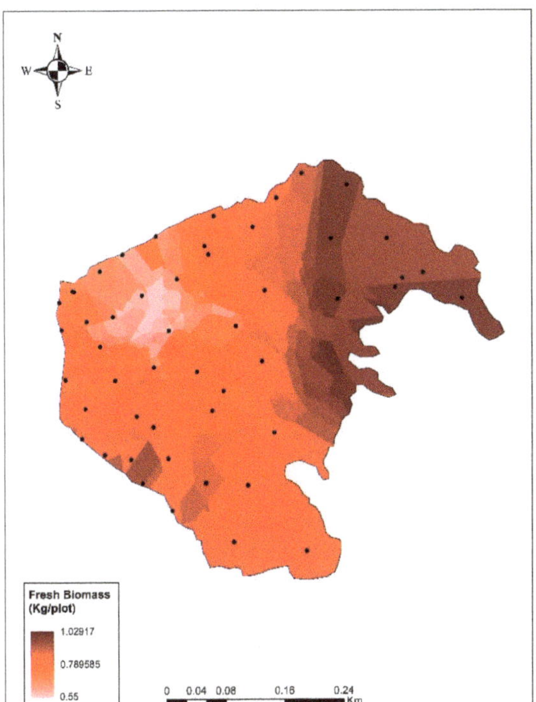

Figure 2. Biomass of *Pistia stratiotes* in the study area. The scale represents the value of fresh biomass (in kilograms) per plot (0.0625 m^2, corresponding with the area of the pixel used).

4. Discussion

This study constitutes the first estimation of the biomass and the carbon captured in a floating aquatic plant population of the Peruvian coast. Other similar studies (e.g., [43,44]) did not consider aquatic vegetation, probably because this requires different equipment (for example, a boat and boat oars were required for this sampling). Interestingly, the carbon stored in the area was not homogeneously distributed (Figure 2), which may have been due to the influence of various activities in other lagoon zones. There was water discharge in the northeastern zone from agricultural fields. In the southeastern zone (adjacent to the reed beds and rushes [12,43]), there was the dumping of construction debris and in the northern and western zones, solid waste and wastewater discharges from urban areas were present, along with waste from grazing and other livestock activities [23]. Unlike the eastern and southeastern zones, the western zone was influenced by the harvesting of *P. stratiotes*. This activity was carried out by volunteers, with the aim of reducing the biomass of the species and decongesting the body of water. This could only be carried out on the edges of the lagoon due to the difficulty of access, while the most significant amounts of biomass and stored carbon were concentrated in the central areas of the lagoon, which were the most difficult to access. Due to their nature, the modeling techniques may have had specific limitations; the dynamics of the species in the lagoon were not considered as a variable (these dynamics could have changed the biomass values since the evaluation was conducted). Also, where access was impossible, sections of the lagoon were not evaluated and there was a margin of error in the estimation presented. Nevertheless, the stabilization of the performance curve and the correlation and error coefficients gave us confidence that the estimation was reliable.

The amount of carbon stored by *P. stratiotes* was lower than that stored by other populations of coastal wetland plant species such as *Typha domingensis*, *Schoenoplectus americanus* and *Sporobolus virginicus*. At the same time, the amount of carbon stored by *P. stratiotes* was higher than the stocks of *Schoenoplectus californicus*, *Scirpus californicus* and other coastal ecosystems such as Lomas de Amancaes (Supplementary Material, Table S6). Certain communities of tall upright plants store more significant reserves than smaller plants such as *P. stratiotes* (and the mixed vegetation of the SRW); the presence of aerenchyma, the lower percentage of dry biomass and the absence of soil in *P. stratiotes* may explain these differences. The population of *P. stratiotes* in the SRW is one of the largest on Peru's central coast with no other similar population in the region, thereby making the SRW the ideal laboratory for this study.

A recent study [27] quantified the carbon stored in other plant communities present in the Santa Rosa wetland such as cattails, reeds and mixed meadows (composed of *Cyperus laevigatus* L., *Eleocharis geniculata* (L) Roem. & Schult., *Bacopa monnieri* (L.) Edwall, *Hydrocotyle ranunculoides* L.F., *Paspalum vaginatum* Sw. and *Distichlis spicata* P.M. Peterson & Romasch). When added to the result of this study, the wetland is estimated to store a total of 3778.27 tC, equivalent to 13,828.47 tCO_2. In this scenario, while the body of water with *P. stratiotes* represents 33% of the total area, it stores 28.46% of the wetland's total carbon. Therefore, taking into account the economic value per ton of sequestered CO_2 (USD 6.39 [45]), the value of the carbon stored in the Santa Rosa wetland amounts to USD 87,494.85. If this store was lost, the carbon would return to the atmosphere. In addition, in recent years, satellite images have revealed changes in the biomass over time, together with a eutrophication process [22]. It is, therefore, recommended that measures be adopted in order to protect this and other similar ecosystems on the Peruvian coast and that further measures be implemented to enable the remuneration for ecosystem services like those studied in this research.

The amount of forage calculated demonstrates the potential of this species for the provision of a resource that would contribute to the surrounding livestock communities. However, given its high water content and high percentage of fiber (20.8% [42]), it would constitute a low-quality forage resource. In addition, on a dry basis, *P. stratiotes* has a protein content of 15.9%, which falls below the average value of 18.21% found across other aquatic plant species such as *Lemna minor*, *Wolffia* spp. and *Hydrocharis laevigata* (Supplementary Material, Table S7). Nonetheless, it is essential that the local Cattle Ranchers Association and the Servicio Nacional de Sanidad Agraria (SENASA-Huaral) be involved in order to encourage further research and the harvesting of this plant for cattle ranching purposes as a way of decolonizing the body of water. In addition, *P. stratiotes* can be used for other purposes such as wastewater treatment [46], medicinal and ornamental uses [47], the removal of heavy metals [48] and in fuel production [14,49]. We encourage, therefore, the further evaluation of alternatives for the use of this interesting aquatic species.

5. Conclusions

Based on the combined use of field measurements and spatial modeling, we estimated the amount of the *P. stratiotes* biomass in the SRW to be 37,809.99 tons, equivalent to 1075.25 tons of stored carbon, 3942.57 tons of CO_2, 2134.41 tons of forage and 339.05 tons of stored protein. These measurements constitute a pioneering development in the study of the carbon stock of Peruvian floating aquatic plants. The methodology applied may form the basis for future studies involving similar species and plant biomasses in other aquatic environments. Subsequent studies should consider focusing on the population dynamics of this species in order to assess changes over time in the biomass and the area's physicochemical conditions. The results of this research may also offer a complementary vision of the role of these species, which are commonly considered to be invasive in nature. These estimates will enable decision-makers to establish plans for the use and management of this plant biomass within the wetland, inspiring new strategies and approaches. The proposal to use plant biomass as forage is just one of the potential uses of this plant and, thanks to this research, it can now be proposed based on more explicit economic values, thereby motivating further support and investment in this area of research.

Supplementary Materials: The following supporting information can be downloaded at: https://www.mdpi.com/article/10.3390/biology13080573/s1, Table S1. Statistical details of spatial modeling. Table S2. Comparison of the mean error with other Kriging models. Table S3. Spherical model settings. Table S4. Normality test for the model values. Table S5. Correlation between measured values and spherical model predictions. Table S6. Carbon stock in this study and in other Peruvian coastal ecosystems and locations. Table S7. Percentage of protein (P) of different aquatic plants on a dry basis. Values taken from [42], with the exception of the value for *Hydrocharis laevigata*, which was taken from [50]. Figure S1. Performance curve of *Pistia stratiotes* biomass in the SRW. The curve was drawn using the accumulated biomass weight (average biomass of all sampling points) during the establishment of the sampling points. References [27,42–44,50–55] are cited in the Supplementary Materials.

Author Contributions: Conceptualization, A.Z.-A. and H.A.; methodology, A.Z.-A. and H.A.; software, A.Z.-A.; validation, A.Z.-A. and H.A.; formal analysis, A.Z.-A. and H.A.; investigation, A.Z.-A.; writing—original draft preparation, A.Z.-A.; writing—review and editing, A.Z.-A.; visualization, A.Z.-A. and H.A..; supervision, A.Z.-A. and H.A.; project administration, A.Z.-A.; funding acquisition, A.Z.-A. and H.A. All authors have read and agreed to the published version of the manuscript.

Funding: This research was partially self-financed and also supported by the Dirección General de Investigación, Desarrollo e Innovación (DGIDI) of Universidad Científica del Sur.

Institutional Review Board Statement: Not applicable.

Informed Consent Statement: Not applicable.

Data Availability Statement: The authors confirm that the data supporting the findings of this study are available within the article and its Supplementary Materials.

Acknowledgments: The authors wish to thank Williams, Jean, Carmen, Nuria and César, members of the Comité de Vigilancia Ambiental of the Santa Rosa wetland, for their support in field data collection; Carlo Orihuela, for his assistance with the ARCGIS work; and Universidad Científica del Sur for the loan of equipment and laboratory facilities. The authors also wish to thank the DGIDI of Universidad Científica del Sur for their support of the project and research (N° 204-2019-PRE11).

Conflicts of Interest: The authors have no conflicts of interest to declare.

References

1. Millennium Ecosystem Assessment. *Ecosystems and Human Well-Being: Synthesis*; Island Press: Washington, DC, USA, 2005. Available online: https://www.researchgate.net/publication/297563785_Millennium_Ecosystem_Assessment_Ecosystems_and_human_well-being_synthesis (accessed on 30 March 2023).
2. Dorado, A. *¿Qué es la Biodiversidad? Una Publicación Para Entender su Importancia, su Valor y Los Beneficios Que Nos Aporta*, 1st ed.; Fundación Biodiversidad: Madrid, España, 2010. Available online: https://ecomilenio.es/wp-content/uploads/2010/10/que-es-la-biodiversidad-web.pdf (accessed on 1 April 2023).
3. Costanza, R.; de Groot, R.; Sutton, P.; Van der Ploeg, S.; Anderson, S.; Kubiszewski, I.; Farber, S.; Kerry Turner, R. Changes in the global value of ecosystem services. *Glob. Environ. Change* **2014**, *26*, 152–158. [CrossRef]
4. Davidson, N.C. How much wetland has the world lost? Long-term and recent trends in global wetland area. *Mar. Freshw. Res.* **2014**, *65*, 934–941. [CrossRef]
5. Gardner, R.; Barchiesi, S.; Beltrame, C.; Finlayson, C.; Galewski, T.; Harrison, I.; Paganini, M.; Perennou, C.; Pritchard, D.; Rosenqvist, A.; et al. *State of the World's Wetlands and Their Services to People: A Compilation of Recent Analyses*; Ramsar Briefing Note No. 7; Ramsar Convention Secretariat: Gland, Switzerland, 2015. [CrossRef]
6. Hoyos Gonzales, A.I.H.; Quinteros Camacho, N.L.; Crisólogo Rodríguez, M.E. Evaluación multitemporal de la superficie del Humedal Huacho-Hualmay-Carquín entre los años 1986 y 2019. *South Sustain.* **2022**, *3*, 1. [CrossRef]
7. Campos Macedo, J.A.; Pozo, J.H.; Núñez, P.C.M. Evolución espacio-temporal de relictos de humedal: El caso del sector de Villa Baja del Sitio Ramsar Los Pantanos de Villa (2003–2019). *South Sustain.* **2021**, *2*, 2. [CrossRef]
8. Morán, B.A.; Bermejo, R.L. Sostenibilidad turística del Santuario Nacional Los Manglares de Tumbes Puerto 25, Provincia de Zarumilla, Tumbes—Perú, 2012. *Manglar* **2016**, *11*, 1. [CrossRef]
9. Gonzáles, S.; Aponte, H. Diversidad taxonómica y patrones de diversidad de la flora en humedales de la costa peruana. *Rev. Acad. Colomb. Cienc. Exactas Físicas Nat.* **2022**, *46*, 730–741. [CrossRef]
10. Apeño, A.; Aponte, H. Caracterización de la diversidad de aves en un humedal altamente intervenido del Pacífico suramericano. *Rev. Acad. Colomb. Cienc. Exactas Físicas Nat.* **2022**, *46*, 380–392. [CrossRef]
11. Castillo Velásquez, R.M.; Huamantinco Araujo, A.A. Spatial variation of the aquatic macroinvertebrates community in the littoral zone of the Santa Rosa Coastal Wetland, Lima, Peru. *Rev. Biol. Trop.* **2020**, *68*, 50–68. [CrossRef]
12. Aponte, H.; Cano, A. Estudio florístico comparativo de seis humedales de la costa de Lima (PERÚ): Actualización y nuevos retos para su conservación. *Rev. Latinoam. Conserv.* **2013**, *3*, 15–27.
13. March-Mifsut, I.; Martínez-Jiménez, M. *Especies Invasoras de alto Impacto a la Biodiversidad: Prioridades en México*; Primera Edición; Instituto Mexicano de Tecnología del Agua: Jiutepec, Mexico, 2007; p. 73. Available online: http://repositorio.imta.mx/handle/20.500.12013/1619 (accessed on 1 April 2023).
14. Thayer, D.D.; Pfingsten, I.A.; Howard, V.; Li, J.; Redinger, J. *Pistia stratiotes* L.: U.S. Geological Survey, Nonindigenous Aquatic Species Database, Gainesville, FL, and NOAA Great Lakes Aquatic Nonindigenous Species Information System, Ann Arbor, MI 2024. Available online: https://nas.er.usgs.gov/queries/greatlakes/FactSheet.aspx?Species_ID=1099 (accessed on 7 December 2023).
15. Neuenschwander, P.; Julien, M.; Center, T.; Hill, M. *Pistia stratiotes* L. (Araceae). In *Biological Control of Tropical Weeds Using Arthropods*; Muniappan, R., Reddy, G.V.P., Raman, A., Eds.; Cambridge University Press: Cambridge, UK, 2009; pp. 332–352.
16. García Murillo, P.; Fernández Zamudio, R.; Cirujano Bracamonte, S. *Habitantes del agua. Macrófitos*, 1st ed.; Junta de Andalucía; Consejería de Medio Ambiente, Agencia Andaluza del agua: Sevilla, España, 2009; 282p. Available online: http://jolube.files.wordpress.com/2011/07/libro_macrofitos_andalucia_2010.pdf (accessed on 1 April 2023).
17. Tavera Escobar, H.A.; Gamba Cubides, N.J. Caracterización de la vegetación de la vía parque Isla de Salamanca, Magdalena-Colombia. *Colomb. For.* **2001**, *7*, 102–115. [CrossRef]
18. Gordon, E. Dinámica de la vegetación y del banco de semillas en un humedal herbáceo lacustrino (Venezuela). *Rev. Biol. Trop.* **2000**, *48*, 25–42.
19. Chaparro, G. Influencia del Nivel Hídrico y de la Macrofitia en los Patrones Espaciales y Temporales del Zooplancton de Una Laguna de Inundación. Ph.D. Thesis, Universidad de Buenos Aires, Buenos Aires, Argentina, 2013. Available online: https://bibliotecadigital.exactas.uba.ar/download/tesis/tesis_n5305_Chaparro.pdf (accessed on 1 April 2023).
20. Jelinski, G.; Sosio, V.; Paredes, S.; Colli, G. Estudio de base de la laguna De Ranchos para la gestión sostenible de un recurso natural bajo presión antrópica en área urbana. In Proceedings of the V Jornadas de Investigación, Transferencia y Extensión de la Facultad de Ingeniería, La Plata, Argentina, 9–11 April 2019; pp. 550–556. Available online: http://sedici.unlp.edu.ar/handle/10915/75506 (accessed on 1 April 2023).
21. Gonzales, S.; Aponte, H.; Cano, A. Actualización de la flora vascular del humedal Santa Rosa—Chancay (Lima, Perú). *Arnaldoa* **2019**, *26*, 867–882. [CrossRef]
22. Loayza, K.; Castillejos, K.; Mestas, R.; Quiliche, J. Estudio de la teledetección y caracterización fisicoquímica del humedal "El Cascajo", Santa Rosa, Chancay-Lima, Perú. *Infinitum* **2017**, *7*, 52–58. [CrossRef]

23. Aponte, H.; Jiménez, R.; Alcántara, B. Challenges for management and conservation of Santa Rosa wetland (Lima-Peru). *Científica* **2012**, *9*, 257–264. Available online: https://www.researchgate.net/publication/344402560_CHALLENGES_FOR_MANAGEMENT_AND_CONSERVATION_OF_SANTA_ROSA_WETLAND_LIMA_PERU (accessed on 4 September 2018).
24. Rosales, A.; Quintero, J.; Buritica, A.; Londoño, A.; Sarria, P.; Leterme, P.; Boudry, C.; Buldgen, A. Valor Nutricional de las Plantas Acuáticas Azolla y Salvinia en Cerdos. In Proceedings of the XII Encuentro Nacional de Zootecnia, Colombia, 2002. Available online: https://www.semanticscholar.org/paper/Valor-Nutricional-de-las-plantas-acuaticas-Azolla-y-Rosales-Quintero/79bb86ea016233aef13e124c8d3320fee8c16719 (accessed on 1 April 2023).
25. Gonzaga, G.; Monteiro, A. Valor nutritivo de macrófitas aquáticas flutuantes (*Eichhornia crassipes, Pistia stratiotes* e *Salvinia molesta*) utilizadas no tramento de efluentes de aqüicultura. *Maringá* **2002**, *24*, 519–526. Available online: http://hdl.handle.net/11449/67094 (accessed on 14 September 2018).
26. Franco, A.M. Evaluación Bromatológica de las Especies Forrajeras Acuáticas del Lago Titicaca. Ph.D. Thesis, Universidad Mayor de San Andrés, La Paz, Bolivia, 1981. Available online: https://repositorio.umsa.bo/bitstream/handle/123456789/5739/T-3.pdf?sequence=1&isAllowed=y (accessed on 30 March 2023).
27. Chavez, R.; Aponte, H. Carbono en el pacífico Sudamericano: Reservas en comunidades vegetales de un humedal costero en Perú. *Rev. De La Acad. Colomb. De Cienc. Exactas Físicas Y Nat.* **2023**, *47*, 962–976. [CrossRef]
28. Lu, W.; Liu, C.; Zhang, Y.; Yu, C.; Cong, P.; Ma, J.; Xiao, J. Carbon fluxes and stocks in a carbonate-rich chenier plain. *Agric. For. Meteorol.* **2019**, *275*, 159–169. [CrossRef]
29. Coutinho, M.E. Ecologia Populacional de Eichhornia Azurea (Kht) e sua Participacao na DINAMIDA da Vegetacao Aquática da Lagoa do Infernao, SP. Master's Thesis, Universidade Federal de Sao Carlos, São Carlos, Brasil, 1989.
30. Camargo, A.F.M.; Florentino, E.R. Population dynamics and net primary production of the aquatic macrophite *Nymphaea rudgeana* C.F. Mey in a lotic environment of the Itanhaém River basin (SP, Brazil). *Rev. Bras. De Biol.* **2000**, *60*, 83–92. [CrossRef]
31. Penha, J.M.F. Ecologia Populational de Pontederia Lanceolata (Nutal) em una área Alágavel do Pantanal Matogrossense, MT. Master's Thesis, Universidade Federal de Sao Carlos, São Carlos, Brasil, 1994.
32. Menezes, C. Biomassa e Producao Primária de Tres Espécies de Macrófitas Aquáticas da Represa do Lobo (Broa). Master's Thesis, Universidade Federal de Sao Carlos, São Carlos, Brasil, 1984.
33. Hernández, M. Suelos de humedales como sumideros de carbono y fuentes de metano. *Terra Latinoam.* **2010**, *28*, 139–147. Available online: https://www.scielo.org.mx/pdf/tl/v28n2/v28n2a5.pdf (accessed on 15 May 2023).
34. Alcántara, B.; Jiménez, V. Expediente Técnico de Creación del área de Conservación Ambiental Humedal Santa Rosa-Sector El Cascajo 2017. Available online: https://www.researchgate.net/publication/344390122_EXPEDIENTE_TECNICO_DE_CREACION_DEL_AREA_DE_CONSERVACION_AMBIENTAL_HUMEDAL_SANTA_ROSA-SECTOR_EL_CASCAJO (accessed on 28 September 2019).
35. Márquez, B.; Díaz-Ramos, J.; Troccoli, L.; Baumar, M.; Ramón, V. Densidad, biomasa y composición del zooplancton, en el estrato superficial de la cuenca de Cariaco, Venezuela. *Rev. Biol. Mar. Oceanogr.* **2009**, *44*, 737–749. Available online: https://www.redalyc.org/articulo.oa?id=47914663019 (accessed on 22 April 2022). [CrossRef]
36. Walkley, A.; Black, I.A. An examination of the Degtjareff method for determining soil organic matter, and a proposed modification of the chromic acid titration method. *Soil Sci.* **1934**, *37*, 29–38. [CrossRef]
37. Eyherabide, M.; Saínz Rozas, H.; Barbieri, P.; Echeverría, E. Comparación de métodos para determinar carbono orgánico en suelo. *Cienc. Suelo* **2014**, *32*, 13–19. Available online: http://www.scielo.org.ar/pdf/cds/v32n1/v32n1a02.pdf (accessed on 22 April 2022).
38. Córdoba, M.; Paccioretti, P.A.; Giannini Kurina, F.; Bruno, C.I.; Balzarini, M.G. *Guía Para el Análisis de Datos Espaciales en Agricultura*, 1st ed.; Editorial Brujas: Córdoba, Argentina, 2019; p. 250. Available online: http://hdl.handle.net/11336/128391 (accessed on 30 March 2023).
39. Majid, I.; Nayik, G.; Dar, S.; Nanda, V. Novel food packaging technologies: Innovations and future prospective. *J. Saudi Soc. Agric. Sci.* **2018**, *17*, 454–462. [CrossRef]
40. Reyes-Cárdenas, O.; Treviño-Garza, E.; Jiménez-Pérez, J.; Aguirre-Calderón, O.; Cuéllar-Rodríguez, L.; Flores-Garnica, J.; Cárdenas-Tristán, A. Modelización de biomasa forestal aérea mediante técnicas deterministas y estocásticas. *Madera Bosques* **2019**, *25*, e2511622. [CrossRef]
41. Romero-Sánchez, M.; Velasco-Bautista, E.; Meza-Juárez, D.; Pérez-Miranda, R. Estimation and analysis of the carbon content in halophilic grasslands from the semi-arid central part of Mexico. *Terra Latinoam.* **2022**, *40*, e1007. [CrossRef]
42. Tacón, A.G. Proteínas unicelulares. In *Nutrición y Alimentación de Peces y Camarones Cultivados. Manual de Capacitación*; FAO: Rome, Italy, 1989.
43. Aponte, H.; Corvacho, M.F.; Lertora, G.; Ramírez, D.W. Reserva de carbono en un humedal del desierto costero de Sudamérica. *Gayana Bot.* **2021**, *78*, 184–190. [CrossRef]
44. Palomino Contreras, D.; Cabrera Carranza, C. Estimación del servicio ambiental de captura del CO_2 en la flora de los humedales de Puerto Viejo. *Rev. Inst. Investig. Fac. Minas Metal. Cienc. Geográficas* **2007**, *10*, 49–59. Available online: https://revistasinvestigacion.unmsm.edu.pe/index.php/iigeo/article/view/494/419 (accessed on 4 April 2023).
45. Jakob, M.; Soria, R.; Edenhofer, O. *Precio al Carbono en América Latina Tendencias y Oportunidades*, Primera Edición; Trinidad, C.; Sociedad Peruana de Derecho Ambiental, 2019. Available online: https://spda.org.pe/wpfb-file/precio-al-carbono-en-al_digital_6nov_2-pdf/ (accessed on 6 April 2023).

46. Mendoza, Y.; Pérez, J.; Galindo, A. Evaluación del aporte de las plantas acuáticas *Pistia stratiotes* y *Eichhornia crassipes* en el tratamiento de aguas residuales municipales. *Inf. Tecnológica* **2018**, *29*, 205–214. [CrossRef]
47. Tripathi, P.; Kumar, R.; Sharma, A.; Mishra, A.; Gupta, R. *Pistia stratiotes*. *Pharmacognosy* **2010**, *4*, 153–160. Available online: https://phcogrev.com/sites/default/files/PhcogRev-4-8-153.pdf (accessed on 14 September 2018).
48. Aguayo, C. Determinación de la Acumulación de los Metales Pesados Plomo, Cadmio y Cromo en la Planta *Pistia stratiotes* Conocida Como Lechuga de Agua. Tesis de Pregrado, Universidad Icesi, Santiago de Cali, 2015. Available online: http://repository.icesi.edu.co/biblioteca_digital/bitstream/10906/78790/1/TG01084.pdf (accessed on 3 April 2022).
49. Mishima, D.; Kuniki, M.; Sei, K.; Soda, S.; Ike, M.; Fujita, M. Ethanol production form candidate energy crops: Water hyacinth (*Eichhornia crassipes*) and wáter letucce (*Pistia stratiotes* L.). *Bioresour. Technol.* **2007**, *99*, 2495–2500. [CrossRef]
50. Ruiz-Merino, M.; Campos-Cuéllar, R.; Germán-Gómez, A.; Aponte, H. Características, historia natural y aplicaciones de *Hydrocharis laevigata*: Una revisión. *Caldasia* **2022**, *44*, 432–441. [CrossRef]
51. Cieza, M. Estimación de la Captura de Dióxido de Carbono Por la Flora Del Área de Conservación Regional Humedales de Ven-tanilla. Tesis de Pregrado, Universidad Nacional Federico Villarreal, Perú, 2014. Available online: https://repositorio.unfv.edu.pe/handle/20.500.13084/716 (accessed on 22 April 2023).
52. Maldonado-Jiménez, I.; Aparicio-Saavedra, M.E. Estimación del almacenamiento de carbono en la biomasa de macrófitas en la Bahía interior de Puno, lago Titicaca. *Ecosistemas Recur. Agropecu.* **2021**, *8*, e2848. [CrossRef]
53. Medrano, Y.; Chupan, M.; Vila, M. Almacenamiento de carbono en especies predominantes de flora en el lago Chinchaycocha. *Apunt. Cienc. Soc.* **2012**, *2*, 110–117. [CrossRef]
54. Arévalo, J.; Aponte, H. Almacenamiento de carbono y agua en *Tillandsia latifolia* Meyen en un sector del Tillandsial de Piedra Campana (Lima/Perú). *Ecol. Apl.* **2020**, *19*, 9–15. [CrossRef]
55. Guerrero-Palomino, V.H.; Malca-Rodríguez, D.; Aponte, H. Reservas de carbono en un ecosistema del desierto sudamericano: El caso de Lomas de Amancaes, Lima, Perú. *Rev. Acad. Colomb. Cienc. Exactas Físicas Nat.* **2022**, *46*, 971–984. [CrossRef]

biology

MDPI

Article

Rotation Culture of Macroalgae Based on Photosynthetic Physiological Characteristics of Algae

Xiaopeng Cheng [1,2,†], Xu Zhao [1,†], Jun Lin [1,*], Shouyu Zhang [1,*], Zhenhua Wang [1], Hong Huang [1], Kai Wang [1] and Jianqu Chen [1]

1 College of Marine Ecology and Environment, Shanghai Ocean University, Shanghai 201306, China; xpcheng@shou.edu.cn (X.C.); xzhao@shou.edu.cn (X.Z.)
2 East China Sea Fisheries Research Institute, Chinese Academy of Fishery Sciences, Shanghai 200090, China
* Correspondence: jlin@shou.edu.cn (J.L.); syzhang@shou.edu.cn (S.Z.)
† These authors have contributed equally to this work.

Simple Summary: Seaweed aquaculture plays an important role in global food supply, but single-species intensive seaweed aquaculture has problems related to simple industrial structure and risk susceptibility. Therefore, there is an urgent need to establish a series of large-scale seaweed rotational culture models to address the current industrial challenges in algae culture. The culture of macroalgae is mainly affected by environmental factors such as water temperature and light. In this study, we constructed a rotational culture model by carrying out experiments with three species of macroalgae, observing in situ their photosynthetic activity responses to different temperature and light conditions. The results of this study can provide a theoretical basis for the establishment of large-scale rotational aquaculture, which can effectively improve the ecological and economic value of macroalgae aquaculture.

Abstract: Seaweed farming has made outstanding contributions to food supply and the restoration of the ecological environment despite the limitations in production and ecological effects due to the current intensive farming of single algae species. These limitations can be overcome by selecting suitable algal species based on their physiological characteristics and by constructing a large-scale seaweed rotation model. This study carried out a trial culture in aquaculture sea areas, and performed in situ monitoring of the environmental conditions and physiological characteristics of *Saccharina japonica*, *Hizikia fusiformis*, and *Gracilariopsis lemaneiformis*. Additionally, a comparative analysis of the three macroalgae at different times was conducted to determine their response characteristics to environmental factors. The results showed that: (1) The three macroalgae had varying light tolerance. The effective quantum yield of *Hizikia fusiformis* and *Gracilariopsis lemaneiformis* remained unchanged during the changes in light environment, while that of *Saccharina japonica* first decreased and then recovered. (2) The relative electron transport rates of the three macroalgae were significantly different under different temperature conditions. *Hizikia fusiformis* and *Saccharina japonica* exhibited the highest relative electron transport rates (70.45 and 106.75, respectively) in May (20.3 °C). Notably, *Gracilariopsis lemaneiformis* demonstrated good growth and exhibited the highest relative electron transport rate (93.07) in September (27.5 °C). These findings collectively support the feasibility of establishing a macroalgae rotation model. Based on the combined environmental conditions of the seas in Shandong, Zhejiang, and Fujian, a macroalgae rotation model was proposed. The application of this model in the construction of artificial seaweed farms in marine ranches can provide a stable output of large-scale seaweed production and ecological benefits.

Keywords: macroalgae; rotation culture; photosynthetic characteristics; ecological effect; physiological suitability

1. Introduction

Algae farming contributes significantly to solving global food demand and ecological challenges. According to statistical reports by FAO [1], the global seaweed production in 2018 was 3180 Mt (wet weight), accounting for 51.3% of the global marine aquacultural production and contributing to about USD13.3 billion in income for farmers [2,3]. In addition to meeting the market food demand, large seaweeds also provide high-quality feeding grounds for the cultivation of invertebrates, such as abalone, and provide abundant raw materials for the extraction of various products, such as fucoidan and mannitol. Macroalgae also act as biofilters, playing an essential role in addressing environmental and ecological problems [4]. At present, many studies have shown that large-scale seaweed culture plays an important ecological role in regulating water eutrophication, alleviating ocean acidification, increasing marine carbon sequestration, and providing a habitat for fishery resources [5–10].

China's large-scale algae culture is dominated by cold-water algae. The total area of algae culture in China in 2018 was 1.44×10^5 hm^2, whose entire output reached 2.34 million tons (wet weight) (China Fishery Statistical Yearbook 2019). For example, the raft *Saccharina japonica* culture structure has solved its demand in light conditions. The technical breakthrough to produce summer seedlings in cooled greenhouses has successfully promoted *Saccharina japonica* culture in the North (N 36°) [11]. Similarly, the high-temperature-resistant kelp varieties, cultivated through gametophyte hybridization technology, have enabled the rapid kelp culture in the south (N 23°) [12]. Currently, the area dedicated to *Saccharina japonica* culture accounts for about 31% of the area available for algae culture. Of note, the culture cycle of large algae is limited by the environmental water temperature. For example, the culture cycle of *Saccharina japonica* is within the time range when the water temperature is below 20 °C [13–15]. The intensive cultivation of a single algae species is marked by a simple industrial structure and limited resilience to risks [16]. In a seaweed cultivation area dominated by cold-water algae species, a common issue is the occurrence of cultivation vacancies during high-temperature conditions in summer, leading to time and space limitations in the yield and ecological utility of large algae. Therefore, it is urgent for researchers to identify suitable algae species, increase the diversity of cultured algae species, and fill the gap in seaweed culture during summer. Establishing large-scale seaweed rotation culture models can help coordinate the supply and demand of individual physiological and environmental rhythms of cultured algae species, improving the viability and stability of the algae culture industry.

The process of photosynthesis is an important target of abiotic stress. Photosynthetic performance is thus widely used as an index to evaluate the physiological adaptability of algae to environmental conditions [17]. The chlorophyll fluorescence technology is a rapid, non-invasive, sensitive, and convenient method for monitoring the photosynthetic performance of autotrophs. It allows us to assess such critical photosynthetic parameters as photosystem II (PSII) light energy absorption, electron transport from PS II to PSI, non-photochemical fluorescence quenching, and others [18,19]. To date, the technology has been widely used in laboratory physiological research. Many researchers have studied the physiological response of large algae to environmental changes, such as global warming and ocean acidification. These studies provide a theoretical basis for understanding the tolerance and ecological suitability of macroalgae to abiotic factors [20–34].

The environmental conditions of large-scale algae cultivation in the field are complex and diverse. Previous physiological experiments primarily focused on observations under controlled indoor conditions, which limits their ability to explain the physiological characteristics of macroalgae in the complex and diverse marine environment. This study used *Saccharina japonica*, *Hizikia fusiformis*, and *Gracilariopsis lemaneiformis* as examples to screen suitable algae species under different temperature conditions and establish an ecological rotation model of macroalgae. The field trial breeding experiment was conducted in Xihu harbor, a typical macroalgae breeding area in the inner bay of Zhejiang Province. The photosynthetic characteristics of the three macroalgae species were observed in situ using a modulated fluorescence instrument, Diving-Pam (Walz GmbH, Effeltrich, Germany), in January (winter), May (spring), and September (summer). The physical and chemical indexes of the environment in the aquaculture area were monitored synchronously. An ecological rotation culture model of large algae was established based on the characteristics of water temperature changes in different culture areas to identify the photosynthetic physiological characteristics of the three kinds of cultured macroalgae in different seasons. The model was also used to explore the possibility of applying a large-scale seaweed rotation culture model in algae culture and ecological restoration engineering.

2. Materials and Methods

2.1. Monitoring Site

Xihu harbor (29°32′20.06″ N, 121°47′28.04″ E) is located in the southeast of Xiangshan Bay, Ningbo City, Zhejiang Province (Figure 1a). The site is a typical subtropical semi-enclosed bay. The average tidal range of Xihu harbor is about 3 m. The large-scale seaweed culture in Xihu harbor mainly includes *Saccharina japonica* and *Pyropia yezoensis*, with the former covering approximately 200 hectares. The seedlings are clamped in December of each year, and the harvest begins in April of the following year. The surface velocity of the *Saccharina japonica* culture area during spring tide is 10.63–82.48 cm/s. Notably, Xihu harbor is extensively eutrophicated due to the low water exchange rate, industrial and agricultural sewage discharge, and excessive mariculture [28].

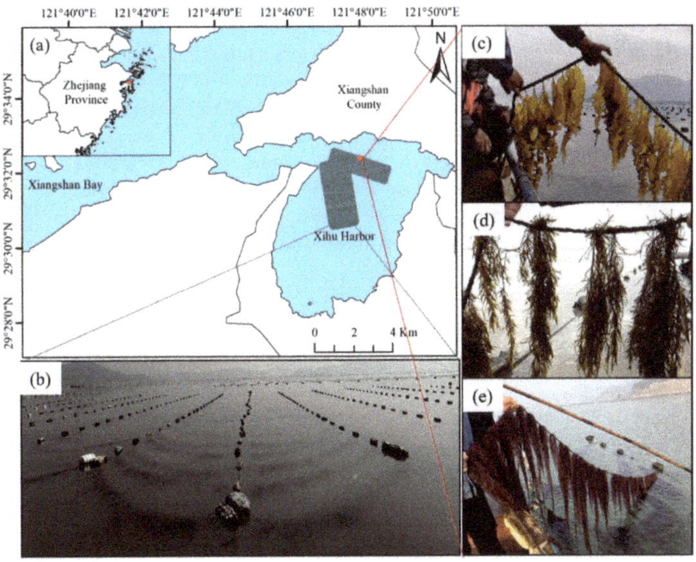

Figure 1. Example of macroalgae rotation in Xihu harbor. (**a**) Locations of example sites; (**b**) kelp of culture; (**c**) *Saccharina japonica*; (**d**) *Hizikia fusiformis*; and (**e**) *Gracilariopsis lemaneiformis*.

2.2. Seaweed Farms

The *Hizikia fusiformis* culture experiment began in 31 October 2019. The seedlings, sourced from the Dongtou *Hizikia fusiformis* culture area in Zhejiang Province measured 24.6 ± 14.93 cm. The *Gracilariopsis lemaneiformis* cultivation experiment began in 25 November 2019. The seedlings were sourced from Qingdao, Shandong Province (provided by Mr. Tao Liu). The *Saccharina japonica* culture experiment began in 20 December 2019, with seedlings sourced from Xiapu, Zhejiang Province, measuring 37.6 ± 0.87 cm long.

The structure of the breeding raft primarily comprises a long-line and cables. The long-lines are 50 m in length and are spaced 3.5 m apart (Figure 1b). The seedling ropes, each measuring 3.8 m in length, are positioned 1.5 m apart from each other. *Hizikia fusiformis*

(Figure 1d) and *Gracilariopsis lemaneiformis* (Figure 1e) have the same cultivation density, with algal bodies spaced 15–20 cm apart. For *Saccharina japonica*, the algal bodies are distanced 10–15 cm from each other (Figure 1c).

The physiological activities of the macroalgae were monitored at the same time (11:00–15:00) on January 9 (winter), May 6 (spring), and September 8 (summer) 2020, respectively. Temperature and light were also monitored synchronously. In situ monitoring of the photosynthetic activity of the macroalgae was carried out from 6 May 2020 to 7 May 2020 to understand the diurnal variation of the photosynthetic activity of the macroalgae.

2.3. Monitoring of Environmental Factors in Aquaculture Water

This study used a calibrated temperature, salinity, and depth multi-probe observation system (SBE 19plus V2, Sea-Bird Electronics, Bellevue, WA, USA) to monitor the surface temperature and photosynthetically active radiation in the aquaculture area of Xihu harbor. The target was to acquire the data at 0.5 m underwater after calibrating the pressure sensor. The surface and bottom water samples were collected to determine the environmental factors based on the "Ocean Monitoring Specification" (GB17378.3-2007) [35]. These samples included dissolved inorganic nitrogen nitrate (NO3-N), nitrite (NO2-N), ammonium salt (NH4-N), and active phosphate (PO4-P). The total nitrogen and total phosphorus were also measured following the "Ocean Monitoring Specification".

2.4. Fluorescence Parameter Determination

The chlorophyll fluorescence parameters of the algae were monitored in situ using an underwater fluorometer, Diving-PAM, and a data acquisition system, WinControl-3 (Walz GmbH, Effeltrich, Germany). Samples (3–5) were randomly selected from the culture water layer, and the surface impurities on the algae were cleaned. An optical fiber probe was fixed with a magnetic blade clip, and the ambient light was adjusted by connecting it to a computer to maintain the light in the range of 200–500 μmol m^{-2}·s^{-1}. The modulation measurement light was first turned on, and the initial fluorescence value ($F_0{}'$) was measured. Then, the saturation pulse was then turned on [4000 μmol m^{-2}·s^{-1}], a time when the fluorescence variable ($F_v{}'$) was the maximum fluorescence value ($F_m{}'$).

$$F_v' = F_m' - F_0' \tag{1}$$

The actual photochemical efficiency ($F_v{}'/F_m{}'$) could be calculated as follows:

$$F_v'/F_m' = (F_m' - F_0')/F_m' \tag{2}$$

The relative electron transport rate (*rETR*) reflects the efficiency of light energy utilization in chloroplasts. The photochemical gradient of the rapid light curve (RLC) of the three algae were 0, 99, 255, 476, 761, 1079, 1453, 2071, and 2762 μmol m^{-2}·s^{-1}. The irradiation time under each photochemical gradient was 10 s, while the interval between two photochemical lights was 20 s. The fluorescence value (F_t) before opening the saturated pulse was recorded after each photochemical irradiation. The measured fluorescence value after opening the saturated pulse was $F_m{}'$, from which the effective quantum yield of photosystem II (PSII) was calculated (effective quantum yield, *Y* (II)):

$$Y(\text{II}) = (F_m\prime - F_t\prime)/F_m\prime \tag{3}$$

$$rETR = Y(\text{II}) \times PAR \times 0.84 \times 0.5 \tag{4}$$

In the formula, *Y* (II) is the effective quantum yield, *PAR* is photosynthetically active radiation, 0.5 is the proportion of light absorbed with PSII, and 0.84 is the absorptivity of the sample [36].

The rapid light curve was fitted using the double exponential attenuation function and the Marquardt–Levenberg regression algorithm [18]:

$$rETR = rETR_m \times \left(1 - e^{-\alpha \times PAR/rETR_m}\right) \times e^{-\beta \times PAR/rETR_m}, \tag{5}$$

$$E_k = rETR_m / \alpha \tag{6}$$

$$E_m = rETR_m / \alpha \times ln[(\alpha + \beta) / \beta] \tag{7}$$

where α is the initial slope of the fast curve, β is the photosynthetic inhibition parameter, $rETR_m$ is the maximum relative electron transport rate, E_k is the half-saturated light intensity, and E_m is the saturated light intensity [18].

2.5. Construct Rotational Models

Based on this study, and considering the species of algae cultured in the major seaweed farming regions in China, we established a model for large-scale seaweed rotation culture. This model takes into account the annual water temperature changes at different latitudes (with observation data provided by the China Ocean Observation Station) and research findings on *Saccharina japonica*, *Undaria pinnatifida*, *Hizikia fusiformis*, *Gracilariopsis lemaneiformis*, and *Eucheuma denticulatum*. It is primarily focused on the provinces of Shandong, Zhejiang, and Fujian.

2.6. Statistical Analysis

All the data obtained from the experimental monitoring were sorted using the Excel software 2016, and the results were expressed as means ± standard deviation (SD). The rapid light curve data were fitted and plotted. The data distribution was tested for normality, and one-way ANOVA was used for data conforming to a normal distribution; otherwise, the Kruskal–Wallis significance test was used. The above analysis was conducted using original software.

3. Results

3.1. Changes in Environmental Factors in Aquaculture Water

The shift in temperature within the breeding area is predominantly noticeable across varying months, while alterations in light become apparent at different times throughout the day. The annual temperature change of the seaweed aquaculture waters of Xihu harbor is shown in Figure 2a. The minimum temperature was recorded in January (12.3 ± 0.15 °C), while the highest temperature was recorded in September (27.5 ± 0.1 °C). The daily trend of photosynthetically active radiation (*PAR*) in the seaweed farming area of Xihu harbor is shown in Figure 2b. The local sunrise time on 6 May 2020 was 5:06 a.m., the midday time was 11:50 a.m., and the sunset time was 18:34. The PAR first increased and then decreased. The maximum *PAR* was detected at around 15:00. Notably, the *PAR* had some errors because of cloud cover and other reasons.

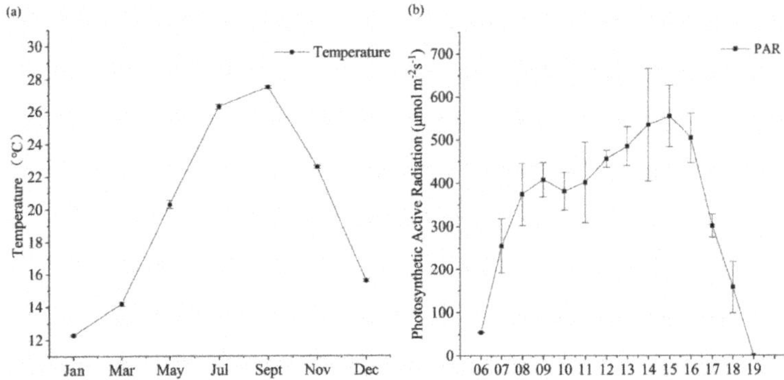

Figure 2. Trends of temperature and light in the trial culture. (**a**) Temperature variation of kelp culture area in Xihu harbor in different months. (**b**) Daily variation of photosynthetically active radiation in the kelp culture area of Xihu harbor.

3.2. Comparison of Photosynthetic Fluorescence Parameters of the Three Macroalgae

Saccharina japonica and *Hizikia fusiformis* tend to wither by September, resulting in a lack of rapid light curves. Comparison of the rapid light curves of *Saccharina japonica, Hizikia fusiformis,* and *Gracilariopsis lemaneiformis* at different time stages revealed that *Hizikia fusiformis* showed the maximum rETRm (106.75 μmol electrons $m^{-2}·s^{-1}$) in May (Figure 3d) and *Saccharina japonica* showed the lowest rETRm (43.58 μmol electrons $m^{-2}·s^{-1}$) in January (Figure 3a). The photosynthetic parameters of all the macroalgae changed significantly at different time stages, except for *Saccharina japonica,* which exhibited significant photo-inhibition in January (Figure 3a). The other two algae did not exhibit photo-inhibition. The rapid light curve fitting of the three macroalgae revealed some differences in the photosynthetic parameters among them. *Hizikia fusiformis* exhibited higher light tolerance than the other two species in light energy use efficiency (January: 0.3; May: 0.29) and semi-saturated light intensity (352.21 μmol ($m^{-2}s^{-1}$)) (Table 1).

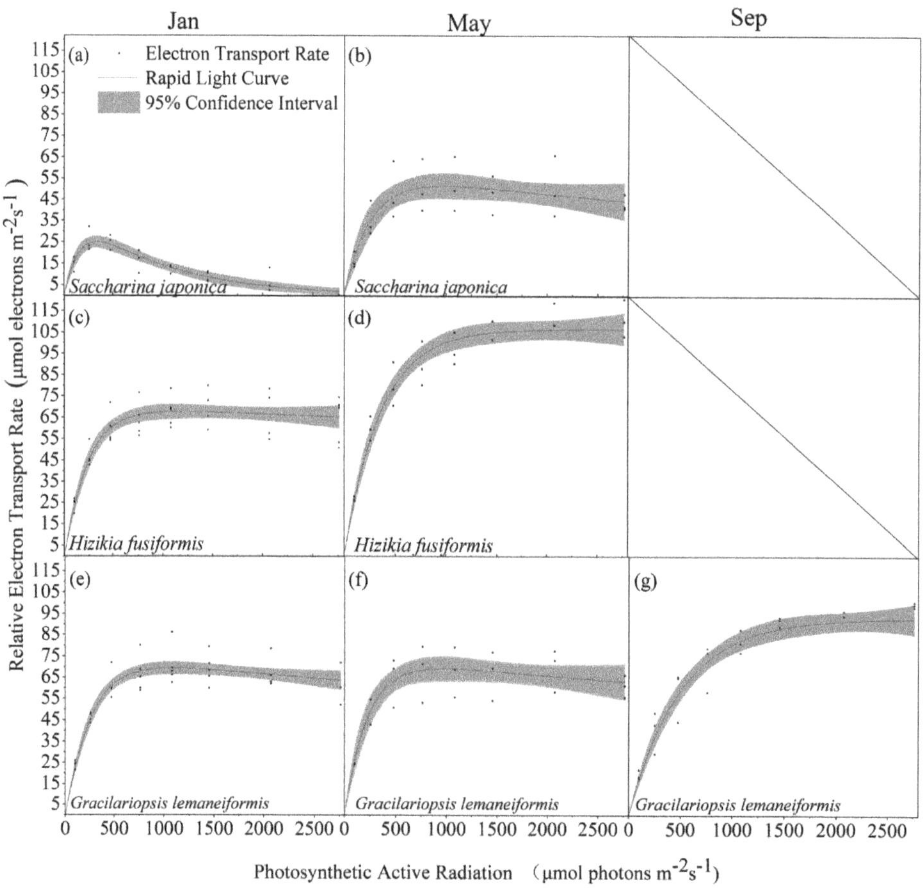

Figure 3. Rapid light curves of three algal species in different seasons. (**a**) *Saccharina japonica,* January; (**b**) *Saccharina japonica,* May; (**c**) *Hizikia fusiformis,* January; (**d**) *Hizikia fusiformis,* May; (**e**) *Gracilariopsis lemaneiformis,* January; (**f**) *Gracilariopsis lemaneiformis,* May; and (**g**) *Gracilariopsis lemaneiformis,* September. ($n \geq 3$).

Table 1. Rapid light curve fitting parameters of three macroalgae.

Species	Parameters	January	May	September
Saccharina japonica	α	0.23 ± 0.04 [a]	0.22 ± 0.04 [a]	-
	β	0.05 ± 0.02 [a]	0.01 ± 0.01 [b]	-
	$rETR_{max}$	43.58 ± 9.59 [b]	58.2 ± 8.78 [a]	-
	E_k	187.33 ± 66.05 [b]	264.91 ± 84.17 [a]	-
	E_m	328.66 ± 39.2 [b]	970.9 ± 148.8 [a]	-
	R^2	0.86	0.79	-
Hizikia fusiformis	α	0.29 ± 0.03 [a]	0.3 ± 0.03 [a]	-
	β	0	0	-
	$rETR_{max}$	70.45 ± 4.48 [b]	106.75 ± 8.62 [a]	-
	E_k	238.85 ± 35.08 [a]	352.21 ± 54.13 [a]	-
	E_m	1201.72 ± 211.15 [a]	-	-
	R^2	0.91	0.96	-
Gracilariopsis lemaneiformis	α	0.29 ± 0.02 [a]	0.33 ± 0.05 [a]	0.19 ± 0.02 [b]
	β	0	0	0
	$rETR_{max}$	75.33 ± 4.37 [b]	74.33 ± 6.72 [b]	93.07 ± 13.21 [a]
	E_k	263.46 ± 32.95 [b]	224.8 ± 47.6 [b]	495.08 ± 111.55 [a]
	E_m	1095.08 ± 84.7 [a]	973.29 ± 130.63 [a]	-
	R^2	0.94	0.89	0.96

Notes: different letters mean significant difference ($p < 0.05$).

3.3. Diurnal Variation of Effective Quantum Yield of the Macroalgae

In May, we conducted measurements every 2 h to record the daily variation of the effective quantum yield (F_v'/F_m') for three species of macroalgae. The outcomes of these measurements are depicted in Figure 4. There were no significant differences in F_v'/F_m' between *Hizikia fusiformis* and *Gracilariopsis lemaneiformis* at different time nodes. However, there were significant differences in F_v'/F_m' of *Saccharina japonica* at other time nodes. Notably, the maximum F_v'/F_m' of *Saccharina japonica* was recorded at 18:30, while the minimum value was recorded at 14:30. In general, there is a pattern of initial decline followed by a subsequent recovery.

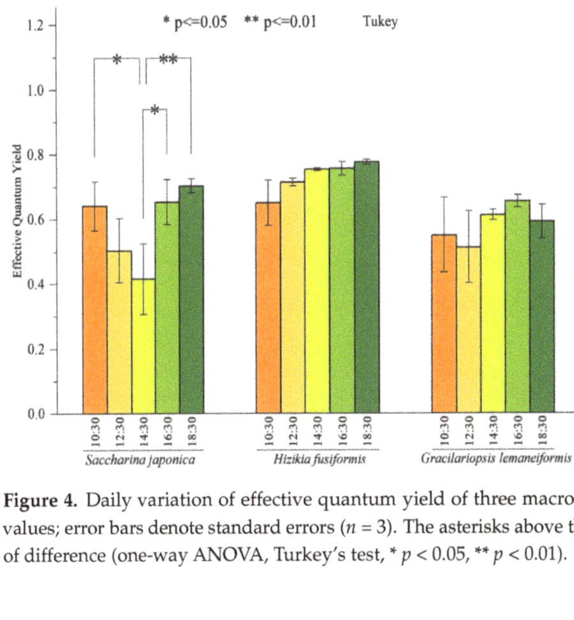

Figure 4. Daily variation of effective quantum yield of three macroalgae. Bars represent average values; error bars denote standard errors ($n = 3$). The asterisks above the bar indicate the significance of difference (one-way ANOVA, Turkey's test, * $p < 0.05$, ** $p < 0.01$).

3.4. Rotation Culture Models of the Macroalgae

The rotation model for macroalgae was developed based on the variation pattern of water temperature across different latitudes, considering the photosynthetic attributes of the primary cultivated algae, as depicted in Figure 5. For instance, Shandong, representing northern China's seaweed rotation cultivation (Figure 5a), is characterized by a considerable water temperature range and an extended duration of low temperature. Here, the principal cultivated algal species are *Saccharina japonica* [37,38], *Undaria pinnatifida* [39], and *Gracilariopsis lemaneiformis*. In contrast, Zhejiang province's macroalgae cultivation mode (Figure 5b) is marked by a brief duration of low temperature. The algae species in rotation predominantly comprises *Saccharina japonica*, *Hizikia fusiformis* [40], *Gracilariopsis lemaneiformis*, and *Eucheuma denticulatum* [41]. In Fujian, as displayed in Figure 5c, the rotation pattern of macroalgae is indicative of a long duration of high temperature, with the rotation algal species mainly consisting of *Saccharina japonica*, *Hizikia fusiformis*, *Gracilariopsis lemaneiformis*, and *Eucheuma denticulatum*.

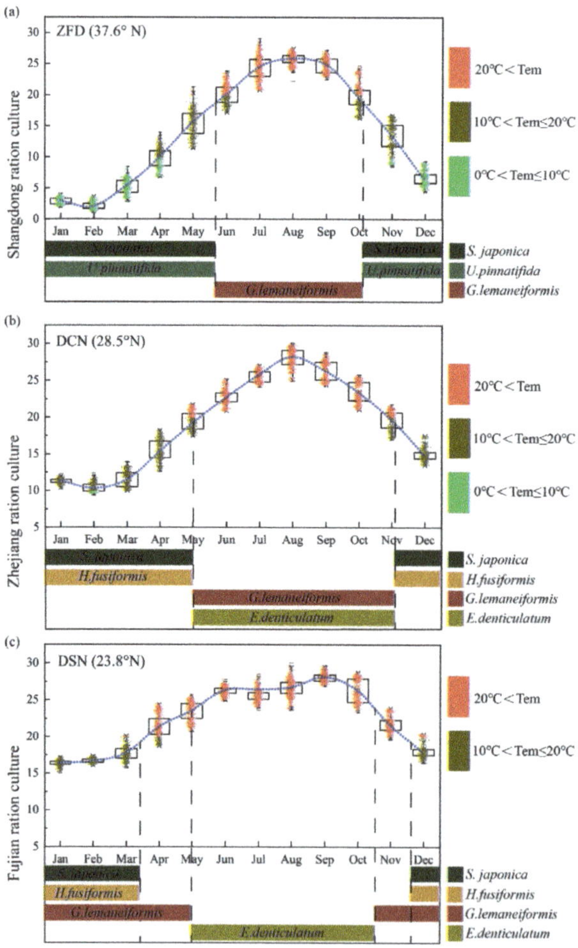

Figure 5. Rotation culture models of macroalgae. Annual water temperature variation at different latitudes in China (observation data in Chinese oceanic stations). **(a)** Shandong ration culture, ZFD: Zhi Fu Dao; **(b)** Zhejiang ration culture, DCN: Da Chen; (c) Fujian ration culture, DSN: Dong Shan.

4. Discussion

4.1. Response of Photosynthetic Physiological Characteristics of the Three Macroalgae Species to Temperature

Seaweeds are heat-sensitive organisms. Environmental temperature profoundly impacts their enzymatic processes and metabolic functions, such as photosynthesis and respiration [30]. Notably, cold-water and warm-water species exhibit different responses to temperature variations. Cold-water species often experience a decreases in photosynthesis rate and a significant increase in respiration rate when ambient temperature exceeds the optimal range, resulting in the net photosynthesis rate of cold-water algae being less than zero under high-temperature conditions [31].

In this study, a large area of withering occurred at the apical zone of the *Saccharina japonica* in late May. Withering became more serious with the increase in temperature. In late June, the *Saccharina japonica* only retained a part of the holdfast, while the rest all withered. Other relevant studies also report that the loss of individual organisms in *Saccharina japonica* culture during the late stage of temperature stress can be between 45% and 88% [42]. Most *Hizikia fusiformis* withered and decomposed in the first ten days of June and were all withered by middle and late July. In contrast, *Gracilariopsis lemaneiformis* showed promising activity in January (12.3 °C), May (20.3 °C), and September (27.5 °C), and did not wither or decompose.

The relative electron transport rates of cold-water algae *Saccharina japonica* varied significantly differently under the ambient temperature conditions in January and May. These results align with previous experimental results which indicated that the relative electron transport rate of *Saccharina japonica* increases with an increase in temperature within the relatively suitable temperature range. However, when the ambient temperature exceeds the appropriate temperature of *Saccharina japonica*, it leads to death. Pang et al. also demonstrated that when the ambient temperature exceeds 25 °C, the photosynthetic fluorescence parameters of *Saccharina japonica* decrease rapidly, leading to death [12].

Comparted to *Saccharina japonica*, *Hizikia fusiformis* exhibited higher photosynthetic characteristics under different temperatures in January and May, especially in May (20.3 °C). Zou et al. reported that *Hizikia fusiformis* enters the sexual maturation stage when the ambient temperature reaches 19–21 °C, with gametophyte release occurring at temperatures of 27.5–30 °C [31]. The energy distribution of *Hizikia fusiformis* at this stage changes from growth to reproduction. Its higher relative electron transport rate indicates that its photosynthesis provides a higher energy supply. The withering and decomposition of *Hizikia fusiformis* during summer may thus be related to its life cycle. *Gracilariopsis lemaneiformis* exhibited a higher relative electron transport rate in January, May, and September, with the highest relative electron transport rate recorded in September (27.5 °C). Remarkably, it maintained high physiological activity during summer. Yang et al. reported that *Gracilariopsis lemaneiformis* has high nitrogen and phosphorus absorption efficiency and is an excellent target algae species for ecological restoration [43]. The results of field trial culture and simultaneous monitoring of the physiological characteristics of *Gracilariopsis lemaneiformis* under different environmental conditions suggest that it is a suitable species to fill the gap of algae culture in high-temperature waters during summer in northern China.

4.2. The Photosynthetic Response of the Three Species of Macroalgae to Light

Macroalgae convert light energy into substances and energy needed for growth, development, and reproduction through the light and dark reaction processes of photosynthesis. Light is the basis of photosynthesis in macroalgae. It affects the light and dark reaction and enzyme activity of macroalgae in the process of photosynthesis [44]. Lack of light limits carbon assimilation and eventually affects the formation of photosynthates in seaweed. In contrast, excessive light causes photo-inhibition to seaweed photosynthesis and even light damage to cells [45]. Different macroalgae have different sensitivity to light because of their morphological differentiation and tissue structure [46].

In this study, the diurnal variation of the effective quantum yield of *Saccharina japonica* decreased at first and then increased. This finding suggested that *Saccharina japonica* has a self-regulation mechanism when the ambient light is less than the saturated light intensity (E_m). Moreover, the photosynthetic reaction center of *Saccharina japonica* does not cause irreversible damage, and its photosynthetic reaction center can recover after adapting to environmental conditions. Previous studies postulate that the photo-regulation mechanism of *Saccharina japonica* is associated with the open state of electron gates in photoreaction center II (PSII) during photosynthesis [47]. Part of the electronic gates of PSII is closed under the continuous irradiation of strong light. PSII transports much of the light energy into heat dissipation to protect the photoreaction, thereby decreasing the actual photochemical efficiency. Notably, the PSII reaction center does not release electrons when the seaweed is in the dark state, and all the electron gates are open, causing the corresponding photochemical efficiency to reach the maximum. In this study, *Hizikia fusiformis* and *Gracilariopsis lemaneiformis* did not show a diurnal changing trend of effective quantum yield. This finding is consistent with the research results of other researchers that report that both algae species have a high tolerance to strong light and that they can adapt to the surface bright-light culture conditions in the actual culture process [23,27,48,49].

Continuous exposure to strong light when the ambient light intensity exceeds the saturated light intensity of macroalgae can cause irreversible photo-damage to the photosystem and result in photo-inhibition ($\beta > 0$). In this study, *Saccharina japonica* exhibited photo-inhibition in the light curves in January and May, with the photo-inhibition phenomenon being most serious during the seedling stage in January ($\beta = 0.05$). Hwang et al. reported that the higher light sensitivity and photo-inhibition phenomenon of *Saccharina japonica* is closely related to the tissue structure of its leaves [50]. In the large-scale seaweed culture process, the transparency and depth of the culture water affect the light intensity received on the surface of the algae. The saturated light intensity and light suppression parameters of *Saccharina japonica* thus provide a reference for farmers to arrange the depth of seedling culture to ensure the survival rate of the seedlings.

Optional arrangement of seaweed culture density can maximize algae production per unit area. Light is one of the main factors restricting the culture density of macroalgae. The differences in light energy utilization efficiency among different algae species provide a reference basis for seaweed culture density. In this study, variations in light energy utilization efficiency among the three macroalgae in different seasons were observed, influenced by complex environmental factors. This suggests that the culture density should be planned based on macroalgae growth and adaptation to environmental conditions to maximize the yield per unit culture area.

4.3. Design and Application of Rotation Culture Model in Large Seaweeds

Seaweed culture in China ranges from N 39° to N 22°, with *Saccharina japonica* culture accounting for about 31% of the total algae culture area (Figure 6, *China Fishery Statistical Yearbook* 2009–2019). The period of *Saccharina japonica* culture is usually when the water temperature is below 20 °C. Consequently, China's algae culture industry thus generally has different degrees of empty window periods from north to south. There are also significant differences between the north and south aquaculture sea areas in the water temperature cycle. It is thus essential to establish a comprehensive algae rotation model based on the water temperature conditions in the south and north. This paper integrates the physiological characteristics of macroalgae with different ecological attributes. Figure 5 outlines the established aquaculture model using Shandong, Zhejiang, and Fujian as examples, based on the environmental and seasonal characteristics of the water temperature of each aquaculture area and the photosynthetic physiological characteristics of each alga. This study focused on Xiapu and other areas in Fujian because of the significant temperature difference between the north and the south of Fujian. There were significant differences in the breeding mode in Xiapu because of the generally high annual water temperature in Dongshan and other southern algae breeding areas in Fujian.

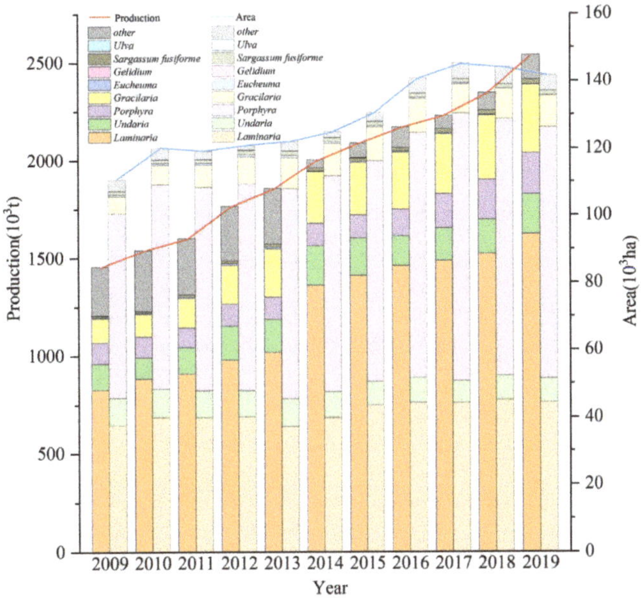

Figure 6. Temporal changes of different aquaculture seaweed productions and farming areas in China between 2009 and 2019 (data from *China Fishery Statical Yearbook*, 2009–2019).

The large seaweed rotation model based on the photosynthetic and ecological characteristics of seaweed is a sustainable development model that uses ecological principles and comprehensively considers the economy, environment, and niche space, thus providing continuous productivity across seasons. The model integrates regulation of the water environment and food web feed supply throughout the year, such as in integrated multi-trophic aquaculture (IMTA) and the construction of seaweed farms. The improvement of water quality also provides bait for fishery resources in aquaculture sea areas. Studies postulate that diversified farming can interrupt the virus transmission cycle and help producers reduce and manage disease risks [51]. The multi-algal species rotation mode and disease control are thus essential for the algae farming industry.

Marine ranches are ecological projects of biological resource conservation and ecological environment restoration. They play a significant role in habitat creation and resource conservation. Seaweed farms are an important part of the marine ranch ecological restoration project. They play a supporting role in the supply of primary productivity, water environment regulation, and habitat creation, and are an indispensable part of the ecological function of marine ranches [52]. Relevant studies also postulate that seaweed farms have made outstanding contributions to the conservation of fishery resources, especially for fish larvae, which directly affect the replenishment of fish populations [5]. However, most sea areas, such as the East China Sea, are characterized by low water transparency and lack of suitable seaweed attachment bases, thus making it challenging to restore natural seaweed farms. The large-scale seaweed raft culture structure provides ideas for solving the challenges of seaweed farm restoration in the East China Sea. The findings of this study, combined with our previous extensive investigation and research work, suggest that the time complementarity should be considered when constructing artificial seaweed farms under the ecological restoration framework of marine ranch. The stability of the project is mainly based on the conservation of biological resources, the restoration of the ecological environment, and the economic benefits for fishermen. The large-scale seaweed rotation mode provides good theoretical support for constructing artificial seaweed farms.

It provides a solution to restoring seaweed farms in the East China Sea and a tool for ecological restoration.

5. Conclusions

This study carried out a trial culture study in aquaculture sea areas and monitored the photosynthetic characteristics of *Saccharina japonica*, *Hizikia fusiformis*, and *Gracilariopsis lemaneiformis* at different times. The study proposes a rotation model of macroalgae, covering the whole year based on the current ecological habits of large cultured algae, combined with the annual water temperature changes in different latitudes in China, taking Shandong, Zhejiang, and Fujian as examples. It provides a solution to the limitations of the production and ecological effects of single algal culture. This study only considers the suitability of macroalgae photosynthetic and physiological characteristics to field temperature and light conditions. Future studies should thus focus on the suitability of macroalgae photosynthetic and physiological activity on environmental factors, such as nutrients, flow field, light quality, and trace elements, to have a holistic conclusion, especially on the rotation model.

Author Contributions: X.C. and X.Z. contributed to the conception and design of the study. X.Z. and J.C. conducted the experiments. X.C. contributed to the analysis and wrote the manuscript. J.L., S.Z., Z.W., H.H. and K.W. contributed to the manuscript revision. All authors have read and agreed to the published version of the manuscript.

Funding: This work was supported by the National Key Research and Development Program of China (Grants 2023YFD2401902, Grants 2023YFD2401905), Southern Marine Science and Engineering Guangdong Laboratory (Zhuhai) (SML2023SP237), China Agriculture Research System (Grants CARS-50), and National Natural Science Foundation of China (Grants 41876191).

Institutional Review Board Statement: Not applicable.

Informed Consent Statement: Not applicable.

Data Availability Statement: The original contributions presented in the study are included in the article, and further inquiries can be directed to the corresponding author.

Acknowledgments: We gratefully acknowledge Tao Liu for providing the seeding of *Gracilariopsis lemaneiformis*. We are very thankful to Longfei Lu for providing assistance with the algae culture. We also give thanks to the National Science & Technology infrastructure—National Marine Science Data Center (http://mds.nmdis.org.cn/, accessed on 2 April 2022) for delivering data support.

Conflicts of Interest: The authors declare no conflicts of interest.

References

1. FAO. *The State of World Fisheries and Aquaculture: Sustainability in Action*; FAO: Rome, Italy, 2020; pp. 1–210.
2. Chopin, T.; Tacon, A.G.J. Importance of Seaweeds and Extractive Species in Global Aquaculture Production. *Rev. Fish. Sci. Aquac.* **2021**, *29*, 139–148. [CrossRef]
3. Duarte, C.M.; Bruhn, A.; Krause-Jensen, D. A seaweed aquaculture imperative to meet global sustainability targets. *Nat. Sustain.* **2021**, *5*, 185–193. [CrossRef]
4. Kang, Y.H.; Kim, S.; Choi, S.K.; Lee, H.J.; Chung, I.K.; Park, S.R. A comparison of the bioremediation potential of five seaweed species in an integrated fish-seaweed aquaculture system: Implication for a multi-species seaweed culture. *Rev. Aquac.* **2020**, *13*, 353–364. [CrossRef]
5. Bertocci, I.; Araújo, R.; Oliveira, P.; Sousa-Pinto, I.; Österblom, H. REVIEW: Potential effects of kelp species on local fisheries. *J. Appl. Ecol.* **2015**, *52*, 1216–1226. [CrossRef]
6. Teagle, H.; Hawkins, S.J.; Moore, P.J.; Smale, D.A. The role of kelp species as biogenic habitat formers in coastal marine ecosystems. *J. Exp. Mar. Biol. Ecol.* **2017**, *492*, 81–98. [CrossRef]
7. Xiao, X.; Agusti, S.; Lin, F.; Li, K.; Pan, Y.; Yu, Y.; Zheng, Y.; Wu, J.; Duarte, C.M. Nutrient removal from Chinese coastal waters by large-scale seaweed aquaculture. *Sci. Rep.* **2017**, *7*, 46613. [CrossRef]
8. Zheng, Y.; Jin, R.; Zhang, X.; Wang, Q.; Wu, J. The considerable environmental benefits of seaweed aquaculture in China. *Stoch. Environ. Res. Risk Assess.* **2019**, *33*, 1203–1221. [CrossRef]

9. Jiang, Z.; Liu, J.; Li, S.; Chen, Y.; Du, P.; Zhu, Y.; Liao, Y.; Chen, Q.; Shou, L.; Yan, X.; et al. Kelp cultivation effectively improves water quality and regulates phytoplankton community in a turbid, highly eutrophic bay. *Sci. Total Environ.* **2020**, *707*, 135561. [CrossRef]
10. Xiao, X.; Agustí, S.; Yu, Y.; Huang, Y.; Chen, W.; Hu, J.; Li, C.; Li, K.; Wei, F.; Lu, Y.; et al. Seaweed farms provide refugia from ocean acidification. *Sci. Total Environ.* **2021**, *776*, 145192. [CrossRef]
11. Tseng, C.K. Algal biotechnology industries and researh activities in China. *J. Appl. Phycol.* **2001**, *12*, 15. [CrossRef]
12. Pang, S.J.; Jin, Z.H.; Sun, J.Z.; Gao, S.Q. Temperature tolerance of young sporophytes from two populations of Laminaria japonica revealed by chlorophyll fluorescence measurements and short-term growth and survival performances in tank culture. *Aquaculture* **2007**, *262*, 493–503. [CrossRef]
13. Zhang, J.; Liu, Y.; Yu, D.; Song, H.; Cui, J.; Liu, T. Study on high-temperature-resistant and high-yield Laminaria variety "Rongfu". *J. Appl. Phycol.* **2011**, *23*, 165–171. [CrossRef]
14. Su, L.; Pang, S.J.; Shan, T.F.; Li, X. Large-scale hatchery of the kelp *Saccharina japonica*: A case study experience at Lvshun in northern China. *J. Appl. Phycol.* **2017**, *29*, 3003–3013. [CrossRef]
15. Shan, T.; Li, Q.; Wang, X.; Su, L.; Pang, S. Assessment of the Genetic Connectivity Between Farmed Populations on a Typical Kelp Farm and Adjacent Spontaneous Populations of *Saccharina japonica* (Phaeophyceae, Laminariales) in China. *Front. Mar. Sci.* **2019**, *6*, 494. [CrossRef]
16. Wu, W.G.; Zhang, J.H.; Liu, Y.; Zhao, Y.X.; Zhang, K.Y.; Wang, L.L. Analysis of the effectiveness of the polyculture of shellfish and algae and influencing factors in the Xuejiadao sea cage aquacultrue area. *Prog. Fish. Sci.* **2022**, *43*, 10. [CrossRef]
17. Borlongan, I.A.; Arita, R.; Nishihara, G.N.; Terada, R. The effects of temperature and irradiance on the photosynthesis of two heteromorphic life history stages of *Saccharina japonica* (Laminariales) from Japan. *J. Appl. Phycol.* **2020**, *32*, 4175–4187. [CrossRef]
18. Ralph, P.J.; Gademann, R. Rapid light curves: A powerful tool to assess photosynthetic activity. *Aquat. Bot.* **2005**, *82*, 222–237. [CrossRef]
19. Bhagooli, R.; Mattan-Moorgawa, S.; Kaullysing, D.; Louis, Y.D.; Gopeechund, A.; Ramah, S.; Soondur, M.; Pilly, S.S.; Beesoo, R.; Wijayanti, D.P.; et al. Chlorophyll fluorescence—A tool to assess photosynthetic performance and stress photophysiology in symbiotic marine invertebrates and seaplants. *Mar. Pollut. Bull.* **2021**, *165*, 112059. [CrossRef] [PubMed]
20. Zou, D.; Xia, J.; Yang, Y. Photosynthetic use of exogenous inorganic carbon in the agarophyte Gracilaria lemaneiformis (Rhodophyta). *Aquaculture* **2004**, *237*, 421–431. [CrossRef]
21. Abdala-Díaz, R.T.; Cabello-Pasini, A.; Pérez-Rodríguez, E.; Álvarez, R.M.C.; Figueroa, F.L. Daily and seasonal variations of optimum quantum yield and phenolic compounds in *Cystoseira tamariscifolia* (Phaeophyta). *Mar. Biol.* **2005**, *148*, 459–465. [CrossRef]
22. Pang, S.J.; Zhang, Z.H.; Zhao, H.J.; Sun, J.Z. Cultivation of the brown alga *Hizikia fusiformis* (Harvey) Okamura: Stress resistance of artificially raised young seedlings revealed by chlorophyll fluorescence measurement. *J. Appl. Phycol.* **2007**, *19*, 557–565. [CrossRef]
23. Zou, D.; Gao, K. Photosynthetic acclimation to different light levels in the brown marine macroalga, *Hizikia fusiformis* (Sargassaceae, Phaeophyta). *J. Appl. Phycol.* **2009**, *22*, 395–404. [CrossRef]
24. Wang, Y.; Xu, D.; Fan, X.; Zhang, X.; Ye, N.; Wang, W.; Mao, Y.; Mou, S.; Cao, S. Variation of photosynthetic performance, nutrient uptake, and elemental composition of different generations and different thallus parts of *Saccharina japonica*. *J. Appl. Phycol.* **2012**, *25*, 631–637. [CrossRef]
25. Zou, D.; Gao, K. Temperature response of photosynthetic light- and carbon-use characteristics in the red seaweed *Gracilariopsis lemaneiformis* (Gracilariales, Rhodophyta). *J. Phycol.* **2014**, *50*, 366–375. [CrossRef] [PubMed]
26. Wang, L.; Zhang, X.; Zou, D.; Chen, W.; Jiang, H. Growth and Fv/Fm in embryos of *Hizikia fusiformis* (Harvey) Okamura (Sargassaceae, Phaeophyta) cultured under different temperature and irradiance conditions. *J. Oceanol. Limnol.* **2018**, *36*, 1798–1805. [CrossRef]
27. Ji, Z.; Zou, D.; Gong, J.; Liu, C.; Ye, C.; Chen, Y. The different responses of growth and photosynthesis to NH4(+) enrichments between *Gracilariopsis lemaneiformis* and its epiphytic alga Ulva lactuca grown at elevated atmospheric CO_2. *Mar. Pollut. Bull.* **2019**, *144*, 173–180. [CrossRef] [PubMed]
28. Jiang, H.; Zou, D.; Lou, W.; Chen, W.; Yang, Y. Growth and photosynthesis by *Gracilariopsis lemaneiformis* (Gracilariales, Rhodophyta) in response to different stocking densities along Nan'ao Island coastal waters. *Aquaculture* **2019**, *501*, 279–284. [CrossRef]
29. Kokubu, S.; Nishihara, G.N.; Watanabe, Y.; Tsuchiya, Y.; Amamo, Y.; Terada, R. The effect of irradiance and temperature on the photosynthesis of a native alga *Sargassum fusiforme* (Fucales) from Kagoshima, Japan. *Phycologia* **2019**, *54*, 235–247. [CrossRef]
30. Van der Loos, L.M.; Schmid, M.; Leal, P.P.; McGraw, C.M.; Britton, D.; Revill, A.T.; Virtue, P.; Nichols, P.D.; Hurd, C.L. Responses of macroalgae to CO_2 enrichment cannot be inferred solely from their inorganic carbon uptake strategy. *Ecol. Evol.* **2019**, *9*, 125–140. [CrossRef]
31. Zou, D.; Ji, Z.; Chen, W.; Li, G. High temperature stress might hamper the success of sexual reproduction in *Hizikia fusiformis* from Shantou, China: A photosynthetic perspective. *Phycologia* **2019**, *57*, 394–400. [CrossRef]
32. Lanctôt, C.M.; Bednarz, V.N.; Melvin, S.; Jacob, H.; Oberhaensli, F.; Swarzenski, P.W.; Ferrier-Pagès, C.; Carroll, A.R.; Metian, M. Physiological stress response of the scleractinian coral *Stylophora pistillata* exposed to polyethylene microplastics. *Environ. Pollut.* **2020**, *263 Pt A*, 114559. [CrossRef] [PubMed]

33. Zhu, Q.; Wu, L.; Li, X.; Li, G.; Li, J.; Li, C.; Zhao, C.; Wang, F.; Du, C.; Deng, C.; et al. Effects of ambient temperature on the redistribution efficiency of nutrients by desert cyanobacteria—*Scytonema javanicum*. *Sci. Total Environ.* **2020**, *737*, 139733. [CrossRef] [PubMed]
34. Wen, J.; Zou, D. Interactive effects of increasing atmospheric CO_2 and copper exposure on the growth and photosynthesis in the young sporophytes of *Sargassum fusiforme* (Phaeophyta). *Chemosphere* **2021**, *269*, 129397. [CrossRef] [PubMed]
35. The State Oceanic Administration of China. *The Specification for Marine Monitoring*; Ocean Press: Beijing, China, 2007.
36. Lideman; Nishihara, G.N.; Noro, T.; Terada, R. Effect of temperature and light on the photosynthesis as measured by chlorophyll fluorescence of cultured *Eucheuma denticulatum* and *Kappaphycus* sp. (Sumba strain) from Indonesia. *J. Appl. Phycol.* **2012**, *25*, 399–406. [CrossRef]
37. Zhang, J.; Wu, W.; Ren, J.S.; Lin, F. A model for the growth of mariculture kelp Saccharina japonica in Sanggou Bay, China. *Aquac. Environ. Interact.* **2016**, *8*, 273–283. [CrossRef]
38. Hu, Z.; Shan, T.; Zhang, J.; Zhang, Q.; Critchley, A.T.; Choi, H.; Yotsukura, N.; Liu, F.; Duan, D. Kelp aquaculture in China: A retrospective and future prospects. *Rev. Aquac.* **2021**, *13*, 1324–1351. [CrossRef]
39. Yu, J.; Kuang, L.; Wang, Q.; Liu, Y.; Gong, Q.; Li, J. Differences in photosynthesis, growth and resource accumulation between drifting alga Sargassum houneri and cultured alga Undaria pinnatifida and their roles in interspecies competition. *Period. Ocean Univ. China* **2021**, *51*, 1–10.
40. Hong, M.; Ma, Z.; Wang, X.; Shen, Y.; Mo, Z.; Wu, M.; Chen, B.; Zhang, T. Effects of light intensity and ammonium stress on photosynthesis in *Sargassum fusiforme* seedlings. *Chemosphere* **2021**, *273*, 128605. [CrossRef] [PubMed]
41. Xing, Y.; Zeng, J.; Wu, X.; Yang, S.; Huang, M.; Tang, X. Cultivation Status and Application Prospect of Three Topical Economic Seaweeds. *Trans. Oceanol. Limnol.* **2019**, *6*, 1–9. [CrossRef]
42. Fieler, R.; Greenacre, M.; Matsson, S.; Neves, L.; Forbord, S.; Hancke, K. Erosion Dynamics of Cultivated Kelp, Saccharina latissima, and Implications for Environmental Management and Carbon Sequestration. *Front. Mar. Sci.* **2021**, *8*, 632725. [CrossRef]
43. Yang, Y.; Chai, Z.; Wang, Q.; Chen, W.; He, Z.; Jiang, S. Cultivation of seaweed Gracilaria in Chinese coastal waters and its contribution to environmental improvements. *Algal Res.* **2015**, *9*, 236–244. [CrossRef]
44. Wu, H.; Jiang, H.; Liu, C.; Deng, Y. Growth, pigment composition, chlorophyll fluorescence and antioxidant defenses in the red alga *Gracilaria lemaneiformis* (Gracilariales, Rhodophyta) under light stress. *S. Afr. J. Bot.* **2015**, *100*, 27–32. [CrossRef]
45. Wu, H. Study on the Responses of the Growth and Photosynthetic Functions of Marine Macroalgae to Diverse Light Environments. Ph.D. Thesis, South China University of Technology, Guangzhou, China, 2016.
46. Cheng, X.; Zhang, S.; Lin, J.; Wang, Z.; Zhao, X.; Huang, L.; Wang, K. Response of photosynthetic activity to different temperature and light lntensities in *Saccharina japonica* sporophyte. *J. Fish. China* **2020**, *44*, 11.
47. Takabayashi, A.; Kishine, M.; Asada, K.; Endo, T.; Sato, F. Differential use of two cyclic electron flows around photosystem I for driving CO_2-concentration mechanism in C4 photosynthesis. *Proc. Natl. Acad. Sci. USA* **2005**, *102*, 16898–16903. [CrossRef] [PubMed]
48. Zou, D.; Gao, K. Photosynthetic characteristics of the economic brown seaweed *Hizikia fusiforme* (Sargassaceae, Phaeophyta), with special reference to its "leaf" and receptacle. *J. Appl. Phycol.* **2005**, *17*, 255–259. [CrossRef]
49. Liu, C.; Zou, D.; Yang, Y.; Chen, B.; Jiang, H. Temperature responses of pigment contents, chlorophyll fluorescence characteristics, and antioxidant defenses in *Gracilariopsis lemaneiformis* (Gracilariales, Rhodophyta) under different CO_2 levels. *J. Appl. Phycol.* **2016**, *29*, 983–991. [CrossRef]
50. Hwang, E.K.; Ha, D.S.; Park, C.S. The influences of temperature and irradiance on thallus length of Saccharina japonica (Phaeophyta) during the early stages of cultivation. *J. Appl. Phycol.* **2018**, *30*, 2875–2882. [CrossRef]
51. Grebe, G.S.; Byron, C.J.; Gelais, A.S.; Kotowicz, D.M.; Olson, T.K. An ecosystem approach to kelp aquaculture in the Americas and Europe. *Aquac. Rep.* **2019**, *15*, 100215. [CrossRef]
52. Zhang, S.; Liu, S.; Zhou, X.; Wang, Z.; Wang, K. Ecological function of seaweed-formed habitat and discussion of its application to sea ranching. *J. Fish. China* **2019**, *43*, 11. [CrossRef]

biology

MDPI

Article

Implications of Environmental Variations on *Saccharina japonica* Cultivation in Xiangshan Bay, China

Yikang Bao and Peng Xu *

College of Oceanography and Ecological Science, Shanghai Ocean University, Shanghai 201306, China; ykb8196@163.com
* Correspondence: p-xu@shou.edu.cn

Simple Summary: Climate change has a significant negative impact on human economic activities, including seaweed mariculture. Kelp cultivation plays an important role in global mariculture due to its significant economic value and important ecological services. Currently, kelp is widely cultivated in China's coastal seas, and the annual yield has increased by approximately 4–5% over the past decade. However, the desired outcomes may not be achieved by blindly increasing the kelp cultivation scale and the cultivation density in one kelp farm, without scientific evaluations of the relationship between kelp cultivation and its environmental variations. Therefore, this study took Xiangshan Bay as an example to elucidate the variation characteristics of the physicochemical environment and its effects on the growth suitability of *Saccharina japonica*. We further explored the key limiting factors restricting the enhancement of kelp cultivation efficiency in Xiangshan Bay. We systematically studied the self-limiting effect in kelp cultivation. These efforts are intended to provide scientific references for improving the cultivation performance.

Abstract: This study took Xiangshan Bay as an example to illustrate the variation characteristics of the physicochemical environments (temperature, salinity, light, nutrients, and currents) during one kelp cultivation cycle. The study was conducted from November 2020 to May 2021 through tracking down observations. Furthermore, the environmental factors were evaluated using suitability functions of kelp growth, aiming to provide references for promoting kelp cultivation in South China. We discussed the self-limiting effect of nutrients in the culture zone. The results showed that the average temperature, salinity, and light intensity during the cruises in Xiangshan Bay kelp farm were characterized by seasonal variations. Temperature was found to be the most critical environmental factor in determining the kelp cultivation window and hence the yield in Xiangshan Bay. The dissolved inorganic nitrogen (DIN) concentrations initially decreased and then increased, while the dissolved inorganic phosphorus (DIP) concentrations remained decreasing along with the kelp cultivation. The surface tide currents were dramatically attenuated by the suspended kelp cultivation, while the quasi-steady circulations which played a key role in nutrient supplementation for kelp cultivation were not weakened by the kelp cultivation. Among the cruises, the suitability indices' ranges for temperature, salinity, light, and nutrients were 0.02–0.94, 0.96–0.99, 0.97–1, 0.96–0.97 (DIN), and 0.92–0.95 (DIP), respectively. The results of the suitability functions demonstrated that the salinity and light conditions in Xiangshan Bay were very suitable for kelp cultivation and would not cause significant cultivation risks. Although the cultivated kelp could greatly absorb nutrients, the suitability index of nutrients remained adequate even during the late stage of cultivation, indicating the present kelp cultivation scale has not reached the carrying capacity of Xiangshan Bay and there is still much potential for development. To this end, further selective breeding of the thermal tolerance variety has become the key to improving the kelp cultivation performance in Xiangshan Bay. Meanwhile, the self-limiting effects in relation to nutrients are not significant in the Xiangshan Bay kelp farm, but it might be more significant in other kelp farms.

Keywords: suspended kelp cultivation; Xiangshan Bay; mariculture; environments; macroalgae

1. Introduction

Changes in environmental characteristics, primarily temperature, affect many physiological processes in different organisms [1,2]. Over time, humans have learned to use the implications of environmental changes in their activities. Macroalgae cultivation holds a prominent position in global mariculture due to its significant economic values and important ecological services [3]. The global yield of cultivated macroalgae reached over 35.1 million tons of wet weight in 2022 [4]. China is the largest producer of cultivated macroalgae, among which the annual yield of kelp was about 1.73 million tons with economic values of about ten billion RMB [5]. Kelp cultivation not only benefits the marine economy but also helps to improve the marine environments such as by mitigating eutrophication and acidification in coastal seas [6,7]. Meanwhile, kelp cultivation contributes to the blue carbon sinks and hence would help China to achieve carbon neutrality [8]. As a consequence, kelp has been widely cultivated in North and South China coastal seas, and the annual yield has increased by approximately 4–5% over the past decade [9].

The rapid development of kelp cultivation calls for scientific guidance, which should be based on a thorough understanding of the relationship between kelp cultivation and the environment. Kelp species cultivated in China is mainly *Saccharina japonica* (Aresch.) C.E. Lane, C. Mayes, Druehl & G.W. Saunders [10] and cold-water [11], and the cultivation period is roughly from the winter to the beginning of the following summer [12]. Kelp seedlings of 2–3 cm need to be temporarily cultured to approximately 20 cm in a horizontal hanging or vertical hanging manner, and then, the kelps are taken for cultivation in the suspended kelp farm [13]. Some researchers argued that the physicochemical factors in the farm could exert significant influences on the kelp (*Saccharina japonica*) growth and hence the yield based on rough observation [11,14,15]. Particularly, they found that a good harvest could be achieved where water exchange was less suppressed by the kelp cultivation [15]. Some researchers obtained quantitative relationships between the kelp growth and the physicochemical factors through laboratory experiments [16–21]. Temperature, salinity, light, nutrients, and carbon dioxide were the main physicochemical factors studied by the researchers in the laboratory [20,22–27]. The suitable water temperature of *Saccharina japonica* ranged from 0.5 °C to 20 °C, and higher or lower temperatures prevented its growth [11,14]. When the salinity was lower than 3 psu, kelp exhibited negative phosphorus uptake [28]. The most suitable salinity for kelp growth was about 29–32 psu [22]. The highest tolerable salinity for kelp was mostly considered to be 40 psu [28]. Excessive and insufficient light intensity can both inhibit the photosynthesis and growth of *Saccharina japonica* [14]. In realistic kelp farms, carbon dioxide would not usually be regarded as a limiting environmental factor to kelp growth since kelps are cultivated near the air–sea interface [29,30]. On the contrary, ocean currents which are not usually considered in the laboratory play an important role in realistic kelp farms due to their transport effects of nutrients [31]. Strong ocean currents directly cause the shedding of kelp withered edges and more brittle parts [32], while some studies have suggested that it is hard to determine the direct impact of currents on the growth rate of macroalgae such as kelp [33]. Nowadays, it is believed that the effects of currents on kelp growth are reflected in indirect mechanisms, such as mixing and material transport in the water column. These hydrodynamic processes can lead to alterations in nutrient concentrations, which may either promote or inhibit algal growth [34]. Therefore, temperature, salinity, light, nutrients, and currents were the main physicochemical factors explored in in-situ studies in kelp farms.

Some scholars summarized some physicochemical factors' functions applicable to kelp growth, which could accurately reflect the quantitative relationship between different ranges of physicochemical factors and kelp growth rate [16–19]. These functions (growth rate curves) could represent the degree of suitability of different ranges of physicochemical factors for kelp growth. Hence, these functions are referred to as suitability functions in the paper. The suitability functions of kelp were integrated into a kelp growth model, which played a fundamental role in modeling studies about kelp cultivation [25,32,35]. However, it should be pointed out that these suitability functions only took the regulation effects of the physicochemical environments on kelp growth into account while ignoring the regulation effects of kelp cultivation on the environments. The regulation effects of kelp cultivation on the physicochemical environments are mainly manifested in two aspects: (1) The cultivated kelp would strongly absorb nutrients and hence decrease the nutrient concentrations [36]; (2) The kelp cultivation would dramatically weaken the hydrodynamic processes dominated by tides in the farm [37]. Some researchers explored assessing the carrying capacity of Sanggou Bay by constructing physical–biological coupling models [38,39], and there were significant self-limiting effects of kelp cultivation in the coupling models, meaning the physicochemical environments regulated by kelp cultivation would limit the performance of kelp cultivation in turn. The research of self-limiting effects during kelp cultivation should not be neglected, especially when scaling up to larger-scale kelp cultivation. In summary, the desired outcomes may not be achieved by blindly increasing the kelp cultivation scale or density in one kelp farm, without scientific evaluations of the effect of physicochemical environment variations on kelp cultivation.

The mature kelp size of the Xiangshan Bay kelp farm is much smaller than that in typical kelp farms of the north [40,41]. It may be attributed to the limitation of some key physicochemical factors. Therefore, this study takes Xiangshan Bay as an example to elucidate the variation characteristics of the physicochemical environment and its effects on the growth suitability of *Saccharina japonica*. Moreover, we explored the key limiting factors restricting the enhancement of kelp cultivation efficiency in Xiangshan Bay and further systematically studied the self-limiting effect in kelp cultivation, to provide scientific references for improving the cultivation performance.

2. Materials and Methods

2.1. Studies Area

Xiangshan Bay (29°23′–29°49′ N, 121°25′–122°00′ E) is located in the southeast of Ningbo City, Zhejiang Province, with a kelp farm of about 3 km^2 (Figure 1a). The floating kelp rafts in Xiangshan Bay consist of floating ropes and seedling ropes (Figure 1c). The floating ropes in Xiangshan Bay are aligned with currents and provide buoyance so that the rafts can be suspended on the sea surface. The seedling ropes perpendicular to the currents are used to hang kelp seedlings (Figure 1c,d). The cultivated kelp species is *Saccharina japonica*, which grows on the surface of Xiangshan Bay. The base of the kelp blades (growing point) is at the water depth of about 0.5 m, and the blades grow downward. The kelp farm in Xiangshan Bay is 4–10 m deep. The hydrodynamics environments in Xiangshan Bay are characterized by strong semi-diurnal tides [42]. Winds and waves are weak in Xiangshan Bay since it is surrounded by hills [43]. The eutrophication in Xiangshan Bay is significant since lots of terrigenous agriculture and industry wastewater flows into the bay during the rainy season [44,45]. The coastlands in South China are dominated by hilly terrain, so there are many small dams to conserve fresh water. A small dam which was built at Xiangshan Bay head (Figure 1a) would release terrigenous fresh water into the bay from time to time when the water level is high, particularly during rainy seasons. The rainy seasons of Xiangshan Bay are from March or April to September [46], and the rest of the time excluding the rainy seasons during one year are dry seasons.

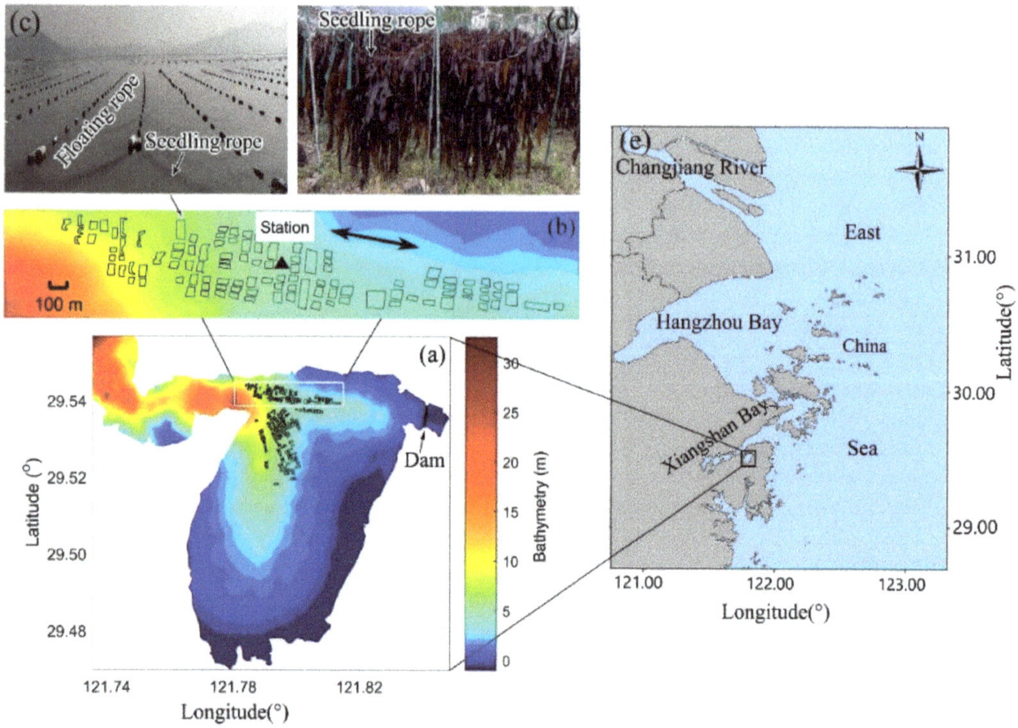

Figure 1. Observation station in Xiangshan Bay. The small black rectangles in (**a**,**b**) represent the suspended kelp farm in Xiangshan Bay. (**b**) is a zoomed-in representation of the white box in (**a**). The observation station is marked with a black triangle in (**b**). The detailed structures of the square suspended rafts are presented in (**c**) consisting of floating and seedling ropes. The seedling ropes in (**c**) are already mounted with kelp seedlings. The harvested kelps are presented in (**d**). Colors in (**a**,**b**) represent bathymetry. The bidirectional arrow in (**b**) represents the flooding and ebbing currents' directions, respectively. The location of Xiangshan Bay in East China Sea is marked in (**e**).

2.2. Data Collection and Processing

The tracking down observations were conducted at a station located in the center of the northern part of Xiangshan Bay kelp farm, where the currents are rectilinear (Figure 1). The tracking down observations consisted of five cruises (Table 1), which covered one complete cultivation cycle. The first and last cruises (202011 and 202105 in Table 1) were conducted before the kelps were planted and after the kelps were harvested, respectively. The remaining three cruises (202101, 202102, and 202104 in Table 1) were conducted during the early, middle and late stages of the kelp cultivation period, respectively. Cruises 202011, 202101 and 202102 were conducted in the dry seasons at Xiangshan Bay, while cruises 202104 and 202105 were conducted in the rainy seasons. During each cruise, we obtained four to six kelps from the same location in the Xiangshan Bay kelp cultivation area, and laid them flat on one laboratory table to measure the size of each kelp blade by using a straightedge (with an accuracy of 1 mm). The average plant size of kelp was 137.4 × 37.1 cm at the late stage (Table 1). Each of the five cruises lasted over seven days which covered spring and neap tides. The observation contents of each cruise were similar and included two main parts. One was continuously sampling throughout each cruise by automatic instruments including an Alec electromagnetic current meter (ECM), which collected the surface currents and temperature every half an hour, and a RDI 1200 kHz Workhorse acoustic Doppler current profiler (ADCP) which could provide bottom currents and tide elevation data every half an hour; the other was spring-tide-day (over 24 h) and neap-tide-day (over 24 h) observations during each cruise including manually launching a SBE 19plus conductivity–temperature–depth sensor (CTD) which collected salinity, temperature and light intensity profiles, manually sampling surface water. Surface water samples were manually collected at the tracking down observations station. During each cruise, approximately 50 surface water samples were manually collected at the depth of about 0.5 m. The water samples (100 mL) were added to about 1 mL of Trichloromethane in the field and placed in an ice box, and then, they were analyzed by a Dechem-Tech CleverChem 200 auto discrete analyzer in the laboratory to obtain the DIN (NO_3^-, NO_2^-, NH_4^+) and DIP concentrations. The molybdenum blue method was used to measure DIP, while NO_3^- and NO_2^- were measured using the cadmium reduction method and diazotization, respectively. NH_4^+ was determined using the indophenol blue method. All measurements were conducted in triplicate, and blanks were analyzed to account for any contamination. The detection limits for DIN and DIP were 0.05 μmol/L and 0.01 μmol/L, respectively. Sample collection and on-site handling, as well as analytical methods, were conducted in accordance with the 'Oceanographic Survey Standards' (GB 17378–2007) [47]. The analysis of data was performed using the software of MATLAB R2018b, EXCEL 2016, and SPSS 24. The means ± standard deviation were derived from the data, and correlations between growth parameters and physicochemical factors were determined with Spearman's correlation test. Unless stated otherwise, significance refers to $p \leq 0.05$.

Table 1. Cruise periods and observed data.

Cruise Periods	Referred as in This Paper (yyyymm)	Observed Data					
		Temperature	Salinity	Light	Nutrients	Currents	Kelp Size: Length × Width (cm)
9~16 November 2020	202011	√	√	√	—	√	Before planting
7~15 January 2021	202101	√	√	√	√	√	(24.2 ± 5.9) × (2.8 ± 0.63)
21 February~2 March 2021	202102	√	√	√	√	√	(108.8 ± 8.1) × (27.0 ± 3.7)
6~14 April 2021	202104	√	√	√	√	√	(137.4 ± 15.5) × (37.1 ± 3.2)
27 May~5 June 2021	202105	√	√	√	—	√	After harvest

The average values of the temperature, salinity, nutrient concentrations, and currents were employed to represent the status of the cruise. The monthly average light intensity was calculated based on the average light intensity during spring-tide-day and neap-tide-day each of cruise, along with the number of sunny, cloudy and rainy days during the month The Eulerian averaging method was employed to filter the periodic tide currents to obtain the quasi-steady residual circulation which was responsible for nutrient transport [48]. The averaging time window was 12 tide periods during each cruise. It should be noted that the ECM was not deployed during 202105, and the surface temperature and currents data were provided by CTD and ADCP, respectively.

2.3. Suitability Assessment Method

The average values of the physicochemical environmental factors collected from each cruise could be substituted into the suitability functions to obtain the suitability indices, which could assess the suitability of the environment for kelp growth [49]. The suitability indices could quantitatively assess whether these physicochemical conditions were favorable or limiting for kelp growth during various cultivation stages. This assessment method enables us to determine the key physicochemical factors that limit kelp cultivation efficiency, as well as the impact of their drastic variations on the suitability of kelp growth. The more the results approached 1, the more suitable the physicochemical conditions were for the kelp growth, and vice versa. The suitability function of temperature is as follows [16]:

$$f(T) = \exp\left[-2.3 \times \left(\frac{T-T_{opt}}{T_x-T_{opt}}\right)^2\right] \quad (1)$$

where T represents sea surface temperature, T_{opt} and T_x are the optimum temperature for kelp growth and the minimum or maximum temperature threshold for kelp growth, respectively. T_{opt} is 10 °C [35], and T_x is 0.5 °C or 20 °C [11,35]. When $T \leq T_{opt}$, Tx = 0.5 °C. When $T > T_{opt}$, T_x = 20 °C. According to the expression for salinity limitation proposed by Martins et al. in the modeling of seaweed, a macroalgae [17], the salinity suitability function for *Saccharina japonica* is adopted when the salinity of seawater is greater than five:

$$f(S) = 1 - \left(\frac{S - S_{opt}}{S_x - S_{opt}}\right)^m \quad (2)$$

where S represents sea surface salinity, S_{opt} and S_x are the optimum salinity for *Saccharina japonica* growth and the minimum and maximum salinity threshold for *Saccharina japonica* growth, respectively. S_{opt} is 30 [22]. S_x is 3 or 40 [28]. When $S < S_{opt}$, S_x = 3, m = 2.5. When $S \geq S_{opt}$, S_x = 40, m = 2. The light suitability function is mostly adopted as Steele's formula (optimal curve of photoinhibition model) [18]:

$$f(I) = \frac{I}{I_{opt}} \times e^{(1-\frac{I}{I_{opt}})} \quad (3)$$

where I and I_{opt} represent the surface light intensity and the optimum light intensity for kelp photosynthesis, respectively. The natural constant e was about 2.72. The optimum light intensity is 180 $\mu mol \cdot m^{-2} \cdot s^{-1}$ [24,35]. Some studies suggested that the absorption characteristics of nitrogen and phosphorus by kelp were in line with the saturation absorption dynamics [19]. Therefore, the Michaelis–Menten equation could be used to analyze the effects of variations in DIN and DIP of the sea surface on the growth suitability of *Saccharina japonica*:

$$f(DIN) = \frac{DIN}{K_{c1} + DIN} \quad (4)$$

$$f(DIP) = \frac{DIP}{K_{c2} + DIP} \quad (5)$$

where DIN and DIP represent the nutrient concentrations. K_{c1} and K_{c2} stand for the half-saturation constant for DIN and DIP, respectively, which were 1.29 µmol/L and 0.13 µmol/L [19].

3. Results

3.1. Characteristics of Prominent Physicochemical Factor Variations

3.1.1. Temperature

The average sea surface temperature initially dropped, then increased with every next cruise (Figure 2). The lowest average surface temperature was 8.40 °C, which was collected during 202101 (deep winter). The average surface temperature during all cruises exceeded the minimum temperature threshold for kelp growth. The average surface temperature of cruises 202011 (late autumn) and 202105 (late spring) were 19.83 °C and

22.86 °C, respectively. The average surface temperature during 202011 and 202105 were, respectively, close and higher than the maximum temperature threshold, determining the time window for kelp cultivation in Xiangshan Bay.

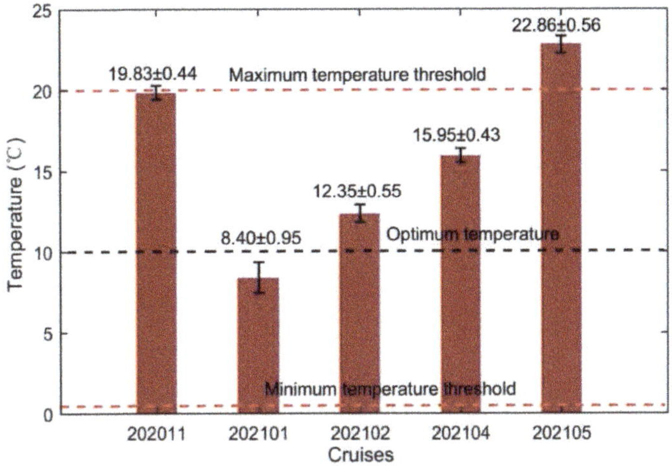

Figure 2. The average sea surface temperature at the observation station during each cruise. The numbers marked above bars are average temperature ± standard deviation. The red dashed lines mark the maximum and minimum temperature threshold for kelp growth, respectively. The black dashed lines mark the optimum temperature for kelp growth.

The standard deviation, representing temperature fluctuations during each cruise, was smaller than the average temperature by one to two orders. Therefore, the most prominent variation in the surface temperature in Xiangshan Bay kelp farm was seasonal variation. The standard deviations of sea surface temperature were mainly contributed by temperature fluctuations at weather events, tide, and daily timescales (Figure 3). A cold wave event during 202101 caused a considerable drop of sea surface temperature by 3 °C in 2 days (Figure 3b). Additionally, the transition from continuous sunny days to continuous cloudy and rainy days during early 202102 caused the sea surface temperature to decrease by about 1.8 °C (Figure 3c). Surface temperature not only fluctuated at tide timescale as bottom temperature, but was also regulated by the processes on the air–sea interface, such as weather events as mentioned above. During continuous cloudy and rainy days as in late 202102 (Figure 3c), no significant deviation was identified between the surface and bottom temperature. In contrast, during continuous sunny days as in late 202101 (Figure 3b), deviations between the surface and bottom temperature varied as day and night, indicating there were fluctuations at the daily timescale in surface temperature.

3.1.2. Salinity

The average sea surface salinity initially increased and then decreased along with the cruises (Figure 4). The highest average surface salinity, which was 25.84, was recorded during 202102, while the two low peaks of the average surface salinity observed during 202011 and 202105 were 22.50 and 24.10, respectively. Across all cruises, the average surface salinity were slightly lower than the optimum salinity for *Saccharina japonica* growth but much higher than the minimum salinity threshold. The standard deviation, representing the salinity fluctuation, was about two orders smaller than the average salinity during each cruise. Therefore, the most prominent variation in the average surface salinity in Xiangshan Bay kelp farm was seasonal variation which was related to rainy seasons and dry seasons.

The salinity continuously increased during dry-season cruises (202011, 202101, 202102) and continuously decreased during rainy-season cruises (202104 and 202105).

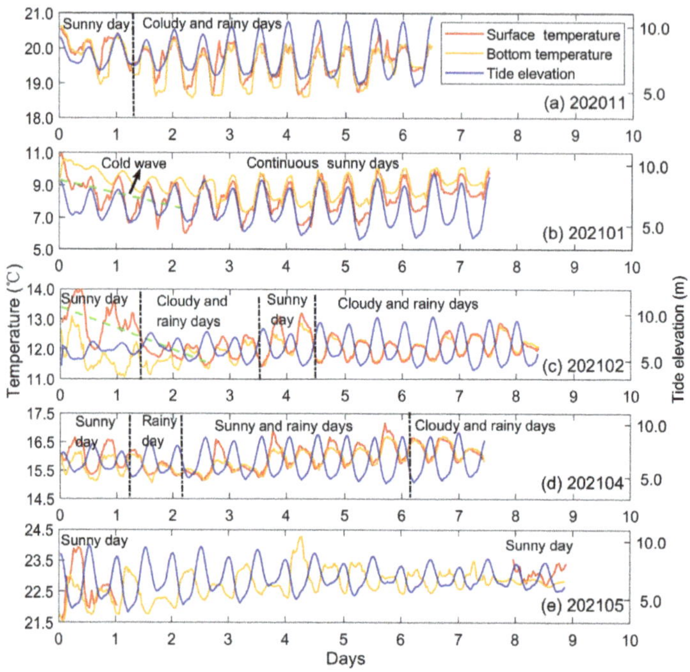

Figure 3. The time series of sea surface temperature (red lines), bottom temperature (yellow lines), and tide elevation (blue lines) during each cruise. The dashed green lines in (**b**,**c**) represent the variation trends of sea surface temperature during that time. The weather conditions are illustrated by words above the time series.

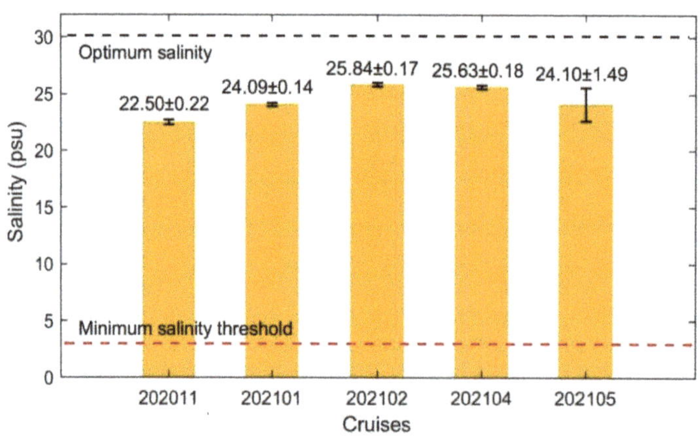

Figure 4. The average sea surface salinity at the observation station during each cruise. The numbers marked above bars are average salinity ± standard deviation. The black and red dashed lines mark the optimum salinity and minimum salinity threshold for kelp growth, respectively.

The standard deviations of the surface salinity, which were generally larger during rainy seasons than during dry seasons (Figure 4). The standard deviations or the salinity fluctuations of one cruise revealed the along-bay salinity gradient. Therefore, the along-bay salinity gradients in Xiangshan Bay kelp farm were stronger during rainy seasons than during dry seasons, and could reverse directions under some conditions such as during 202011. The fluctuations of the surface salinity during the 202105 neap tides were particularly larger than those during the 202105 spring tides. Considering there was no heavy rain during 202105, the dramatic drop of the surface salinity during 202105 neap tides was attributed to the release of terrestrial water from the dam at the bay head (Figure 1), giving rise to larger fluctuations (Figures 4 and 5).

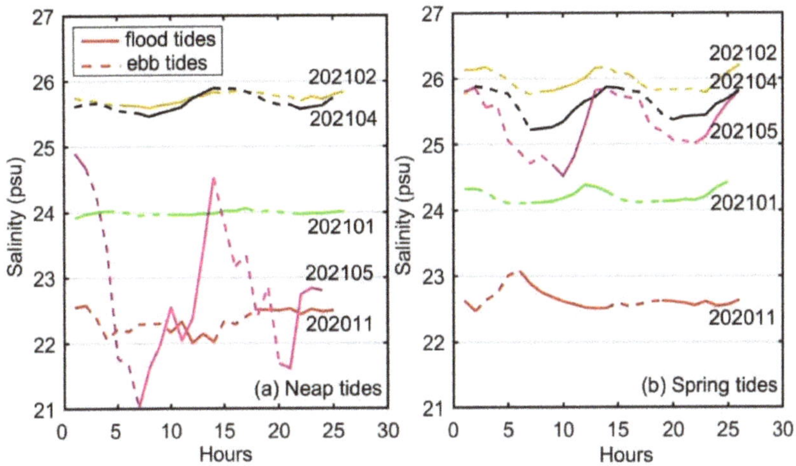

Figure 5. The time series of sea surface salinity during neap-tide-day and spring-tide-day. Different colored lines were used to distinguish various cruises. The solid and dashed lines represent the flood and ebb tides, respectively.

3.1.3. Light

The average surface light intensity during 202011, 202101, and 202102 which were conducted during dry season (when the ratio of cloudy days in the month ranged between 29 and 54% according to our observations) presented a trend of initially decreasing and then increasing along with the cruises (Figure 6). The weakest surface light intensity was found during 202101 with a value of 173.12 $\mu mol \cdot m^{-2} \cdot s^{-1}$. The surface light intensity during 202011 and 202102 was stronger, with values of 178.79 and 187.18 $\mu mol \cdot m^{-2} \cdot s^{-1}$, respectively. When the rainy season came, the ratio of cloudy days increased. The surface light intensity during 202104 was strongest among the cruises, reaching about 225.86 $\mu mol \cdot m^{-2} \cdot s^{-1}$. The surface light intensity during 202105 was a little lower than 202104 since the ratio of cloudy days became higher.

3.1.4. Nutrients

The average DIN concentrations initially decreased and then increased during the kelp farming period (Figure 7). The highest average DIN concentration was 42.36 μmol/L collected during 202101, which then decreased to 31.81 μmol/L during 202102 and finally increased to 38.19 μmol/L during 202104, which was still lower than that during 202101. Unlike DIN, the average concentrations of DIP remained decreasing throughout the kelp farming period, with values of 2.51 μmol/L, 1.71 μmol/L, and 1.37 μmol during 202101, 202102, and 202104, respectively. The different variation trends between DIN and DIP during the second half-cultivation period might be attributed to the terrigenous agriculture and industry wastewater loaded into the bay as characterized by a high N/P ratio [50]. The standard deviations of the DIN and DIP concentrations during each cruise accounted for approximately 27.70% and 56.60% of the average concentrations on average (Figure 7), respectively. The standard deviations of DIN and DIP concentrations could reveal the along-bay gradient of nutrient concentrations under the advection of tide currents as salinity. Therefore, the considerable standard deviations of DIN and DIP concentrations revealed the great absorption effects of nutrients by the cultivated kelp.

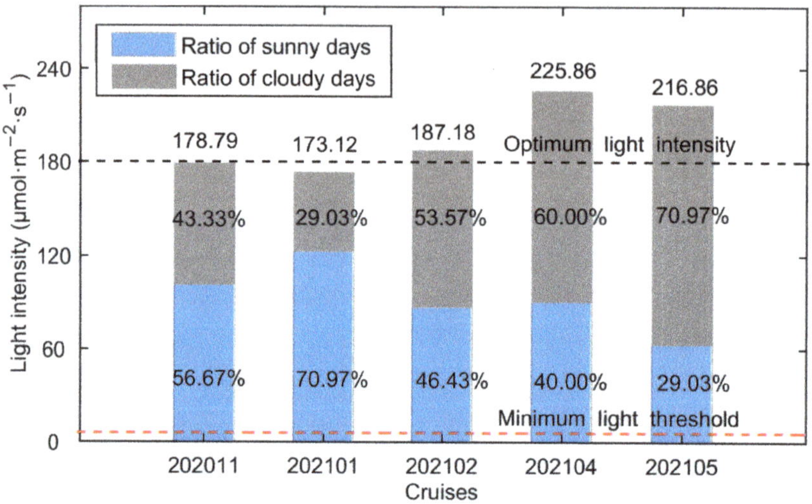

Figure 6. The average light intensity at the observation during each cruise. The numbers marked above bars are the average light intensity. The gray and blue parts in a bar represent the time ratios of sunny days, and cloudy and rainy days in the cruise month, which are marked by the percent numbers. The black and red dashed lines represent the optimum light intensity and minimum light threshold (6.21 $\mu mol \cdot m^{-2} \cdot s^{-1}$) [14] for kelp growth, respectively.

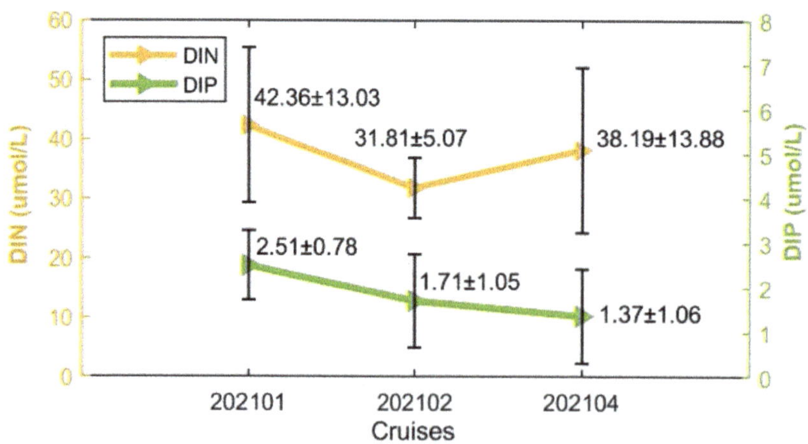

Figure 7. The concentrations of DIN and DIP during cruises when kelps were cultivated. The numbers marked close to the triangles are average nutrient concentrations ± standard deviation.

3.1.5. Currents

The average magnitude of surface tide currents for each cruise in Xiangshan Bay kelp farm is shown in Figure 8. The average magnitude of the surface tide currents was 0.27 m/s during 202011 without the suspended kelp cultivation. When the kelp seedlings were planted in 202101, the average magnitude of surface tide currents decreased by about 22.23% to 0.21 m/s. When the kelps grew to about mature size in 202102 and 202104, the average magnitudes of surface tide currents further decreased by about 44.44% and 51.86% compared to 202011, to the values of 0.15 m/s and 0.13 m/s, respectively. Meanwhile, the standard deviation of surface tidal currents decreased along with cruises (Figure 8), indicating the damping effect of tide waves caused by the suspended kelp cultivation.

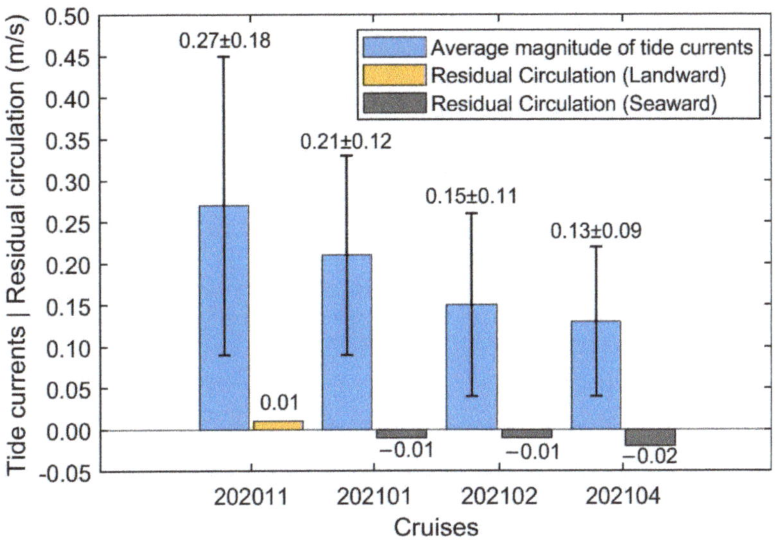

Figure 8. The average magnitudes of the surface tide currents (blue bars) and the surface circulations (yellow and gray bars). The numbers marked above blue bars are the average ± standard deviation of magnitudes of the surface tidal currents. The numbers marked above yellow and gray bars represent the circulation during each cruise. Note the positive circulation is landward while the negative ones are seaward.

The surface circulations after filtering periodic tide currents based on the Eulerian averaging method are presented in Figure 8. The surface circulations in 202011, 202101, 202102 and 202104 were about 0.01 m/s, −0.01 m/s, −0.01 m/s and −0.02 m/s, respectively. The surface circulation during 202105 was not calculated since ADCP fell over 1.5 days after the deployment. It should be noted that the surface circulation reversed during the tracking down observations. The surface circulation was landward (positive) in 202011, while it was seaward (negative) in 202101, 202102 and 202104. The turning of the circulation directions from 202011 to 202101 might be caused by the reverse of the along-bay salinity gradient which drove the gravity circulation. Among the cruises, the surface circulation was strongest during 202104 in the rainy season and the *Saccharina japonica* grew to its mature size.

As shown in Table 2, the length of kelp exhibited significant positive correlations with surface temperature and salinity, while it showed significant negative correlations with surface DIP, surface DIN, and surface currents. Similarly, the width of kelp was significantly positively correlated with surface temperature and salinity and significantly negatively correlated with surface DIP, while the correlations with surface light intensity and surface current were not significant. These results suggested that surface temperature, salinity, DIN, DIP, and currents in Xiangshan Bay had significant impacts on kelp growth. It was worth noting that the lack of significant correlation between surface light intensity and kelp growth parameters was mainly due to the nonlinear effect of light intensity on kelp growth.

Table 2. Spearman correlation coefficient of kelp size and physicochemical factors.

Parameters	Surface Temperature	Surface Salinity	Surface Light	Surface DIN	Surface DIP	Surface Currents
Kelp length	0.629 *	0.902 **	0.343	−0.629 *	−0.930 **	−0.853 **
Kelp width	0.888 **	0.594 *	0.455	−0.14	−0.671 *	−0.531

Note: **, * represent the significance levels of $p \leq 0.01$ and $p \leq 0.05$, respectively.

3.2. Suitability of Physicochemical Factors for *Saccharina japonica* Growth in Different Cultivation Periods

The average surface temperature, salinity, light, DIN and DIP concentrations were substituted into their suitability functions (Equations (1)–(5)). The results showed that the temperature suitability indices during different periods ranged from 0.02 to 0.94 (Figure 9). The suitability indices of the temperature during 202011 and 202105 were as low as 0.11 and 0.02, respectively. Among the three cruises when kelps were cultivated, the suitability index of 202101 was highest with a value of 0.94. Since the standard deviation of temperature was much smaller than the average temperature (Figure 2), the fluctuations of sea surface temperature would not cause considerable drops of the suitability index of temperature during most of the farming period. However, the cold wave during 202101 which caused a drop of the sea surface temperature reduced the suitability index of temperature by about 0.31. The suitability indices of the surface salinity during all cruises varied in the range of 0.96–0.99 (Figure 10). The standard deviations of surface salinity were so small compared to the average salinity that they would not make great drops in the suitability index of salinity (Figure 4). Although the release of terrestrial water from the dam caused a drop of surface salinity by 2.50, the suitability index of salinity only decreased by 0.02. The obtained suitability indices of light intensity varied in the range of 0.97–1.00 (Figure 11). The suitability indices of light intensity during 202011, 202101 and 202102 were all about 1.00. The suitability indices of 202104 and 202105 were decreased to 0.97 and 0.98, respectively. The suitability indices of DIN of each cruise decreased and then increased, ranging from 0.96 to 0.97, while the suitability indices of DIP continued to decrease, ranging from 0.92 to 0.95 (Figure 12).

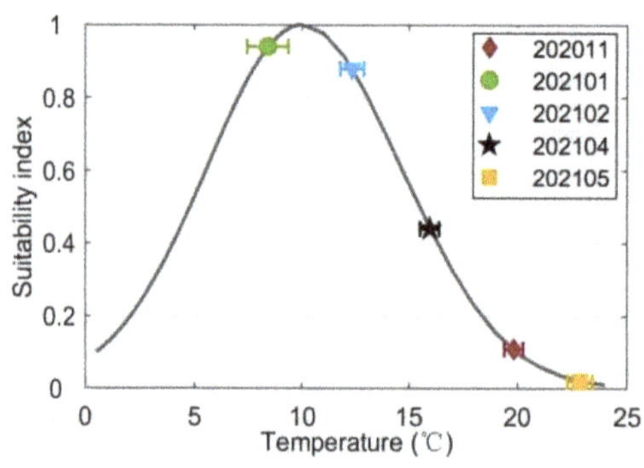

Figure 9. The suitability indices of the average surface temperature collected during each cruise. The horizontal bars represent the standard deviations of the temperature during each cruise.

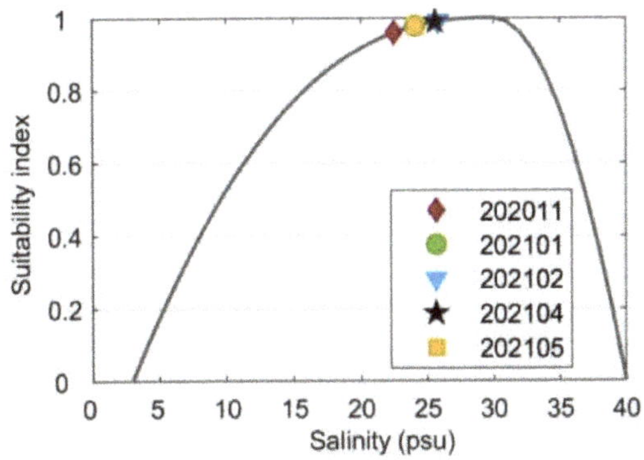

Figure 10. The suitability indices of the average surface salinity collected during each cruise. Note the standard deviations of the surface salinity during each cruise are not plotted in the figure since they are about two orders smaller than the average values.

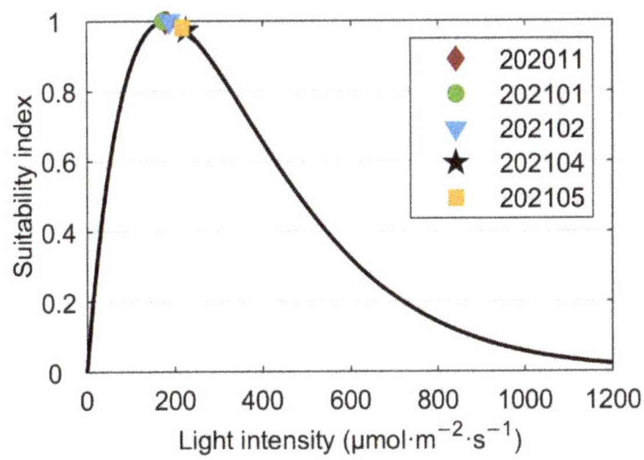

Figure 11. The suitability indices of the average surface light intensity collected during each cruise. Note the standard deviations of the surface light intensity are not calculated as illustrated in the texts.

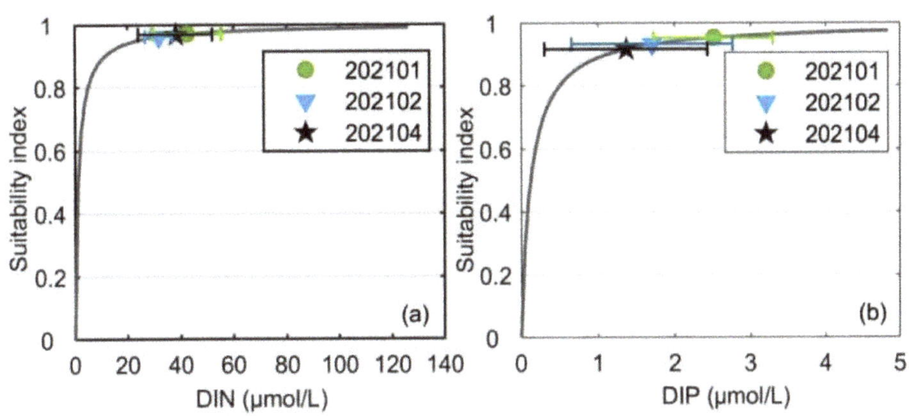

Figure 12. The suitability indices of the DIN (**a**) and DIP (**b**) concentrations during cruises when kelps were cultivated. The horizontal bars represent the standard deviations of the DIN and DIP concentrations during each cruise.

4. Discussion

4.1. Implications of the Variation in Prominent Physicochemical Factors for Saccharina japonica Cultivation

The lowest suitability index of the average surface temperature among all cruises was 0.02 (Figure 9), while the suitability indices of other factors during all cruises exceeded 0.92, which showed that temperature was the most critical environmental factor that determined the kelp cultivation time window in Xiangshan Bay. The cultivation time window revealed by the observed temperature in this study was consistent with the kelp cultivation practice in Xiangshan Bay (personal communication with local kelp farmers). The cultivation time window of Sanggou Bay was about eight months [32], when the climate was colder. The time window length in Xiangshan Bay was about three months shorter than that in Sanggou Bay. Therefore, it might be the reason why the kelp harvested in Xiangshan Bay (137.4×37.1 cm, Table 1) was much smaller than that in Sanggou Bay (313.0×45.0 cm) [41]. Therefore, it is critical to further explore selective breeding of the thermal tolerance variety to increase the kelp yield in South China such as in Xiangshan Bay. Although the size of mature kelp (118.41 ± 16.57 cm) measured by Yang [40] in the same region was similar to our measurement result, our sample size of kelp individuals in each cruise was relatively small. Future studies could increase the sample size to provide more robust results. Some severe weather events during the early and late stages of the farming period when the suitability indices of temperature were inherently low might cause considerable cultivation risks. For example, the cold wave during 202101 caused the suitability index of sea surface temperature to be reduced by about 0.31. In addition, cold waves during the early stage of cultivation and heat waves during the late stage should also be investigated [51,52]. Especially when the water temperature exceeds the maximum physiological threshold of kelp due to marine heat waves, the mortality rate of kelp increases significantly [53]. Therefore, it is necessary to develop high-precision climate models to predict the severe weather events to avoid kelp cultivation risks in South China, and more attention should be paid to the timely harvesting of kelp according to temperature variations during the late stage of kelp cultivation.

The decline of the average DIN concentrations from 202101 to 202102 might be mainly due to the strong absorption of nutrients by the cultivated kelp [36]. The concentrations of DIN and DIP in Xiangshan Bay were affected greatly by terrigenous runoff [44], and the increase in average DIN concentrations from 202102 to 202104 was due to the transition from dry to rainy seasons, during which the terrigenous nutrients loaded into the kelp farm overwhelmed the nutrients absorbed by the cultivated kelp. Unlike DIN, the average DIP concentrations remained decreasing throughout the kelp farming period. In addition to the influence of the high N/P ratio of terrestrial water [50], both phytoplankton and kelp have a high demand for phosphorus in spring, which could cause the DIP concentration to decrease significantly [54]. Therefore, the variation in average DIP concentration in kelp farm might be attributed to the greater consumption by kelp and phytoplankton than supplementation. However, the suitability indices of DIN and DIP during all cruises exceeded 0.92 (Figure 12), even during the late stage of the kelp cultivation, indicating the nutrient conditions were suitable for kelp and would not become a limiting environmental factor at the present scale. To this end, the current kelp cultivation scale did not reach the carrying capacity of Xiangshan Bay, and there is still much potential for development.

The along-bay salinity gradients reversed in direction during 202011, and the reverse of the along-bay salinity gradient was not very unusual in the bay junction as pointed out by Warner et al. [55]. Although there were some variations in the average sea surface salinity between the cruises, the suitability indices of average surface salinity data during all cruises exceeded 0.96. The standard deviations of surface salinity and the dramatic fluctuations during cruise 202105 would not cause great cultivation risks. The average surface light intensity suitability indices were high for all cruises, with the lowest still very close to 1.00. Moreover, the standard deviation of the light intensity was not calculated since the average light intensity was far more important for kelp growth than the fluctuations [56]. In summary, the salinity and light intensity in Xiangshan Bay was very suitable for kelp growth and would not limit the extension of the time window for kelp cultivation. It should be noted that the current research was based on a complete cultivation cycle, which might limit the robustness and generalizability of our findings. Future studies could consider repeating the experiments across more cultivation cycles to further elucidate the effect of environmental conditions on the growth of *Saccharina japonica*.

4.2. Assessment of Self-Limiting Effects of Nutrients

The regulation effects of ocean currents on kelp cultivation were usually indirect except when the ocean currents were extremely strong to tear apart kelp blades, such as under typhoon weather [32]. There were few strong winds and waves in Xiangshan Bay [43], so the direct loss of kelp cultivation caused by ocean currents ripping off kelp blades was less. The indirect regulation effects of ocean currents took place through influencing other environmental factors such as nutrients [57] and further caused the self-limiting effects of kelp cultivation. The self-limiting effects due to nutrients could be caused in two ways: (1) the nutrient concentrations became too low to support the proper growth of the abundant kelps during the late stage of the cultivation since the kelps absorbed many nutrients as they grew; (2) the nutrient supplementation from the outer sea to the kelp farm decreased as a result of the weakening circulation caused by the high-density suspended kelp cultivation [39], in turn restricting the further growth of kelp [46].

The ocean currents in coastal seas where kelps were planted usually consisted of periodic tide currents and the quasi-steady circulations. Previous studies observed that the self-limiting effect of kelp cultivation due to nutrients was caused by weakening the circulation and decreasing the related nutrient supplementation in Sanggou Bay [49]. The water exchange was relatively weak due to the obstruction of cultivation facilities and macroalgae in Sanggou Bay kelp farm, and the nutrients were absorbed and could not be replenished in time, so the *Saccharina japonica* yield was significantly affected [58]. Although the surface tide currents in Xiangshan Bay kelp farm were significantly attenuated along with the kelp growth as in Sanggou Bay, the nutrient concentration during the late stage of kelp farming period in Xiangshan Bay was not significantly affected, and the nutrient suitability index remained high. On the one hand, it could be attributed to the fact that the terrigenous water which was rich in nutrients would enhance the nutrient supplements to Xiangshan Bay kelp farm during the rainy season [45]. On the other hand, the offshore residual current had a vital transport effect on environmental factors such as playing a key role in the offshore transportation of nutrients [59], which was crucial for nutrient supplements in the kelp farm. The surface circulation in Xiangshan Bay kelp farm was not weakened by the suspended kelp cultivation. Therefore, the self-limiting effects of the kelp cultivation, which were due to nutrients through weakening circulation and the related nutrient supplements in Xiangshan Bay kelp farm, were not significant.

5. Conclusions

The variation characteristics of the five physicochemical environmental factors were analyzed based on the tracking down observations in Xiangshan Bay kelp farm. We found that temperature was the most critical environmental factor that determined the kelp cultivation time window in Xiangshan Bay. The key to increasing the kelp yield in Xiangshan Bay was further selective breeding of the thermal tolerance variety. In addition, the remarkable temperature fluctuations caused by severe weather events during the early and late stages of cultivation would cause great drops in the suitability indices and hence cause cultivation risks. Therefore, developing high-precision climate models to predict the severe weather events is essential to avoid kelp cultivation risks caused by remarkable temperature fluctuations in South China, and more attention should be paid to the timely harvesting of kelp according to temperature variations during the late stage of kelp cultivation. The salinity and light conditions in Xiangshan Bay were very suitable for kelp cultivation. The average salinity and light intensity mainly varied at the season timescale, and their conditions were very suitable for kelp cultivation. Therefore, the salinity and light also did not limit the expansion of the present kelp cultivation time window. The suitability indices of nutrients remained adequate even during the late stage of the kelp cultivation, indicating the self-limiting effects subject to nutrients was not significant in Xiangshan Bay kelp farm. To this end, the present scale of kelp cultivation in Xiangshan Bay did not reach the carrying capacity of the bay, and there is still much potential for development.

The surface tide currents in Xiangshan Bay kelp farm were dramatically attenuated by the suspended kelp cultivation. However, the surface circulation in Xiangshan Bay kelp farm was not suppressed as the kelp growth. The self-limiting effect subject to nutrient supplemented by circulation and the terrigenous water was not significant in Xiangshan Bay.

Author Contributions: Conceptualization, Y.B. and P.X.; Methodology, Y.B.; Software, Y.B.; Validation, Y.B. and P.X.; Formal Analysis, Y.B. and P.X.; Investigation, Y.B. and P.X.; Resources, P.X.; Data Curation, Y.B.; Writing—Original draft, Y.B.; Writing—Review and Editing, Y.B. and P.X.; Visualization, Y.B.; Supervision, P.X.; Project Administration, P.X.; Funding Acquisition, P.X. All authors have read and agreed to the published version of the manuscript.

Funding: This work was supported by [Youth Foundation of Natural Science Foundation of China] with grant number [42106002].

Institutional Review Board Statement: Not applicable.

Informed Consent Statement: Not applicable.

Data Availability Statement: The original contributions presented in the study are included in the article, and further inquiries can be directed to the corresponding author.

Conflicts of Interest: The authors declare no conflicts of interest.

References

1. Amin, M.R.; Zhang, J.; Yang, M. Effects of climate change on the yield and cropping area of major food crops: A case of Bangladesh. *Sustainability* **2014**, *7*, 898–915. [CrossRef]
2. Andreychev, A.V. Daily and seasonal feeding activity of the greater mole-rat (*Spalax microphthalmus*, Rodentia, Spalacidae). *Biol. Bull.* **2019**, *46*, 1172–1181. [CrossRef]
3. Moreira, A.; Cruz, S.; Marques, R.; Cartaxana, P. The underexplored potential of green macroalgae in aquaculture. *Rev. Aquac.* **2022**, *14*, 5–26. [CrossRef]
4. Farghali, M.; Mohamed, I.M.; Osman, A.I.; Rooney, D.W. Seaweed for climate mitigation, wastewater treatment, bioenergy, bioplastic, biochar, food, pharmaceuticals, and cosmetics: A review. *Environ. Chem. Lett.* **2023**, *21*, 97–152. [CrossRef]
5. Fisheries and Fisheries Administration Bureau of Ministry of Agriculture and Rural Affairs, National Fisheries Technology Extension Center. *China Fisheries Statistical Yearbook*; China Agriculture Press: Beijing, China, 2021.
6. Xu, S.; Yu, Z.; Zhou, Y.; Yue, S.; Liang, J.; Zhang, X. The potential for large-scale kelp aquaculture to counteract marine eutrophication by nutrient removal. *Mar. Pollut. Bull.* **2023**, *187*, 114513. [CrossRef] [PubMed]
7. Xiao, X.; Agustí, S.; Yu, Y.; Huang, Y.; Chen, W.; Hu, J.; Li, C.; Li, K.; Wei, F.; Lu, Y.; et al. Seaweed farms provide refugia from ocean acidification. *Sci. Total Environ.* **2021**, *776*, 145192. [CrossRef] [PubMed]
8. Fang, J.; Wang, F.; Strand, Ø.; Liu, D. The ecological function of mariculture. *Front. Mar. Sci.* **2022**, *9*, 1070683. [CrossRef]

9. Hu, Z.M.; Shan, T.F.; Zhang, Q.S.; Liu, F.L.; Jueterbock, A.; Wang, G.; Sun, Z.M.; Wang, X.Y.; Chen, W.Z.; Critchley, A.T.; et al. Kelp breeding in China: Challenges and opportunities for solutions. *Rev. Aquac.* **2024**, *16*, 855–871. [CrossRef]
10. Li, X.; Zhang, Z.; Qu, S.; Liang, G.; Zhao, N.; Sun, J.; Song, S.; Cao, Z.; Li, X.; Pan, J.; et al. Breeding of an intraspecific kelp hybrid Dongfang no. 6 (*Saccharina japonica*, Phaeophyceae, Laminariales) for suitable processing products and evaluation of its culture performance. *J. Appl. Phycol.* **2016**, *28*, 439–447. [CrossRef]
11. Tseng, C.K.; Wu, C.Y. *Manual of Haidai (Laminaria japonica Aesech) Cultivation*; Science Press: Beijing, China, 1962; pp. 1–71.
12. Li, L.; Xu, J.; Kong, G.; Li, P.; Ren, Y.; Wang, H. Hydrodynamics in the tidal flat in semi-enclosed Xiangshan Bay. *Front. Mar. Sci.* **2023**, *10*, 1073254. [CrossRef]
13. Li, X.D.; Su, L.; Li, X.J.; Li, J.; Xu, Y.J.; Chang, L.R.; Yu, R.C.; Yang, D.Z.; Feng, S.J. Comprehensiveanalysis of large-scale *saccharina japonica* damage in the principal farming area ofrongcheng in Shandong province from 2021 to 2022. *J. Agric. Sci. Technol.* **2023**, *25*, 206–222.
14. Zhang, Q.X. Relationship between kelp growth and light. *China Fish.* **1994**, *6*, 34–35.
15. Zhang, D.M.; Miao, G.R.; Yang, Q.M. Research on the relationship between the coastal current and *Laminaria japonica* raising. An experiment of comparision between the current speed and *Laminaria japonica* growth. *J. Shandong Coll. Oceanol.* **1986**, *16*, 180–186.
16. Bowie, G.L.; Mills, W.B.; Porcella, D.B.; Campbell, C.L.; Pagenkopf, J.R.; Rupp, G.L.; Johnson, K.M.; Chan, P.W.; Gherini, S.A.; Chamberlin, C.E. *Rates, Constants, and Kinetics. Formulations in Surface Water Quality Modeling*, 2nd ed.; EPA (Environmental Protection Agency): Washington, DC, USA, 1985; p. 455.
17. Martins, I.; Marques, J.C. A model for the growth of opportunistic macroalgae (*Enteromorpha* sp.) in tidal estuaries. *Estuar. Coast. Shelf Sci.* **2002**, *55*, 247–257. [CrossRef]
18. Steele, J.H. Environmental control of photosynthesis in the sea. *Limnol. Oceanogr.* **1962**, *7*, 137–150. [CrossRef]
19. Kitadai, Y.; Kadowaki, S. The growth process and N, P uptake rates of *Laminaria japonica* cultured in coastal fish farms. *Aquac. Sci.* **2003**, *51*, 15–23.
20. Kirihara, S.; Fujikawa, Y.; Notoya, M. Effect of temperature and day length on the zoosporangial sorus formation and growth of sporophytes *Laminaria japonica* Areschoug (Laminariales, Phaeophyceae) in tank culture. *Suisan Zoshoku* **2003**, *51*, 385–390.
21. Mizuta, H.; Ogawa, S.; Yasui, H. Phosphorus requirement of the sporophyte of *Laminaria japonica* (Phaeophyceae). *Aquat. Bot.* **2003**, *76*, 117–126. [CrossRef]
22. Chen, G.L.; Wang, D.S. Experience in pilot production management of kelp farming. *China Fish.* **1958**, *4*, 10.
23. Chen, D.Y.; Wang, J.X. Observation of the relationship between the growth and water temperature of kelp in southern Zhejiang. *Zhejiang Agric. Sci.* **1964**, *2*, 89–93.
24. Yao, N.Y.; Li, J.Z. A study on the photosynthesis traits of *Laminaria japonica*. *Plant Physiol. Commun.* **1981**, *4*, 18–20.
25. Duarte, P.; Meneses, R.; Hawkins, A.J.S.; Zhu, M.; Fang, J.; Grant, J. Mathematical modelling to assess the carrying capacity for multi-species culture within coastal waters. *Ecol. Model.* **2003**, *168*, 109–143. [CrossRef]
26. Xu, D.; Wang, D.S.; Li, B.; Fan, X.; Zhang, X.W.; Ye, N.H.; Wang, Y.T.; Mou, S.L.; Zhuang, Z.M. Effects of CO_2 and seawater acidification on the early stages of *Saccharina japonica* development. *Environ. Sci. Technol.* **2015**, *49*, 3548–3556. [CrossRef] [PubMed]
27. Zhang, X.S.; Xu, D.; Guan, Z.; Wang, S.H.; Zhang, Y.; Wang, W.; Zhang, X.W.; Fan, X.; Li, F.; Ye, N.H. Elevated CO_2 concentrations promote growth and photosynthesis of the brown alga *Saccharina japonica*. *J. Appl. Phycol.* **2020**, *32*, 1949–1959. [CrossRef]
28. Ji, Z.Q. Study on Nearshore Ecological Remediation of Nitrogen and Phosphorous Pollution and Bioenergy Extraction from Macroalgae. Master's Thesis, Zhejiang University, Hangzhou, China, 2011; pp. 71–73.
29. Beer, S.; Björk, M.; Beardall, J. *Photosynthesis in the Marine Environment*; John Wiley Sons: Hoboken, NJ, USA, 2014.
30. Hill, R.; Bellgrove, A.; Macreadie, P.I.; Petrou, K.; Beardall, J.; Steven, A.; Ralph, P.J. Can macroalgae contribute to blue carbon? An Australian perspective. *Limnol. Oceanogr.* **2015**, *60*, 1689–1706. [CrossRef]
31. Stephens, T.A.; Hepburn, C.D. A kelp with integrity: *Macrocystis pyrifera* prioritises tissue maintenance in response to nitrogen fertilisation. *Oecologia* **2016**, *182*, 71–84. [CrossRef]
32. Cai, B.Y.; Zhu, C.B.; Liu, H.; Chang, L.R.; Xiao, L.Y.; Sun, Q.W.; Lin, F. Model simulated growth of the kelp *Saccharina japonica* in Sanggou Bay. *Prog. Fish. Sci.* **2019**, *40*, 31–41.
33. Kregting, L.T.; Hurd, C.L.; Pilditch, C.A.; Stevens, C.L. The relative importance of water motion on nitrogen uptake by the subtidal macroalga *adamsiella chauvinii* (rhodophyta) in winter and summer 1. *J. Phycol.* **2008**, *44*, 320–330. [CrossRef]
34. Wang, Y.B. Study on mechanism of action of hydrodynamic on algae. *J. Anhui Agric. Sci.* **2013**, *41*, 5000–5001.
35. Wu, R.J.; Zhang, X.L.; Zhu, M.Y.; Zheng, Y.F. A model for the growth of Haidai (*Laminaria japonica*) in aquaculture. *Mar. Sci. Bull.* **2009**, *28*, 34–40.
36. Li, R.; Liu, S.; Zhang, J.; Jiang, Z.; Fang, J. Sources and export of nutrients associated with integrated multi-trophic aquaculture in Sanggou Bay, China. *Aquac. Environ. Interact.* **2016**, *8*, 285–309. [CrossRef]
37. Zeng, D.; Huang, D.; Qiao, X.; He, Y.; Zhang, T. Effect of suspended kelp culture on water exchange as estimated by in situ current measurement in Sanggou Bay, China. *J. Mar. Syst.* **2015**, *149*, 14–24. [CrossRef]

38. Shi, J.; Wei, H.; Zhao, L.; Yuan, Y.; Fang, J.; Zhang, J. A physical–biological coupled aquaculture model for a suspended aquaculture area of China. *Aquaculture* **2011**, *318*, 412–424. [CrossRef]
39. Lin, F.; Du, M.R.; Liu, H.; Fang, J.G.; Asplin, L.; Jiang, Z.J. A physical-biological coupled ecosystem model for integrated aquaculture of bivalve and seaweed in sanggou bay. *Ecol. Model.* **2020**, *431*, 109181.
40. Yang, L. Carbon, Nitrogen and Phosphorus Stoichiometric Ratios of *Laminaria japonica* in Xiangshan Harbor. Master's Thesis, Shanghai Ocean University, Shanghai, China, 2019.
41. Fang, J.H.; Jiang, Z.J.; Lin, F.; Gao, Y.P.; Fang, J.G.; Zhang, P.; Guo, Z.L.; Du, M.R.; Meng, S. Analysis on the advantages of standard kelp long line culture in Sanggou Bay. *Prog. Fish. Sci.* **2020**, *41*, 134–140.
42. Fan, X.; Wei, H.; Yuan, Y.; Zhao, L. Vertical structure of tidal current in a typically coastal raft-culture area. *Cont. Shelf Res.* **2009**, *29*, 2345–2357. [CrossRef]
43. Gao, S.; Xie, Q.C.; Feng, Y. Fine grained sediment transport and sorting by tidal exchange in Xiangshan Bay, Zhejiang, China. *Estuar. Coast. Shelf Sci.* **1990**, *31*, 397–409.
44. Jiang, Z.B.; Chen, Q.Z.; Zeng, J.N.; Liao, Y.B.; Shou, L.; Liu, J. Phytoplankton community distribution in relation to environmental parameters in three aquaculture systems in a Chinese subtropical eutrophic bay. *Mar. Ecol. Prog. Ser.* **2012**, *446*, 73–89. [CrossRef]
45. Ye, R.; Liu, L.; Wang, Q.; Ye, X.; Cao, W.; He, Q.; Cai, Y. Identification of coastal water quality by multivariate statistical techniques in two typical bays of northern Zhejiang Province, East China Sea. *Acta Oceanol. Sin.* **2017**, *36*, 1–10. [CrossRef]
46. Huang, Z.Z.; Lou, F.Q.; Ding, B.L.; Lv, Y.G.; Xu, H.Y.; Wu, Z.M. Spatial variation analysis of precipitation characteristics and statistical parameters in Zhejiang Province. *Water Resour. Prot.* **2022**, *38*, 117–123.
47. State Oceanic Administration. *Marine Monitoring (Seawater Analysis GB 17378-2007)*; Ocean Press: Beijing, China, 2007.
48. Xu, P.; Mao, X.; Jiang, W. Mapping tidal residual circulations in the outer Xiangshan Bay using a numerical model. *J. Mar. Syst.* **2016**, *154*, 181–191. [CrossRef]
49. Shi, J.; Wei, H.; Zhao, L.; Wan, J.; Zhang, J. Study on ecosystem model of multi-species culture in Sanggou Bay: III Numerical study on the kelp culture carrying capacity. *Prog. Fish. Sci.* **2010**, *31*, 43–52.
50. Zhang, J.; Liu, S.M.; Ren, J.L.; Wu, Y.; Zhang, G.L. Nutrient gradients from the eutrophic Changjiang (Yangtze River) Estuary to the oligotrophic Kuroshio waters and re-evaluation of budgets for the East China Sea Shelf. *Prog. Oceanogr.* **2007**, *74*, 449–478. [CrossRef]
51. Nepper-Davidsen, J.; Andersen, D.T.; Pedersen, M.F. Exposure to simulated heatwave scenarios causes long-term reductions in performance in *Saccharina Latissima*. *Mar. Ecol. Prog. Ser.* **2019**, *630*, 25–39. [CrossRef]
52. Jiang, M.; Gao, L.; Huang, R.; Lin, X.; Gao, G. Differential responses of bloom-forming *Ulva intestinalis* and economically important *Gracilariopsis lemaneiformis* to marine heatwaves under changing nitrate conditions. *Sci. Total Environ.* **2022**, *840*, 156591. [CrossRef] [PubMed]
53. Wernberg, T. *Marine Heatwave Drives Collapse of Kelp Forests in Western Australia. Ecosystem Collapse and Climate Change*; Springer International Publishing: Cham, Switzerland, 2021; pp. 325–343.
54. Yang, Z.; Ran, L.H.; Xu, X.Q.; Ji, Z.Q.; Zhu, Y.; Chen, Q.N.; Chen, J.F. The behavior of phosphate in the Xiangshan Bay and its potential impact on harmful algal blooms. *Haiyang Xuebao* **2018**, *40*, 61–70.
55. Warner, J.; Schoellhamer, D.; Burau, J.; Schladow, G. Effects of tidal current phase at the junction of two straits. *Cont. Shelf Res.* **2002**, *22*, 1629–1642. [CrossRef]
56. Wing, S.R.; Denny, L.M.W. A dynamic model for wave-induced light fluctuations in a kelp forest. *Limnology and Oceanography* **1993**, *38*, 396–407. [CrossRef]
57. Millar, R.V.; Houghton, J.D.; Elsäβer, B.; Mensink, P.J.; Kregting, L. Influence of waves and currents on the growth rate of the kelp *Laminaria digitata* (Phaeophyceae). *J. Phycol.* **2020**, *56*, 198–207. [CrossRef]
58. Fang, J.G.; Sun, H.L.; Kuang, S.H.; Sun, Y.; Zhou, S.L.; Song, Y.L.; Cui, Y.; Zhao, J.; Yang, Q.; Li, F.; et al. Assessing the carrying capacity of Sanggou Bay for culture of kelp *Laminaria japonica*. *Mar. Fish. Res.* **1996**, *17*, 7–17.
59. Sigaúque, P.J.; Schettini, C.A.; Valentim, S.S.; Siegle, E. The role of tides, river discharge and wind on the residual circulation of Maputo Bay. *Reg. Stud. Mar. Sci.* **2021**, *41*, 101604. [CrossRef]

biology

MDPI

Article

Effects of Ocean Acidification and Temperature Coupling on Photosynthetic Activity and Physiological Properties of *Ulva fasciata* and *Sargassum horneri*

Kai Wang [1,2], Xiang Tao [1], Shouyu Zhang [1,2] and Xu Zhao [1,2,*]

1 College of Oceanography and Ecological Science, Shanghai Ocean University, Shanghai 201306, China; kwang@shou.edu.cn (K.W.); 18957451910@163.com (X.T.); syzhang@shou.edu.cn (S.Z.)
2 Research Center of Marine Ranching, Shanghai Ocean University, Shanghai 201306, China
* Correspondence: xzhao@shou.edu.cn

Simple Summary: Macroalgae in natural marine areas play an important role in mitigating ocean climate change. The complexity of natural conditions also makes it necessary to study macroalgae not only by considering the effects of changes in a single factor but also by exploring the coupled effects of different environmental conditions on macroalgae. Therefore, in this study, two species of macroalgae were used as experimental subjects to observe their growth processes under different co-treatments of temperature and CO_2 concentration. The results of this study can provide a reference for how natural macroalgae can cope with future changes in ocean climate.

Abstract: To investigate the ecological impacts of macroalgae in the framework of shifting global CO_2 concentrations, we conducted a study utilizing *Ulva fasciata* and *Sargassum horneri* specimens sourced from the Ma'an Archipelago in Zhejiang Province on how ocean acidification (OA) and temperature changes interact to affect the photosynthetic physiological responses of macroalgae. The results of the study showed that OA reduced the tolerance of *U. fasciata* to bright light at 20 °C, resulting in more pronounced photoinhibition, while 15 °C caused significant inhibition of *U. fasciata*, reducing its growth and photosynthetic activity, but OA alleviated the inhibition and promoted the growth of the alga to a certain extent. The tolerance of *S. horneri* to bright light was also reduced at 20 °C; the inhibition was relieved at 15 °C, and the OA further improved the algal growth. The Relative Growth Rate (*RGR*), photosynthetic pigment content, and the release of the dissolved organic carbon (DOC) of *U. fasciata* were mainly affected by the change in temperature; the growth of the alga and the synthesis of metabolites were more favored by 20 °C. A similar temperature dependence was observed for *S. horneri*, with faster growth and high metabolism at 15 °C. Our results suggest that OA reduces the tolerance of macroalgae to high light at suitable growth temperatures; however, at unsuitable growth temperatures, OA effectively mitigates this inhibitory effect and promotes algal growth.

Keywords: macroalgae; ocean acidification; greenhouse effect; chlorophyll fluorescence

1. Introduction

In recent years, with rapid economic development, the global environment has undergone swift transformations, particularly with the concerning escalation of CO_2 levels in the atmosphere, posing a critical challenge to the sustainable progress of humanity. Gattuso anticipates a potential surge in atmospheric CO_2 concentrations to 1000 μatm by the century's conclusion [1]. Primarily absorbed by oceans, escalating atmospheric CO_2 quantities will lead to a concurrent uptick in oceanic absorption, catalyzing a pH decline in surface oceanic waters—a phenomenon known as ocean acidification [2]. Moreover, the mounting atmospheric CO_2 load will further intensify global warming, elevating the temperatures of oceanic surface waters [3,4]. Notably, within coastal ecosystems, intertidal macroalgae exhibit heightened sensitivity to fluctuating oceanic CO_2 levels. This heightened sensitivity stems from the fact that macroalgae typically complete their entire life cycle within their habitats, thus rendering them particularly susceptible to enduring environmental vicissitudes in adjacent waters [5,6]. In addition, different species of macroalgae in the intertidal zone often show different responses to changes in CO_2 concentration: no change [7], growth promotion [8], and growth inhibition [9]. The different responses are mainly related to macroalgae photosynthesis, as well as environmental adaptation [6].

Macroalgae thriving in coastal intertidal waters not only yield economic benefits within coastal seafood aquaculture but also serve as efficient absorbers of a myriad of pollutants abundant in elements such as N and P [10,11]. These macroalgae play a critical role in the marine ecosystem's carbon cycle, effectively mitigating eutrophication while contributing positively to the regulation of marine ecological equilibrium and the moderation of atmospheric CO_2 concentration increases. They stand as prime candidates for environmental rehabilitation in near-shore aquaculture marine regions [12]. Their growth and development primarily rely on the absorption and utilization of CO_2 and HCO_3^- from seawater to synthesize organic compounds. Ocean acidification can enhance the photosynthetic and growth rates of macroalgae to a certain extent [13]; however, it may diminish their photosynthetic efficiency when seawater pH falls below a specific threshold [14].

The response of macroalgae to temperature variations has been extensively scrutinized. Some research indicates that temperature impacts the growth, biochemical composition, and physiological processes of macroalgae [15,16]. Furthermore, changes in CO_2 levels within seawater can influence how macroalgae adapt to external temperature shifts [17–19]. Investigations reveal that alterations in CO_2 concentrations prompt biochemical changes in macroalgae, facilitating enhanced adaptation to temperature fluctuations in the external environment [19]. Numerous studies have delved into the photosynthetic physiology of macroalgae concerning temperature and CO_2 alterations, examining facets like photosynthetic rates, pigments, and organic matter metabolism. However, these studies oftentimes isolate either factor individually, neglecting the combined effects of these influential variables. Gordillo proposed that the impact of CO_2 concentrations becomes substantial when interacting with other factors [18]. Hence, it becomes imperative to explore the consequences of temperature and CO_2 concentration fluctuations on macroalgae in a more integrated manner.

Ulva fasciata [11] and *Sargassum horneri* [20] stand out as the predominant macroalgae that naturally thrive in the coastal waters of southeastern China. These nutrient-rich species play pivotal roles across diverse sectors, including food, feed, medicine, and chemicals [17]. Possessing substantial biomass and thriving expansively, these key species not only serve crucial ecological functions as biological habitats but also prove pivotal in the restoration of macroalgae beds, a process of paramount significance for near-shore marine ecological initiatives [17,20]. Nestled in the northeastern region of Shengsi, Zhejiang Province, China, Ma'an Archipelago finds itself enveloped by a multitude of islands and abundant macroalgae. This investigation delves into the variations in photosynthetic activity exhibited by these two macroalgae species amidst the combined impacts of acidification and temperature shifts. Leveraging chlorophyll fluorescence techniques, we scrutinized changes in the biochemical compositions and physiological metabolism of *U. fasciata* and *S. horneri* in the waters surrounding Ma'an Archipelago, Zhejiang Province as the chosen test subjects. It is expected that insights into the effects of acidification and temperature synergism on the photosynthesis and physiological properties of *U. fasciata* and *S. horneri* will provide valuable data for further exploration in this area.

2. Materials and Methods

2.1. Sample Collection and Processing

The samples were collected from the intertidal waters of the Ma'an Archipelago, with *U. fasciata* originating from the coastal rocky reef zone and *S. horneri* from the mussel culture raft area. Subsequently, the collected macroalgae were cleansed in situ using seawater to eliminate any contaminants, such as floating debris or attached organisms. The indoor adaptive cultivation process lasted for 24 h. The acclimation was carried out using sterile seawater with salinity (28 ± 1‰) and temperatures (15 ± 1 °C) consistent with that of in situ seawater. The cultivation environment maintained a light intensity of 100 ± 5 μmol·m^{-2}·s^{-1}, determined through a field light intensity survey and the photosynthetic requirements of the macroalgae, alongside a photoperiod of L:D = 12 h:12 h.

2.2. Experimental Condition

After completing the acclimation phase, compliant samples (2 ± 0.005 g) were chosen from the collected macroalgae for subsequent experimentation. These selected samples were positioned in 2 L round-bottomed flasks, into which 1.2 L of sterile seawater was introduced along with an appropriate quantity of nutrient solution to avert any potential nutrient deficiencies (NH_4: 150 mmol/L, NO_3: 2100 mmol/L, P: 150 mmol/L, K: 900 mmol/L, Ca: 600 mmol/L, Mg: 300 mmol/L, and S: 300 mmol/L, adding about 2 mL). The flasks were then transferred to a CO_2 incubator for the cultivation process. Acidification levels were precisely calibrated based on the present atmospheric CO_2 concentration and Gattuso's projections for CO_2 levels by the century's conclusion [1]: 400 μL·L^{-1} (blank treatment/LC) and 1000 μL·L^{-1} (acidification treatment/HC). Temperature settings were based on the seawater temperature at the time of sample collection (April and May) and the contribution of rising CO_2 concentrations to global warming: 15 °C (natural temperature/LT) and 20 °C (warming temperature/HT). Each of the four cultivation conditions involved three sample groupings to study the combined effects of CO_2 concentration and temperature, while a control group contained sterile seawater without any samples under each cultivation condition. Lighting conditions and photoperiod remained consistent with the acclimatization phase throughout the 7-day cultivation duration. Parameters of algae and seawater were measured during the early (day 1), middle (days 3 and 5), and later (day 7) of cultivation.

2.3. Rapid Light Curve and Fluorescence Induction Parameters

The Rapid Light Curve (*RLC*) of the samples was determined for each cultivation condition 1 h after the start of light exposure at a set number of cultivation days. A chlorophyll fluorometer (WALZ DIVING PAM, Effeltrich, Germany) was used to provide the samples sequentially with a total of 8 gradients of 0–2000 μmol·m^{-2}·s^{-1} of photochemical light, with an interval of 20 s between two photochemical lights. After the determination, the RLC of each sample was derived from the obtained Relative Electron Transfer Rate (*rETR*) fitted by the exponential function formula, with reference to the following exponential function formula:

$$rETR = rETRm(1 - e^{-\alpha \cdot PAR/rETRm})\, e^{-\beta \cdot PAR/rETRm} \quad (1)$$

where PAR is the corresponding light intensity at the time of the measurement of the sample, *rETR*. Other relevant parameters included the maximum relative electron transfer rate (*rETRm*), macroalgae light energy utilization efficiency (α), photoinhibition parameter (β), and half-saturated light intensity (E_k) were obtained with the *RLC*.

The samples were dark-adapted for 1–2 h after the end of light exposure, and the fluorescence induction parameters were determined using a chlorophyll fluorometer. The maximum quantum yield (Fv/Fm) of photosystem II (PSII) was calculated by the following equation:

$$Fv/Fm = (Fm - Fo)/Fm \quad (2)$$

where *Fv* is the dark-adapted variable fluorescence value, *Fm* is the maximum fluorescence value at the dark-adapted saturating light intensity, and *Fo* is the dark-adapted initial fluorescence value. The effective quantum yield (Fv'/Fm') of PSII was calculated by the following equation:

$$Fv'/Fm' = (Fm' - Fo')/Fm' \quad (3)$$

where Fm' is the maximum chlorophyll fluorescence at a preset level of photochemical light, and Fo' is the minimum chlorophyll fluorescence at a preset level of photochemical light.

Photochemical quenching (*qP*) and non-photochemical quenching (*NPQ*) were calculated by the following equations:

$$qP = (Fm' - F)/(Fm' - Fo) \tag{4}$$

$$NPQ = (Fm - Fm')/Fm' \tag{5}$$

2.4. Relative Growth Rate

After the determination of the chlorophyll fluorescence parameters of the samples, the surface of the samples was dried with absorbent paper, the weight (fresh weight, Fw) was determined by using an electronic balance with 3 decimal places, and the Relative Growth Rate (RGR) of the macroalgae was calculated using the following formula:

$$RGR = [(W_t/W_o) - 1]/\mathrm{t} \tag{6}$$

where W_o is the initial weight of the sample (Fw), and W_t is the weight of the sample on day t (Fw).

2.5. Dissolved Organic Carbon Release

After 1 h of light initiation at the set number of cultivation days, the culturing seawater in each group of flasks was measured, and the DOC content of the seawater was determined. Seawater from 50 mL flasks of each group was aspirated and filtered using cauterized GF/F (Whatman, Maidstone, Kent, UK) glass fiber filtration membranes, after which, the resulting filtrate was subjected to DOC concentration determination with a Total Organic Carbon analyzer (Shimadzu TOC-L, Otsu, Japan). Each set of experiments was replicated 3 times.

2.6. Photosynthetic Pigment Content

After the growth parameters were determined, 0.2 ± 0.005 g samples (Fw) were accurately weighed by using an electronic balance with 3 decimal places, cut with scissors, and put into a mortar, adding appropriate amounts of quartz sand with 80% acetone, grinding thoroughly, followed by fixing them to 15 mL and then keeping them at 4 °C for 24 h in the dark; after that, the samples were centrifuged with a freezing centrifuge at 5000 r/min for 10 min at 4 °C. Then, the supernatant was taken, and the absorbance value was determined with a UV spectrophotometer (Mapada UV-3200, Shanghai, China). Each set of experiments was replicated 3 times. The chlorophyll a (Chl-a) and carotenoid (Car) contents were calculated according to Hellebustand Craigie [21]:

$$\text{Chl-a (mg/g)} = (11.85 \times A_{665} - 1.54 \times A_{647} - 0.08 \times A_{639}) \times V/W \tag{7}$$

$$\text{Car (mg/g)} = 7.6 \times (A_{480} - 1.49 \times A_{510}) \times V/W \tag{8}$$

where A denotes the absorbance value of the supernatant at different wavelengths of light, V denotes the volume of the fixed volume, and W denotes the fresh weight of the macroalgae.

2.7. Statistical Analysis

The Excel 2019 software was used for experimental data processing, and Origin 2023 was used for plotting; one-way ANOVA analysis was performed using Tukey's test, and two-way ANOVA analysis was performed to analyze the interaction of different CO_2 concentrations and temperature changes in the macroalgae (Signifcance levels were set at $p < 0.05$). The results of the experimental measurements are expressed as the mean and standard deviation.

3. Results

3.1. Changes in Chlorophyll Fluorescence Parameters

The acidification effect induced by elevated CO_2 concentration on the *RLC* of *U. fasciata* varied significantly at different temperatures. At 15 °C, the effect of the acidification treatment (1000 $\mu L \cdot L^{-1}$) was not significant during the cultivation period ($p > 0.05$), and relative to the blank treatment (400 μL·L^{-1}), the *rETR* of the acidification treatment group showed a decrease in the early stage of cultivation, but in the middle and later stages of cultivation, the CO_2 concentration did not have a significant effect on the *rETR* of *U. fasciata* ($p < 0.05$). The effect of the acidification treatment on *U. fasciata* was more obvious at 20 °C. The *rETR* of the algae decreased with the increase in cultivation time and was lower than that of the blank treatment group at the same cultivation time; moreover, the acidification treatment further aggravated the photoinhibitory effect of the algae, which led to a larger decrease in the *rETR* (Figure 1, left side).

At different temperatures, the acidification effect induced by the elevated CO_2 concentration did not have a significant effect on the *RLC* of *S. horneri* ($p < 0.05$). At 15 °C, the *rETR* of *S. horneri* increased in both groups, but the increase was larger in the acidification treatment group. At 20 °C, elevated CO_2 concentrations inhibited the growth of *S. horneri* and exacerbated the photoinhibition of the alga in low light (Figure 1, right side).

Acidification treatments generally suppressed the maximum relative electron transfer rate (*rETRm*) of *U. fasciata* at different temperatures in the early stage of cultivation, but in the later stage of cultivation, the acidification treatments presented a promotional effect on the *rETRm* of *U. fasciata* at 20 °C; on the other hand, the *rETRms* of the blank treatment group at different temperatures all presented a continuous decreasing trend (Figure 2a). Relative to changes in CO_2 concentrations, the light energy utilization efficiency (α) and half-saturated light intensity (E_k) of *U. fasciata* were mainly affected by changes in temperature (Figure 2b). *U. fasciata* cultured at different CO_2 concentrations at 20 °C had higher α than those cultured at 15 °C during the cultivation period. The acidification treatment had no significant effect on the E_k of *U. fasciata* at different temperatures ($p > 0.05$). The E_k was higher at 15 °C for the early stage of cultivation and at 20 °C for the later stage of cultivation (Figure 2c). Temperature changes had an effect on the trend of the photoinhibition parameter (β) of *U. fasciata*. At 20 °C, the acidification treatment resulted in a sustained increase in β with cultivation time, which was higher than that of the blank treatment group at the same cultivation time; at 15 °C, the CO_2 concentration had no significant effect on β (Figure 2d).

Likewise, the variation trend of *rETRm* in *S. horneri* closely paralleled its *RLC*. In the early stage of cultivation, *rETRm* was not affected by changes in temperature or CO_2 concentration, but in the middle stage of cultivation, the *rETRm* of *S. horneri* at 15 °C decreased and then finally returned to normal levels in the later stage of cultivation (Figure 2e). The α, E_k, and β of *S. horneri*, on the other hand, were mainly affected by temperature changes during the cultivation period ($p < 0.05$), and the effect of CO_2 concentration changes on them was not significant. At 15 °C, the α of *S. horneri* cultured with different CO_2 concentrations showed a significant decreasing trend in the middle and later stages of cultivation, while the α at 20 °C was more stable (Figure 2f). The E_k showed a continuous increase at 15 °C, which was higher than that of *S. horneri* at 20 °C during the same cultivation time throughout the cultivation period, but the E_k was more stable at 20 °C (Figure 2g). The acidification treatment had no significant effect on the β of *S. horneri* ($p > 0.05$). The β of the algae was higher at 15 °C in the early stage of cultivation and at 20 °C in the later stage of cultivation (Figure 2h).

The results show that the maximum quantum yield (Fm/Fv) of *U. fasciata* was more stable in the early and middle stages of cultivation and showed a decreasing trend in the later stage of cultivation under different cultivation conditions, but the statistical analyses showed that the temperature and the CO_2 concentration did not significantly affect the Fm/Fv of *U. fasciata* during the same cultivation time ($p > 0.05$) (Figure 3a). The effective quantum yield (Fv'/Fm') of *U. fasciata* showed a decreasing trend under different cultivation conditions, and the Fv'/Fm' was mainly affected by the temperature throughout the whole cultivation period; on the other hand, the change in the CO_2 concentration did not have a significant effect on it. The Fv'/Fm' at 20 °C was higher than that of the *U. fasciata* at 15 °C during the same cultivation time (Figure 3b). The trend of the photochemical quenching parameter (qP) of *U. fasciata* was similar to that of Fm/Fv, and qP was more stable throughout the cultivation period under different cultivation conditions, with less of an effect from temperature and CO_2 concentration on *qP* (Figure 3c). In addition, the non-photochemical quenching parameter (*NPQ*) of the blank treatment group showed a continuous increase throughout the cultivation period at 15 °C, whereas the changes in *NPQ* were not significant under other cultivation conditions. Temperature changes significantly affected the *NPQ* of *U. fasciata* at different CO_2 concentrations ($p < 0.05$), but the acidification treatment only had a more significant effect on *NPQ* at 15 °C (Figure 3d).

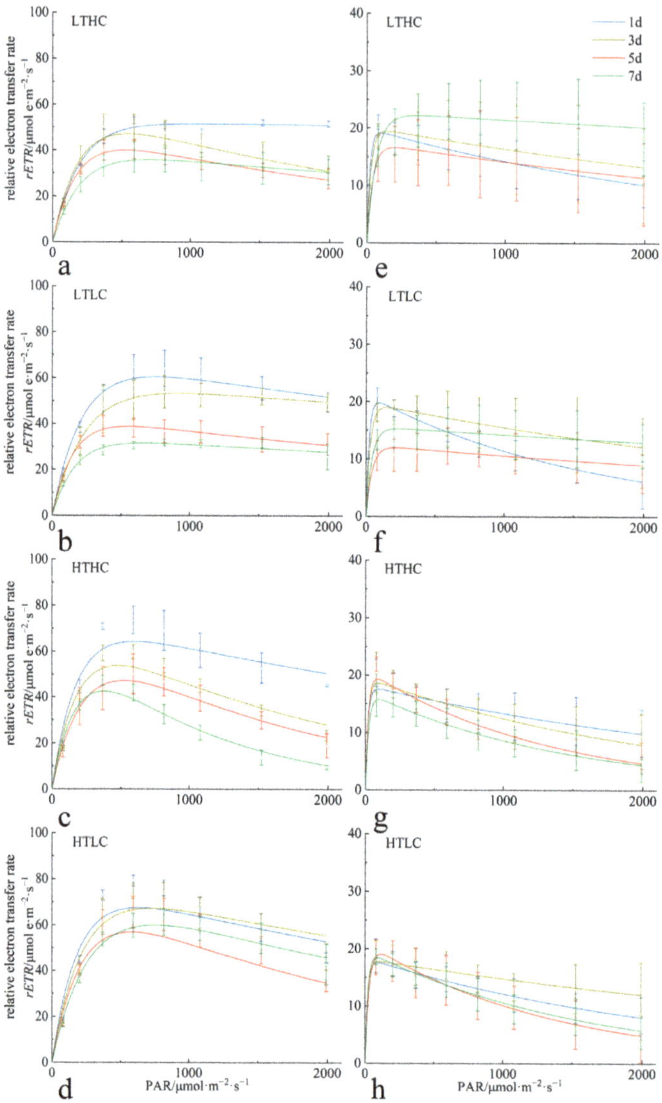

Figure 1. Changes in *RLC* of *U. fasciata* (**a–d**) and *S. horneri* (**e–h**) during cultivation in different conditions. Cultivation conditions for each group: LTLC (15 °C and 1000 $\mu L \cdot L^{-1}$), LTLC (15 °C and 400 $\mu L \cdot L^{-1}$), HTHC (20 °C and 1000 $\mu L \cdot L^{-1}$), and HTLC (20 °C and 400 $\mu L \cdot L^{-1}$). The error line is the SD.

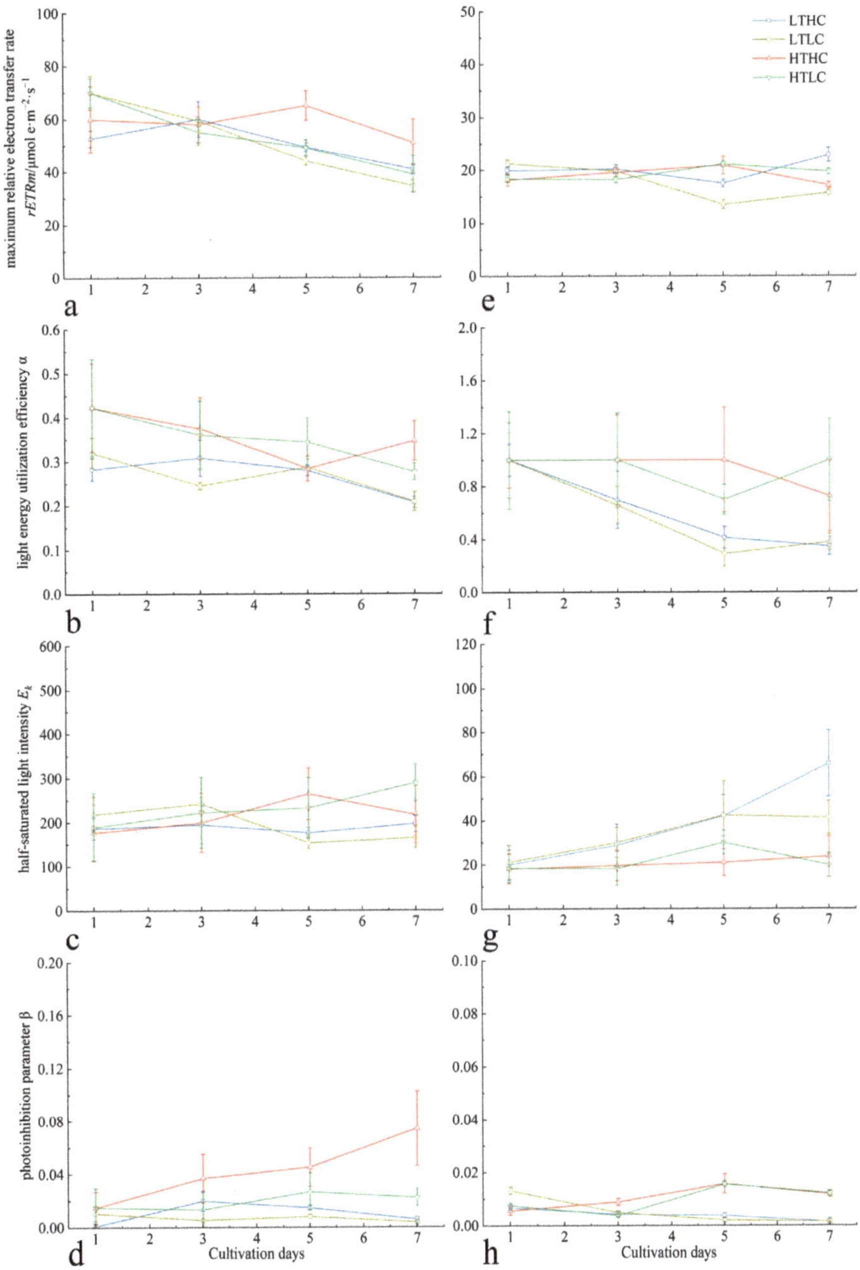

Figure 2. Changes in *RLC*-related parameters of *U. fasciata* (**a–d**) and *S. horneri* (**e–h**) during cultivation under different conditions. Cultivation conditions for each group: LTLC (15 °C and 1000 μL·L^{-1}), LTLC (15 °C and 400 μL·L^{-1}), HTHC (20 °C and 1000 μL·L^{-1}), and HTLC (20 °C and 400 μL·L^{-1}). The error line is the SD.

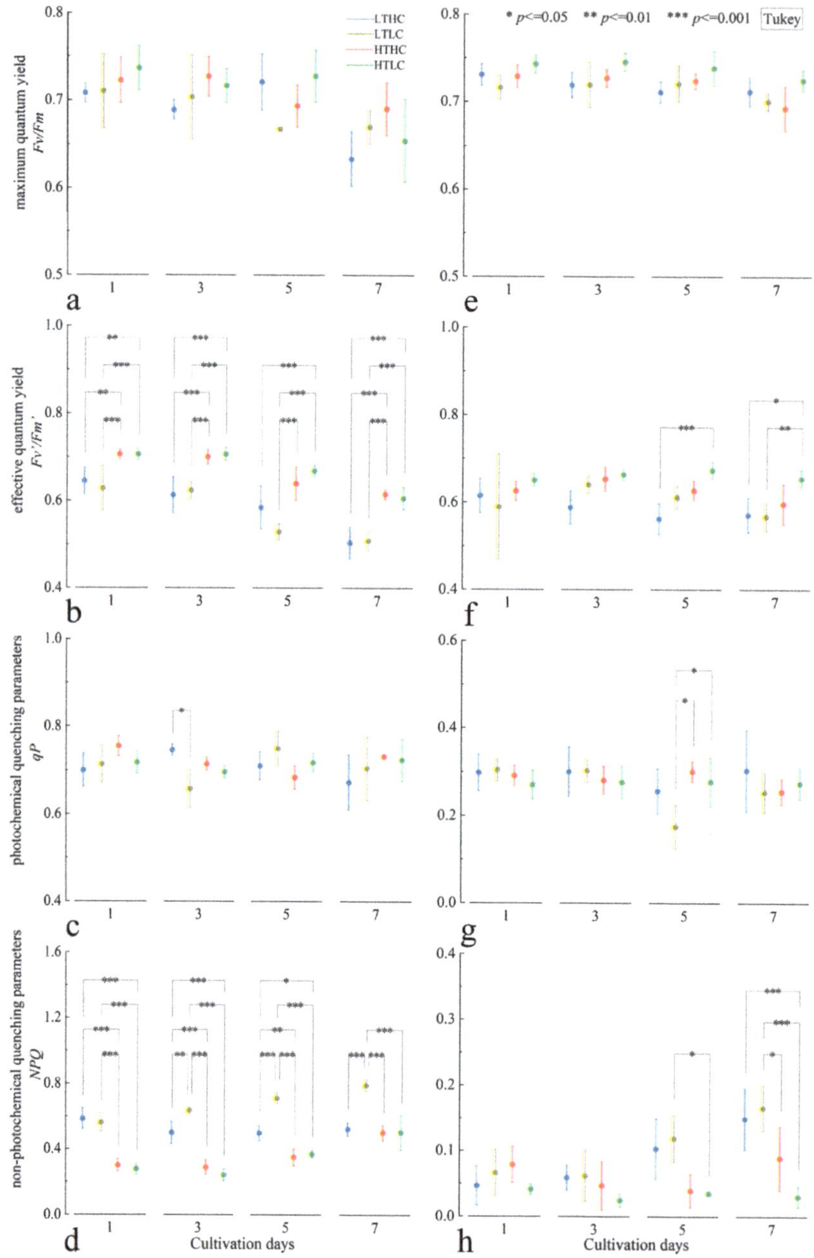

Figure 3. Changes in fluorescence induction parameters of *U. fasciata* (**a–d**) and *S. horneri* (**e–h**). Cultivation conditions for each group: LTLC (15 °C and 1000 μL·L^{-1}), LTLC (15 °C and 400 μL·L^{-1}), HTHC (20 °C and 1000 μL·L^{-1}), and HTLC (20 °C and 400 μL·L^{-1}). The error line is the SD.

Under different cultivation conditions, the Fm/Fv of *S. horneri* was relatively stable throughout the cultivation period, and the effects of temperature and CO_2 concentration on Fm/Fv were not significant (Figure 3e). The Fv'/Fm' of *S. horneri* was relatively stable in the early and middle stages of cultivation and was not affected by the cultivation conditions.

However, as the cultivation progressed, the Fv'/Fm' of the blank treatment group at 20 °C showed a significant difference from that of *S. horneri* at 15 °C ($p < 0.05$), and throughout the whole cultivation period, the Fv'/Fm' of *S. horneri* at 20 °C was higher than that of *S. horneri* at 15 °C in the same cultivation time (Figure 3f). The trend in qP changes in *S. horneri* was similar to that of *U. fasciata*, with qP being more stable throughout the cultivation period under different cultivation conditions, while the effects of temperature and CO_2 concentration on qP were smaller (Figure 3g). The NPQ of *S. horneri* at 15 °C showed a high level and a continuous increase with the cultivation time, while the NPQ at 20 °C was lower and did not change significantly, showing some differences compared to 15 °C in the middle and later stages of cultivation (Figure 3h).

3.2. Changes in RGR and DOC Release

There were some differences in the effect of the acidification treatments on the *RGR* of *U. fasciata* at different cultivation temperatures. At 20 °C, the acidification treatment inhibited the growth of *U. fasciata*. At 15 °C, the acidification treatment promoted growth. In addition, the *RGR* at 20 °C was higher than that at 15 °C during the same cultivation time in both the early and middle stages of cultivation but was slightly lower than that at 15 °C in the later stages of cultivation (Figure 4a). At the end of cultivation, the interaction of different cultivation conditions did not significantly affect the *RGR* of *U. fasciata* ($p > 0.05$). The effect of the acidification treatments on the *RGR* of *S. horneri* was more similar to that of *U. fasciata*. Acidification treatments at 15 °C promoted growth in the early and middle stages of cultivation. The blank treatment group at 20 °C had a higher *RGR* in the early stage of cultivation, but the *RGR* of *S. horneri* at different CO_2 concentrations converged as the cultivation time progressed. In addition, the *RGR* of *S. horneri* at 15 °C was higher than that at 20 °C during the same cultivation time in both the early and middle stages of cultivation, but in the later stage of cultivation, the *RGR* under each cultivation condition was more similar and did not show a significant difference ($p > 0.05$) (Figure 4b).

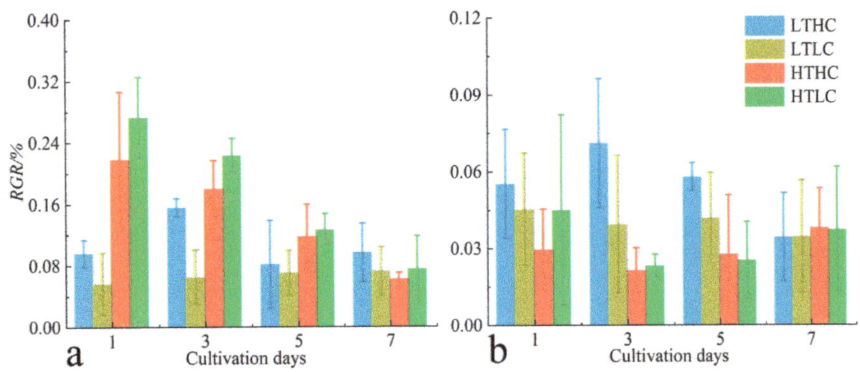

Figure 4. Changes in Relative Growth Rate (*RGR*) of *U. fasciata* (**a**) and *S. horneri* (**b**) during cultivation under different conditions. Cultivation conditions for each group: LTLC (15 °C and 1000 $\mu L \cdot L^{-1}$), LTLC (15 °C and 400 $\mu L \cdot L^{-1}$), HTHC (20 °C and 1000 $\mu L \cdot L^{-1}$), and HTLC (20 °C and 400 $\mu L \cdot L^{-1}$). The error line is the SD.

The DOC release of *U. fasciata* showed a similar trend to its *RGR*. At 20 °C, the DOC release of the acidified treatment group was lower than that of the blank treatment group during the same cultivation time throughout the cultivation period. At 15 °C, the acidification treatment promoted the release of DOC from *U. fasciata* to external seawater in the early and middle stages of cultivation, but the DOC release from *U. fasciata* treated with different CO_2 concentrations was more similar in the later stages of cultivation as the cultivation time progressed. At the end of cultivation, the DOC release from *U. fasciata* was significantly affected by the different cultivation conditions, as well as by the interaction

($p > 0.05$). In addition, the DOC release at 20 °C was higher than that of *U. fasciata* at 15 °C during the same cultivation time (Figure 5a). At 15 °C, the DOC release of *S. horneri* was higher in the acidified treatment group than in the blank treatment group during the same cultivation time throughout the cultivation period. At 20 °C, the DOC release of the acidified treatment group was higher than that of the blank treatment group at the same cultivation time in the early and middle stages of cultivation, but as the cultivation time progressed, the DOC release amounts of the two groups of *S. horneri* become closer in the later stages of cultivation. Changes in CO_2 concentration and temperature did not significantly affect the DOC release from *S. horneri* at the end of the cultivation ($p > 0.05$). In addition, the DOC release was higher at 15 °C than at 20 °C for the same cultivation time for *S. horneri* (Figure 5b).

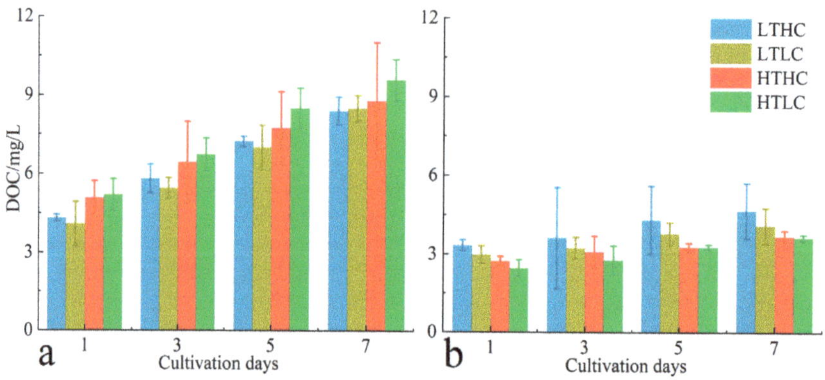

Figure 5. Changes in the release of dissolved organic carbon (DOC) from seawater during cultivation of *U. fasciata* (**a**) and *S. horneri* (**b**) under different conditions. Cultivation conditions for each group: LTLC (15 °C and 1000 $\mu L \cdot L^{-1}$), LTLC (15 °C and 400 $\mu L \cdot L^{-1}$), HTHC (20 °C and 1000 $\mu L \cdot L^{-1}$), and HTLC (20 °C and 400 $\mu L \cdot L^{-1}$). The error line is the SD.

3.3. Changes in Photosynthetic Pigment Content

Under different cultivation conditions, the Chl-a content of *U. fasciata* mostly showed a decreasing trend. At 20 °C, the Chl-a content of the acidification treatment group was mostly higher than that of the blank treatment group during the whole cultivation period; the Chl-a content in the acidification treatment group showed a decreasing and then increasing trend, but the blank treatment group showed a continuous decreasing trend throughout the cultivation period. At 15 °C, the CO_2 concentration did not cause significant differences in the Chl-a content of *U. fasciata*; the acidification treatment group showed a continuous decreasing trend in Chl-a content, while the blank treatment group did not show any significant changing trend. In addition, the Chl-a content at 20 °C was mostly higher than that of *U. fasciata* at 15 °C during the same cultivation time (Figure 6a). The trend in the Car content of *U. fasciata* under each cultivation condition was more similar to that of its Chl-a content, which was mainly affected by temperature changes throughout the cultivation period, and the Car content at 20 °C was mostly higher than that of *U. fasciata* at 15 °C during the same cultivation time (Figure 6b).

The trend in the photosynthetic pigment content of *S. horneri* differed somewhat from that of *U. fasciata*. At 20 °C, the Chl-a content of the acidification treatment group was lower than that of the blank treatment group in the same cultivation time throughout the cultivation period; the Chl-a content of the acidification treatment group showed a tendency to increase and then decrease, whereas the Chl-a content of the blank treatment group showed a tendency to slowly decrease. At 15 °C, the acidification treatment caused the Chl-a content of *S. horneri* to be lower than that of the blank treatment group in the early stage of cultivation, but the effect of CO_2 concentration on the Chl-a content gradually decreased with the cultivation time; the Chl-a content of *S. horneri* incubated with different

CO_2 concentrations showed a slow decreasing tendency throughout the whole cultivation period. In addition, the Chl-a content at 20 °C was mostly higher than that of *S. horneri* at 15 °C during the same cultivation time (Figure 6c). The trend in the Car content of *S. horneri* was similar to that of its Chl-a content under all cultivation conditions. The highest levels of Chl-a and Car were found in the blank treatment group at 20 °C and were at high levels throughout the cultivation period (Figure 6d). As with *U. fasciata*, the interaction between CO_2 concentration and temperature changes did not significantly affect the photosynthetic pigment content of the two macroalgae at the end of the cultivation ($p > 0.05$).

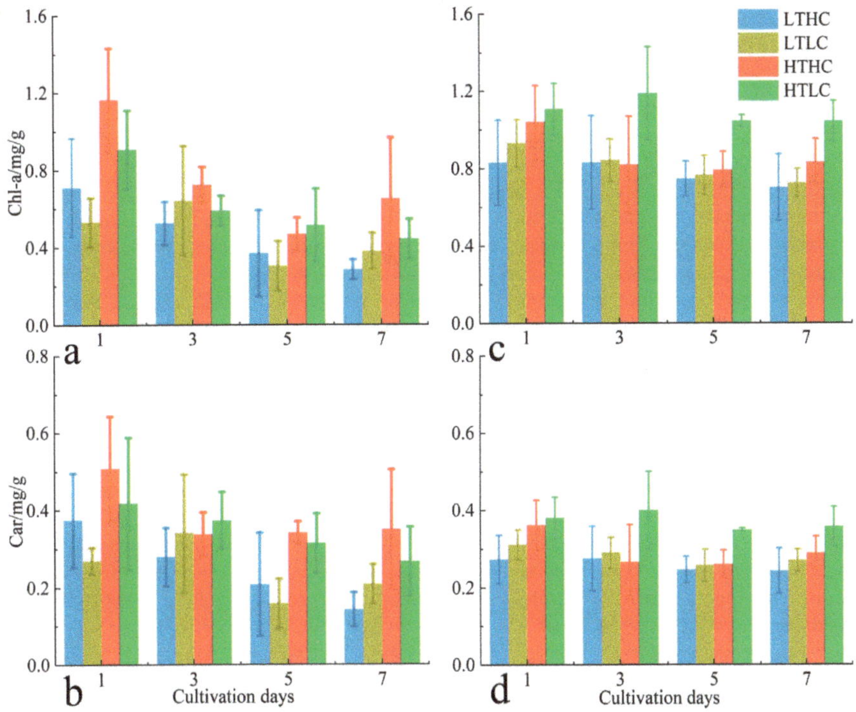

Figure 6. Changes in chlorophyll a (Chl-a) and carotenoids (Car) of *U. fasciata* ((**a**) Chl-a; (**b**) Car) and *S. horneri* ((**c**) Chl-a; (**d**) Car) during cultivation in different conditions. Cultivation conditions for each group: LTLC (15 °C and 1000 $\mu L \cdot L^{-1}$), LTLC (15 °C and 400 $\mu L \cdot L^{-1}$), HTHC (20 °C and 1000 $\mu L \cdot L^{-1}$), and HTLC (20 °C and 400 $\mu L \cdot L^{-1}$). The error line is the SD.

4. Discussion

4.1. Changes in Fluorescence Parameters and Growth of Macroalgae

The continued rise in atmospheric CO_2 concentrations has led to ocean acidification and global warming, which, in turn, affects the lives of most marine organisms. Among them, organisms in offshore intertidal waters show stronger environmental adaptability than those in pelagic waters [22], and intertidal macroalgae are often used as research models when exploring the effects of environmental changes on intertidal marine organisms [23]. The results of this study showed that the Fv/Fm and Fv'/Fm' of *U. fasciata* at 15 °C were lower than those at 20 °C during the same cultivation time throughout an entire cultivation period, suggesting that *U. fasciata* was subjected to the stressful effects of low temperature. In addition, the rise in *NPQ* and the decrease in *rETRm* in the blank treatment group at 15 °C suggests that *U. fasciata* protects its PSII structure by increasing the heat released from non-photochemical processes as a means of depleting captured light energy, allowing it to maintain a high photosynthetic capacity and tolerance [24].

However, under an acidification treatment at 15 °C, the *NPQ* of *U. fasciata* was more stable throughout the cultivation period, and the *RGR* was also higher than that of the blank treatment group, suggesting that the acidification treatment could alleviate the stressful effects of low temperature. This may be due to the fact that the acidification treatment increased the amount of CO_2 in the cultured seawater [25]. Under high CO_2 concentrations, *U. fasciata* regulates the energy required for CO_2 concentration mechanisms (CCMs) [4] to better cope with the stressful effects of low temperatures. Higher cultivation temperatures promoted the growth and photosynthesis of *U. fasciata* relative to 15 °C. It has been shown that the suitable temperature for the growth of *U. fasciata* under natural conditions is around 25 °C [20]. This experiment also found that 20 °C was more favorable for the growth of *U. fasciata*, and the fluorescence parameters were at a higher level, which Fu et al. attributed to the higher activity of the key enzymes of photosynthesis in macroalgae at higher temperatures [26]. However, at this temperature, the acidification treatment did not significantly affect the growth and photosynthetic use efficiency of *U. fasciata*. In the present study, it was found that the *RGR* and the fluorescence parameters of the acidified treatment group at 20 °C did not show any significant difference from the blank treatment group. According to Wu et al. [6], the supply of CO_2 promotes the growth and photosynthesis of macroalgae at suitable temperatures, but their photosystems also carry out a coordinated action to integrate the carbon assimilation process affected by temperature. Furthermore, it has also been shown that an appropriate increase in seawater CO_2 concentrations does not significantly affect macroalgae grown under saturated light conditions [26]. Excessive CO_2 concentrations at 20 °C increased the *rETRm* and β of *U. fasciata* during long-term cultivation, also indicating that acidification treatments increase the photosynthetic potential of *U. fasciata* but decrease the tolerance of algae to high light.

At 15 °C, the *RGR* of *S. horneri* was mostly higher than that of *S. horneri* at 20 °C in the early stage of cultivation, but the *RGR* of *S. horneri* at 15 °C decreased rapidly with cultivation time in the later stage of cultivation. Therefore, subsequent studies found that *S. horneri* increased its demand for light energy in order to maintain normal growth, and the E_k and *NPQ* of *S. horneri* at 15 °C gradually increased with cultivation time, indicating that *S. horneri* enhanced its protective mechanism to prevent excessive light energy from destroying the structure of PSII while improving its ability to capture light energy [13]. In addition, the effect of the acidification treatment on the fluorescence parameters of *S. horneri* was not significant at 15 °C, but the *RGR* was significantly higher than that of the blank treatment group in the early and middle stages of cultivation. This may be because the CO_2 concentration in the seawater of the blank treatment was low, which failed to meet the demands of photosynthesis for *S. horneri*, while the acidification treatment allowed *S. horneri* to obtain more CO_2, which, in turn, promoted the growth of *S. horneri* [27], but with the time of cultivation, excess CO_2 produced a negative feedback regulation on *S. horneri* and inhibited its growth. [28]. It has been pointed out that 11–16 °C is the optimal range for *S. horneri* growth and reproduction [29,30]. The results of the present study showed that 20 °C did not significantly affect the levels of the fluorescence parameters *rETRm*, Fv/Fm, and Fv'/Fm' in *S. horneri*, suggesting that higher temperatures do not diminish the photosynthetic activity of *S. horneri*. However, the β of *S. horneri* continued to rise at higher temperatures with cultivation time, suggesting that 20 °C may have attenuated light energy capture and light acclimation in *S. horneri*. In addition, at 20 °C, the acidification treatment increased the *NPQ* of *S. horneri*, causing it to release excess energy in the form of heat dissipation, alleviating the stressful effects of high temperature to some extent [6].

4.2. Changes in Biochemical Characterization Parameters of Macroalgae

The 15 °C cultivation temperature resulted in lower DOC release from *U. fasciata*, possibly due to lower temperatures inhibiting the activity of key enzymes in photosynthesis [24]. The acidification treatment had a mitigating effect on the stressful effects of low temperature, resulting in a higher level of DOC release in the acidification treatment group at 15 °C. In addition, the macroalgae adapted to the low-temperature environment over the cultivation time, resulting in a convergence of DOC release from *U. fasciata* at different CO_2 concentrations. Rogers also found that too much CO_2 inhibits the activity of Rubisco enzymes at suitable temperatures, which, in turn, inhibits the rate of assimilation of inorganic carbon by macroalgae [31]. Therefore, the activity of *U. fasciata* photosynthesis enzymes is saturated under acidification treatments at 20 °C, and an excessive CO_2 concentration decreases the rate of synthesis of metabolites such as proteins [32] and lipids [33]. The DOC release from *S. horneri* was at a higher level at 15 °C, which was attributed by Marañón [34] to the higher *RGR* of *S. horneri* at 15 °C. The acidification treatment increased the CO_2 concentration in the cultured seawater, which promoted the growth of *S. horneri* and also increased the synthesis rate of its own metabolites, and the DOC release from *S. horneri* reached its highest level under the acidification treatment at 15 °C. In addition, different CO_2 concentrations did not produce a significant effect on the DOC release under 20 °C cultivation conditions, probably because under the stress of high temperature *S. horneri* allocated more energy to the synthesis of photosynthetic pigments, and thus, the acidification treatment did not significantly increase the activity of metabolic enzymes in the cells [35].

The photosynthetic pigment content reflects changes in the ability of macroalgae to regulate their own physiological responses [36]. It has been shown that macroalgae are able to self-regulate the content of pigments to adapt to changes in the external growth environment [37]. In this study, we showed that the photosynthetic pigment content of *U. fasciata* at 15 °C was lower than that of *U. fasciata* at 20 °C during the same cultivation period. This may be due to *U. fasciata* cells at 15 °C reducing the synthesis of photosynthetic pigments to decrease the absorption area of Photosystem I (PSI) and the activity ratio of PSI to PSII. This reduction leads to a decrease in the synthesis of high-energy compound adenosine triphosphate (ATP) to conserve energy for maintaining normal life activities [27]. Under acidification treatment conditions at 15 °C, low temperatures inhibit the photosynthesis of *U. fasciata*, so even though the amount of dissolved inorganic carbon (DIC) available in cultured seawater is increased, the photosynthetic capacity of *U. fasciata* is essentially saturated so that it regulates its own biochemical reactions without significantly altering photosynthetic pigment content [38]. However, a cultivation temperature of 20 °C promoted the activity of enzymes, as well as other physiological components, in *U. fasciata*, resulting in the need to synthesize more Chl-a and Car in the 20 °C acidification treatment group in order to maintain a higher photosynthetic rate for more light energy. *S. horneri* is a light-adapted macroalga that grows well under higher seawater transparency and shallow cultivation conditions; thus, its growth is significantly affected under lower seawater transparency and unfavorable light conditions, which may even lead to slow growth or death [29]. Where cultivation light is not the maximum light intensity required for *S. horneri*, the stressful effects of higher cultivation temperatures can force *S. horneri* to enhance its own light energy harvesting capacity by increasing the synthesis of photosynthetic pigments [39]. The present study also showed that the acidification treatment provided some relief from the effects of high-temperature stress on *S. horneri*. In addition, the high concentration of CO_2 saves the energy required for *S. horneri*'s CCMs [40], allowing it to maintain normal levels of vital activity without synthesizing excessive photosynthetic pigments under low-temperature cultivation conditions [41].

5. Conclusions

At 20 °C, the growth metabolism of *U. fasciata* is at a relatively high level. The acidification treatment further enhances the photosynthesis of *U. fasciata* but at the same time weakens its tolerance to high light intensity. In contrast to *U. fasciata*, *S. horneri* is more adapted to low-temperature environments. At 15 °C, both the growth and adaptability of *S. horneri* are improved. Although the acidification treatment promotes the growth metabolism of *S. horneri*, it suppresses the synthesis of its photosynthetic pigments. Long-term cultivation results show that compared with *U. fasciata*, *S. horneri* exhibits greater adaptability to environmental changes. With the passage of cultivation time, the regulation of growth metabolism and energy required for synthesizing substances in *S. horneri* becomes more rational. This enables *S. horneri* to grow and develop without significant inhibition under different stress conditions, with no significant differences in its physiological characteristic parameters.

Author Contributions: K.W. and X.Z. contributed to the conception and design of the study. K.W., X.T. and X.Z. conducted the experiments. X.T. and X.Z. contributed to the analysis and wrote the manuscript. K.W. and S.Z. contributed to the manuscript's revision. All authors have read and agreed to the published version of the manuscript.

Funding: This work was supported by the Southern Marine Science and Engineering Guangdong Laboratory (Zhuhai) (Grants Number SML2023SP23); the Open Fund Project of the Key Laboratory of Marine Ecological Monitoring and Restoration Technology, the Ministry of Natural Resources of China (Grants Number MEMRT202113); and the National Natural Science Foundation of China (Grants Number 41876191).

Institutional Review Board Statement: Not applicable.

Informed Consent Statement: Not applicable.

Data Availability Statement: The original contributions presented in this study are included in the article, and further inquiries can be directed to the corresponding author.

Acknowledgments: Sincere thanks to those who have helped me on my research journey.

Conflicts of Interest: The authors declare no conflicts of interest.

References

1. Gattuso, J.P.; Magnan, A.; Bille, R.; Cheung, W.W.L.; Howes, E.L.; Joos, F.; Allemand, D.; Bopp, L.; Cooley, S.R.; Eakin, C.M.; et al. Contrasting futures for ocean and society from different anthropogenic CO_2 emissions scenarios. *Science* **2015**, *349*, aac4722. [CrossRef] [PubMed]
2. Delille, B. CO_2 in Seawater: Equilibrium, Kinetics, Isotopes. *J. Mar. Syst.* **2002**, *36*, 269–270. [CrossRef]
3. Scheffer, M.; Brovkin, V.; Cox, P.M. Positive feedback between global warming and atmospheric CO_2 concentration inferred from past climate change. *Geophys. Res. Lett.* **2006**, *33*, 10702-1–10702-4. [CrossRef]
4. Feely, A.R.; Doney, C.S.; Cooley, R.S. Ocean Acidification: Present Conditions and Future Changes in a High-CO_2 World. *Oceanography* **2009**, *22*, 36–47. [CrossRef]
5. Gao, G.; Clare, A.S.; Rose, C.; Caldwell, G.S. Intrinsic and extrinsic control of reproduction in the green tide-forming alga, Ulva rigida. *Environ. Exp. Botany* **2017**, *139*, 14–22. [CrossRef]
6. Wu, H.L.; Feng, J.C.; Li, X.S.; Zhao, C.Y.; Liu, Y.H.; Yu, J.T.; Xu, J.T. Effects of increased CO_2 and temperature on the physiological characteristics of the golden tide blooming macroalgae *Sargassum horneri* in the Yellow Sea, China. *Mar. Pollut. Bull.* **2019**, *146*, 639–644. [CrossRef] [PubMed]
7. Liu, C.X.; Zou, D.H. Do increased temperature and CO_2 levels affect the growth, photosynthesis, and respiration of the marine macroalga *Pyropia haitanensis* (Rhodophyta)? An experimental study. *Hydrobiologia* **2015**, *745*, 285. [CrossRef]
8. Suárez-Alvarez, S.; Gómez-Pinchetti, J.L.; García-Reina, G. Effects of increased CO_2 levels on growth, photosynthesis, ammonium uptake and cell composition in the macroalga *Hypnea spinella* (Gigartinales, Rhodophyta). *J. Appl. Phycol.* **2012**, *24*, 815–823. [CrossRef]
9. Mercado, J.M.; Javier, F.; Gordillo, L.; Niell, F.X.; Figueroa, F.L. Effects of different levels of CO_2 on photosynthesis and cell components of the red alga *Porphyra leucosticta*. *J. Appl. Phycol.* **1999**, *11*, 455–461. [CrossRef]
10. Coutinho, R.; Zingmark, R. Interactions of light and nitrogen on photosynthesis and growth of the marine macroalga *Ulva curvata* (Kützing) De Toni. *J. Exp. Mar. Biol. Ecol.* **1993**, *167*, 11–19. [CrossRef]
11. Chakraborty, K.; Lipton, A.P.; Raj, R.P.; Vijayan, K.K. Antibacterial labdane diterpenoids of *Ulva fasciata* Delile from southwestern coast of the Indian Peninsula. *Food Chem.* **2010**, *119*, 1399–1408. [CrossRef]
12. He, P.M.; Liu, Y.Y.; Zhang, J.W.; Wu, H.L.; Yu, K.F.; Huo, Y.Z.; Zhang, J.H. Research progress on the effects of macroalgae on carbon sink. *J. Fish. Sci. China* **2015**, *22*, 588–595.
13. Li, M.Y.; Xu, K.; Tang, X.H.; Zhang, L.R.; Xu, Y.; Wang, W.L.; Ji, D.H.; Chen, D.S.; Xie, C.T. Effects of ocean acidification on the light and temperature adaptations of commercial seaweed *Pyropia haitanensis*. *J. Appl. Oceanogr.* **2021**, *40*, 388–394. [CrossRef]
14. Xu, X.T.; Li, Y.H.; Wang, D.; Xu, N.J. Effects of simulated acid rain on the photosynthetic physiological characteristics in *Ulva fasciata* under salt stress. *J. Fish. China* **2016**, *40*, 731–739. [CrossRef]

15. Huang, Y.H.; Yang, R.; Sun, Q.H. Effect of Temperature on Growth and Chlorophyll Fluorescence of *Ulva fasciata*. *Biotechnol. Bull.* **2016**, *32*, 99–105. [CrossRef]
16. Zhang, T.; Shen, Z.; Xu, P.; Zhu, J.Y.; Lu, Q.Q.; Shen, Y.; Wang, Y.; Yao, C.Y.; Li, J.F.; Wang, Y.X.; et al. Analysis of photosynthetic pigments and chlorophyll fluorescence characteristics of different strains of *Porphyra yezoensis*. *J. Appl. Phycol.* **2012**, *24*, 881–886. [CrossRef]
17. Russell, B.D.; Passarelli, C.A.; Connell, S.D. Forecasted CO_2 modifies the influence of light in shaping subtidal habitat. *J. Phycol.* **2011**, *47*, 744–752. [CrossRef]
18. Gordillo, F.J.L.; Aguilera, J.; Wiencke, C.; Jiménez, C. Ocean acidification modulates the response of two Arctic kelps to ultraviolet radiation. *J. Plant Physiol.* **2015**, *173*, 41–50. [CrossRef]
19. Gordillo, F.J.L.; Carmona, R.; Viñegla, B.; Wiencke, C.; Jiménez, C. Effects of simultaneous increase in temperature and ocean acidification on biochemical composition and photosynthetic performance of common macroalgae from *Kongsfjorden* (Svalbard). *Polar Biol.* **2016**, *39*, 1993–2007. [CrossRef]
20. Choi, C.G.; Kim, H.G.; Sohn, C.H. Transplantation of Young Fronds of *Sargassum horneri* for Construction of Seaweed Beds. *Korean J. Fish. Aquat. Sci.* **2003**, *36*, 469–473. [CrossRef]
21. Zhao, X. *Photosyntheitc Physiology and Ecological Effects of Macroalgae in Gouqi Island*; Shanghai Ocean University: Shanghai, China, 2022.
22. Cai, W.J.; Hu, X.; Huang, W.J.; Murrell, M.C.; Lehrter, J.C.; Lohrenz, S.E.; Chou, W.C.; Zhai, W.D.; Hollibaugh, J.T.; Wang, Y.C.; et al. Acidiffcation of subsurface coastal waters enhanced by eutrophication. *Nat. Geosci.* **2011**, *4*, 766–770. [CrossRef]
23. Kim, J.K.; Yarish, C.; Pereira, R. Tolerances to hypo-osmotic and temperature stresses in native and invasive species of *Gracilaria* (Rhodophyta). *Phycologia* **2016**, *55*, 257–264. [CrossRef]
24. Fu, F.X.; Warner, M.E.; Zhang, Y.; Feng, Y.Y.; Hutchins, D.A. Effects of increased temperature and CO_2 on photosynthesis, growth, and elemental ratios in marine *Synechococcus* and *Prochlorococcus* (cyanobacteria). *J. Phycol.* **2007**, *43*, 485–496. [CrossRef]
25. Ji, Y.; Jiang, Y.; Jin, P.; Xia, J.R. Interactive effects of ocean acidification and nitrate on *Ulva lactuca*. *J. Appl. Phycol.* **2024**. [CrossRef]
26. Rautenberger, R.; Fernandez, P.A.; Strittmatter, M.; Heesch, S.; Cornwall, C.E.; Hurd, C.L.; Roleda, M.Y. Saturating light and not increased carbon dioxide under ocean acidification drives photosynthesis and growth in *Ulva rigida* (Chlorophyta). *Ecol. Evol.* **2015**, *5*, 874–888. [CrossRef] [PubMed]
27. Bao, O.N.; Shi, D.G.; Guan, W.C.; Sun, M.; Zhang, P.; Peng, X.; Wang, T.G.; Chen, S.B.; Qiu, J.B. Effect of CO_2 concentration and irradiance on the growth of *Sargassum horneri* (Phaeophyceae) in Nanji Archipelago. *Acta Agric. Zhejiangensis* **2014**, *26*, 649–655. [CrossRef]
28. Pederson, M.F.; Paling, E.I.; Walker, D.I. Nitrogen uptake and allocation in the seagrass Amphibolis Antarctica. *Aquat. Botany* **1997**, *56*, 105–117. [CrossRef]
29. Sun, J.Z.; Zhuang, D.G.; Sun, Q.H.; Pang, S.J. Artificial cultivation trials of *Sargassum horneri* at Nanji island of China. *South China Fish. Sci.* **2009**, *5*, 41–46. [CrossRef]
30. Dai, W.; Wang, X.R.; Zhuang, M.M.; Sun, J.Y.; Shen, Y.F.; Xia, Z.Y.; Wu, T.J.; Jiang, R.T.; Li, A.Q.; Bi, F.L.; et al. Responses of photosynthesis-related genes in *Sargassum horneri* to high temperature stress. *Mar. Pollut. Bull.* **2024**, *199*, 115944. [CrossRef]
31. Rogers, A.; Fischer, B.U.; Bryant, J.; Frehner, M.; Blum, H.; Raines, C.A.; Long, S.P. Acclimation of Photosynthesis to Elevated CO_2 under Low-Nitrogen Nutrition Is Affected by the Capacity for Assimilate Utilization. Perennial Ryegrass under Free-Air CO_2. *Enrich. Plant Phys.* **1998**, *118*, 683–689. [CrossRef]
32. Chen, B.B.; Xia, J.R.; Zou, D.H.; Zhang, X. Responses to ocean acidification and diurnal temperature variation in a commercially farmed seaweed, *Pyropia haitanensis* (Rhodophyta). *Eur. J. Phycol.* **2019**, *54*, 184–192. [CrossRef]
33. Barakat, K.M.; El-Sayed, H.S.; Khairy, H.M.; El-Sheikh, M.A.; Al-Rashed, S.A.; Arif, I.A.; Elshobary, M.E. Effects of ocean acidification on the growth and biochemical composition of a green alga (*Ulva fasciata*) and its associated microbiota. *Saudi J. Biol. Sci.* **2021**, *28*, 5106–5114. [CrossRef]
34. Marañón, E.; Cermeño, P.; Fernändez, E.; Rodríguez, J.; Zabala, L. Significance and Mechanisms of Photosynthetic Production of Dissolved Organic Carbon in a Coastal Eutrophic Ecosystem. *Limnol. Oceanogr.* **2004**, *49*, 1652–1666. [CrossRef]
35. Luan, Q.; Lv, F.; Wu, H.Y.; Ding, G.; Zhan, D.M. Effects of culture conditions on nutrient composition of *Sargassum horneri*. *Prog. Fish. Sci.* **2019**, *40*, 123–130. [CrossRef]
36. Fu, W.T.; Liu, J.; Zhang, J.L.; Feng, T.W.; Su, Y.M.; Zhu, X.L.; Li, Y.Y.; Liu, J.; Liu, Y.; Cao, S.Q.; et al. Effects of elevated atmospheric CO_2 concentration on growth and pigment contents of macroalga *Ulva pertusa*. *J. Dalian Ocean Univ.* **2013**, *28*, 481–486. [CrossRef]
37. Miller, S.M.; Wing, S.R.; Hurd, C.L. Photoacclimation of *Ecklonia radiata* (Laminariales, Heterokontophyta) in Doubtful Sound, Fjordland, Southern New Zealand. *Phycologia* **2006**, *45*, 44–52. [CrossRef]
38. Mercado, J.M.; Gordillo, F.J.L.; Figueroa, F.L.; Niell, F.X. External carbonic anhydrase and affinity for inorganic carbon in intertidal macroalgae. *J. Exp. Mar. Biol. Ecol.* **1998**, *221*, 209–220. [CrossRef]
39. Wang, Q.H. *Effects of Diel Fluctuating Environmental Factors on Germination, Growth of Macroalage and Their Ecophysiological Mechanisms*; Ocean University of China: Qingdao, China, 2008.

40. Wu, Y.; Gao, K.; Riebesell, U. CO_2-induced seawater acidification affects physiological performance of the marine diatom *Phaeodactylum tricornutum*. *Biogeosciences* **2010**, *7*, 2915–2923. [CrossRef]
41. James, I.L.M.; Michael, D.M. *Plant Growth and Climate Change*; Blackwell Publishing: Oxford, UK, 2006; pp. 17–47. ISBN 79780470994184.

biology

MDPI

Article

Protective Effect of Polysaccharides Isolated from *Sargassum horneri* against H_2O_2-Induced Oxidative Stress Both In Vitro, in Vero Cells, and In Vivo in Zebrafish

Shuangyan Wei [1,†], Li Wang [1,†], Jia Yang [1], Ruihang Xu [1], Rui Jia [1,2,*] and Peimin He [1,2,*]

1 College of Oceanography and Ecological Science, Shanghai Ocean University, Shanghai 201306, China; wsy97221@163.com (S.W.); orange709wl@163.com (L.W.); yangjia0808@163.com (J.Y.); xrh_cn@163.com (R.X.)

2 Marine Biomedical Science and Technology Innovation Platform of Lingang Special Area, Shanghai 201306, China

* Correspondence: rjia@shou.edu.cn (R.J.); pmhe@shou.edu.cn (P.H.); Tel./Fax: +86-021-61900449 (R.J.)

† These authors contributed equally to this work.

Simple Summary: The environment within which human beings exist is progressively becoming more complex. A multitude of external factors prompt the body to produce excessive amounts of reactive oxygen species (ROS), thereby leading to heightened levels of oxidative stress, which is detrimental to health. In recent years, researchers have discerned the potential of secondary metabolites derived from brown algae in combating diseases. Additionally, *Sargassum horneri* has witnessed a large-scale outbreak in China, exerting a severe negative influence on the local ecology and economic development. We extracted polysaccharides (SHPs) from *S. horneri* and discovered that SHPs possess outstanding antioxidant properties and can be utilized as a natural antioxidant. Our work can alleviate the environmental stress caused by algae and achieve the utilization of algae resources, which hold considerable social value and significance.

Abstract: The extensive outbreak of *Sargassum horneri* in China has not merely imposed a severe threat to the ecological environment and human life in coastal waters but also impeded the development of waterway transportation and the local economy. Consequently, we isolated polysaccharides from *S. horneri*, designated as SHP, and evaluated the antioxidant activity of SHP both in vitro and in vivo by investigating the effect of SHP on H_2O_2-induced African green monkey kidney cells (Vero cells) and zebrafish. The results demonstrated that SHP can enhance the activities of superoxide dismutase, catalase, and glutathione peroxidase in zebrafish. It also effectively inhibits micro malondialdehyde and ROS levels in Vero cells and zebrafish to mitigate the oxidative damage caused by H_2O_2, thereby achieving the protective effect of SHP on Vero cells and zebrafish. In conclusion, SHP holds the potential as a natural antioxidant. SHP can be contemplated for utilization as a natural antioxidant in the biomedical, cosmetic, and food industries, thereby alleviating the environmental stress caused by *S. horneri* and achieving resource utilization.

Keywords: *Sargassum horneri*; polysaccharides; oxidative stress; Vero cell; zebrafish

1. Introduction

With the swift advancement of industry, science, and technology, human beings inhabit an increasingly intricate environment. Reactive oxygen species (ROS) will escalate in response to external stimuli (including ultraviolet light, cigarette smoke, air pollution, etc.) [1]. Generally, ROS produced by normal metabolism can be eliminated by endogenous antioxidant systems. However, excessive environmental stress will result in the abnormal generation of ROS, thereby causing oxidative stress [1,2]. Oxidative stress, a series of reactions caused by an imbalance between reactive oxygen species and the body's antioxidant system, is extensively implicated in the progression of numerous chronic diseases such as neurodegenerative disorders, diabetes, cancer, inflammation, cardiovascular diseases, aging, and many geriatric diseases [3,4]. Elevated levels of superoxide dismutase (SOD), catalase (CAT), and glutathione peroxidase (GSH-PX) can down-regulate the production of ROS. Antioxidants can enhance the activity of these enzymes, not only possessing a potent free radical scavenging capacity but also regulating signaling pathways like MAPK, NF-κB, and Nrf2, lowering the level of ROS production, protecting cells from oxidative stress damage, and thereby mitigating the related diseases caused by oxidative stress [5,6]. Natural antioxidants are safer, more moderate, and have fewer side effects compared to chemically synthesized antioxidants [7]. Hence, natural bioactive substances with robust ROS scavenging ability will be ideal therapeutics for oxidative stress diseases.

Polysaccharides isolated from seaweed have been reported to possess numerous health benefits, including antioxidant, UV protection, liver protection, anti-inflammatory, antibacterial, and anticancer activities [8–15]. Furthermore, the polysaccharide content in seaweed is particularly abundant, which constitutes one of the most crucial bioactive substances in seaweed [16,17]. The research suggests that extracts such as polysaccharides in *Fucus vesiculosus* and *Turbinaria conoides* exhibit antioxidant properties and that they could be employed to prevent skin aging and diseases [18]. The polysaccharides isolated from *Ecklonia cava* demonstrated outstanding antioxidant activity in both in vivo and in vitro experiments [19]. The study indicated that the polysaccharides extracted from *Padina boryana* can significantly reduce oxidative stress in Vero cells and zebrafish [20]. The polysaccharides isolated from *Codium fragile* through enzyme-assisted extraction have been shown to inhibit H_2O_2-induced oxidative stress both in vitro and in vivo [8], which indicates that seaweed polysaccharides are highly promising natural antioxidants. Previous studies have unveiled the capacity of *Sargassum horneri* polysaccharides to scavenge free radicals, and the in vitro antioxidant effects of *S. horneri* polysaccharides have been evaluated at the cellular level [21–24]. Additionally, the anti-inflammatory [25–28], anti-tumor [29], immunomodulatory [30], whitening [24], and moisturizing effects [21,31] of *S. horneri* polysaccharides have been explored, suggesting a wide array of potential applications.

S. horneri is an edible brown alga, predominantly distributed in the warm temperate sea regions of China. It exhibits rapid growth and possesses a substantial biomass. Consequently, it often floats in patches on the sea surface, subsequently forming an extensive seaweed field, which is even referred to as the "forest in the sea" [32]. Widespread outbreaks of *S. horneri* have occurred in China, inflicting severe damage on tourism, aquaculture, and waterways, and even posing a threat to the ecological environment and human health [33–35]. Hence, it is highly significant to develop *S. horneri* polysaccharides as antioxidants, as this can not only reduce the environmental pressure of offshore waters but also achieve resource utilization. We chose to employ hydrogen peroxide (H_2O_2) to induce oxidative stress to investigate the antioxidant effect of the polysaccharide extract of *S. horneri* in Vero cells and zebrafish.

2. Materials and Methods

2.1. Chemicals and Regents

The African green monkey kidney cell (Vero) was purchased from Procell Life Science and Technology Co., LTD (Wuhan, China). The wild-type zebrafish was provided by the College of Fisheries and Life Sciences of Shanghai Ocean University. Minimum essential medium (MEM), dimethyl sulfoxide (DMSO), penicillin/streptomycin (P/S), and trypsin/EDTA were purchased from Nanjing BioChannel Biotechnology Co., LTD (Nanjing, China). Fetal bovine serum (FBS) was obtained from Gibco (Thermo Fisher, Mulgrave, VIC, Australia). 2′,7′-dichlorodihydrofurescin diacetate (DCFH-DA), cell counting kit-8 (CCK-8), and protein assay kit (BCA kit) were purchased from Beyotime Biotechnology Co., LTD (Shanghai, China). Catalase activity (CAT) assay kit, superoxide dismutase (SOD) assay kit, micro malondialdehyde (MDA) assay kit, and glutathione peroxidase (GSH-PX) assay kit were bought from Nanjing Jiancheng Bioengineering Institute (Nanjing, China). All other chemicals and reagents were of analytical grade.

2.2. Plant Material and Sample Preparation

S. horneri was collected in October 2020 from Gouqi Island, Zhejiang Province. The preparation of polysaccharides has been described in previous work [36]. The polysaccharide of *S. horneri* obtained was named SHP. Moreover, the SHP was characterized by FT-IR measurements within the frequency range of 500–4000 cm^{-1}.

2.3. The Cell Cultures

The Vero cells were cultured in a medium comprising 10% FBS and 1% penicillin/streptomycin, and placed in a cell culture chamber with 5% CO_2 and a temperature of 37 °C. Once the cells reached a growth density ranging from 80% to 90%, they were digested using 0.25% trypsin and re-culture. Subsequently, cells in the logarithmic phase were selected for the subsequent experiments.

2.4. The Assay for Cytotoxicity

Vero cells were seeded into 96-well plates at a density of 1×10^5 cells mL^{-1}. After 24 h of cell culture in the cell incubator, the cells were exposed to various concentrations of H_2O_2 (0.1, 0.2, 0.3, 0.4, 0.5, 0.6 mmol L^{-1}) to establish the H_2O_2-induced damage model. Alternatively, Vero cells were treated with different concentrations of SHP (0, 25, 50, 75, 100, 150, 200 μg mL^{-1}), and then the effect of SHP on the survival rate of Vero cells was determined. Alternatively, the same volume of SHP (25, 50, and 100 μg mL^{-1}) was pretreated for 1 h, followed by the treatment of Vero cells with H_2O_2 (0.4 mmol L^{-1}), while the control group was treated with medium alone to determine the protective effect of SHP on the viability of Vero cells induced by H_2O_2. The cells were further cultured for another 24 h, and then CCK-8 solution was added, and the absorbance was measured at 450 nm (iBIO-RAD Mark, Vallejo, CA, USA).

2.5. Measurement of Intracellular ROS Levels

The DCFH-DA probe per se has no fluorescence. Once DCFH-DA enters cells, it will be oxidized by ROS and transformed into a green, fluorescent substance. Therefore, it is frequently employed to detect the level of ROS within cells [37]. Vero cells were seeded into 6-well plates at a cell density of 1×10^5 cells/mL; the same volume of SHP (25, 50, and 100 μg mL^{-1}) was pretreated for 1 h, and subsequently, the Vero cells were treated with H_2O_2 (0.4 mmol L^{-1}), while the control group was treated merely with the medium. After 24 h incubation, DCFH-DA with a final concentration of 10 μg mL^{-1} was added, and the incubation was carried out for 30 min in the dark. The medium was cleaned twice. The cells were collected, the fluorescence intensity within the Vero cells was measured, and the observations were conducted under a fluorescence microscope (Nikon, Tokyo, Japan).

2.6. The Effect of SHP on Intracellular MDA Content and SOD Activity

When the body is stimulated by the external environment, ROS are abnormally produced. However, SOD can convert superoxide into hydrogen peroxide and oxygen, while maintaining ROS within the body within an appropriate range to reduce the damage caused by oxidative stress [38]. MDA is recognized as the ultimate decomposition product of lipid peroxidation. The concentration of MDA within a biological sample can be utilized as an indicator of the degree of damage sustained by the cell due to oxidative stress [39]. Briefly, Vero cells were seeded as described above. After a 24 h culture period, the medium was discarded, washed twice with PBS, and the supernatant was discarded after centrifugation. Subsequently, 200 μL of deionized water was added to the cells, and the cells were disrupted by the repeated freeze–thaw method. The contents of SOD and MDA in the cells were measured in accordance with the kit instructions.

2.7. Maintenance of Zebrafish

The maintenance of zebrafish was as described in one study with minor modifications [17]. Briefly, fifteen wild-type adult zebrafish were fed three times daily in a tank at

28 ± 0.5 °C, under a 14/10 h light/dark cycle. A pair of adult zebrafish were mated in spawning tanks, and healthy embryos were collected within 120 min. The use of zebrafish adheres to the document of the Animal Ethics Committee of Shanghai Ocean University (SHOUDW-2016-003).

2.8. The Survival Rate of Zebrafish Embryos

The survival rate of zebrafish embryos was determined in accordance with the method described by Kim et al. [40]. Briefly, 20 zebrafish embryos fertilized were selected, cultured to 7–9 h post-fertilization (hpf), and various concentrations of H_2O_2 (0, 1, 3, 5, 7, 9, 11, 13, 15 mmol L^{-1}) were added to induce zebrafish embryos for the establishment of an oxidative stress model. The zebrafish embryos were placed in a 28 °C incubator for further culturing for 96 h, while the survival rate and development of zebrafish embryos were recorded. A control group (culture medium), an induction group (H_2O_2), and SHP groups with different doses (25, 50, and 100 µg mL^{-1}) were established to assess the protective effect of SHP on H_2O_2-induced oxidative stress of zebrafish embryos.

2.9. Measurement of Oxidative Stress-Related Indicators (SOD, CAT, GSH-PX, and MDA) in Zebrafish

The activities of SOD, CAT, and GSH-PX constitute a primary antioxidant defense system, which plays a crucial and fundamental role in the overall defense mechanisms and strategies within biological systems [41]. The control group (medium), induction group (H_2O_2), and SHP groups (25, 50, and 100 µg mL^{-1}) were established to treat zebrafish embryos. Several zebrafish that developed to 96 hpf were collected and placed into a 1.5 mL centrifuge tube, and 2 to 3 magnetic beads were added to prepare a 10% tissue homogenate, followed by centrifugation to obtain the supernatant. The SOD activity, CAT activity, GSH-PX content, and MDA content were measured in accordance with the kit instructions.

2.10. Determination of ROS in Zebrafish

Zebrafish embryos were treated in accordance with the above description. Several zebrafish that had developed to 96 hpf were randomly selected. Subsequently, DCFH-DA was diluted in the embryo culture medium (with a final concentration of 10 µmol L^{-1}), incubated for 30 min in the dark, and washed three times to eliminate DCFH-DA that had not entered the zebrafish body. The fluorescence microscope was utilized for observation and photography, and the fluorescence intensity in zebrafish was quantified (using Image J software V 1.8.0).

2.11. Statistical Analysis

The data are expressed as the mean ± standard error (S.E.), and the ANOVA test was employed for statistically comparing the mean values (with SPSS 26.0 statistical software). All experiments were performed in triplicate.

3. Results

3.1. FTIR Spectrum of S. horneri Polysaccharide (SHP)

The extraction and structural characterization of SHP have been elaborated in detail in our previous study [36]. The FTIR of SHP exhibited the presence of characteristic peaks specific to polysaccharides at 3432 cm^{-1}, 2938 cm^{-1}, 1625 cm^{-1}, 1200 cm^{-1}, and 842 cm^{-1} and indicated the presence of sulfate, pyranoside, α-glycosidic bond, and other features. In summary, the results demonstrate that SHP exhibits the characteristic features of sulfate polysaccharides.

3.2. Determination of the H_2O_2 Damage Model

Currently, H_2O_2 has been extensively employed as an inducer in the assessment of antioxidant studies. Thus, in this study, Vero cells were exposed to diverse concentrations of H_2O_2 to construct a model of oxidative damage. The activity of Vero cells declined as the concentration of H_2O_2 increased (Figure 1A). In comparison with the control group, when the concentration of H_2O_2 was 0.4 mmol/L, the activity of Vero cells was 55.48%, which was approximately the lethal concentration 50 (LC_{50}). Consequently, the H_2O_2-induced concentration of Vero cells was 0.4 mmol/L. Our results demonstrate that at SHP concentrations (25–200 μg mL^{-1}), Vero cells were more than 90% active, indicating that SHP had no significant toxicity to Vero cells (Figure 1B). Therefore, three doses of 25, 50, and 100 μg/mL were chosen in the subsequent experiment to examine the antioxidant capacity of SHP in Vero cells.

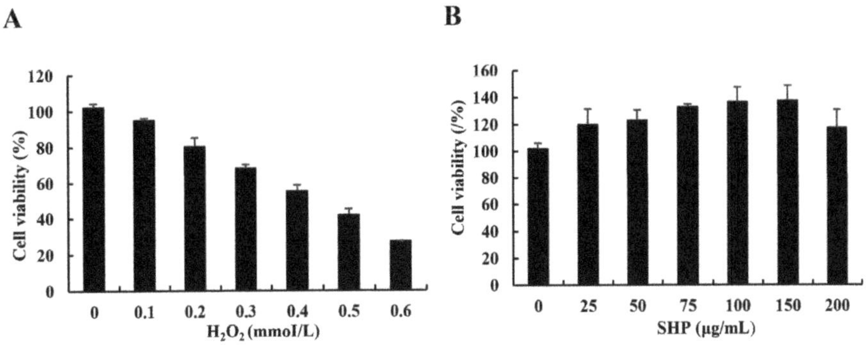

Figure 1. Vero cytotoxicity assay. (**A**) The effect of H_2O_2 on Vero cell viability. (**B**) The effect of SHP on Vero cell viability. The experiments were performed in triplicate and data are expressed as mean ± SE.

3.3. The Effect of SHP against Oxidative Stress Induced by H_2O_2 in Vero Cells

In contrast to the control group, Vero cells treated with H_2O_2 (0.4 mmol L^{-1}) exhibited a significant decrease in cell viability, reaching merely 55.09% ($p < 0.01$) (Figure 2). However, in reality, the cell viability increased following treatment with SHP at concentrations of 25, 50, and 100 μg mL^{-1} in a dose-dependent manner. After treatment with 100 μg mL^{-1} SHP, the activity of Vero cells was restored to 88.62% ($p < 0.01$) (Figure 2). The results showed that SHP had a protective effect on H_2O_2-induced injury on Vero cells.

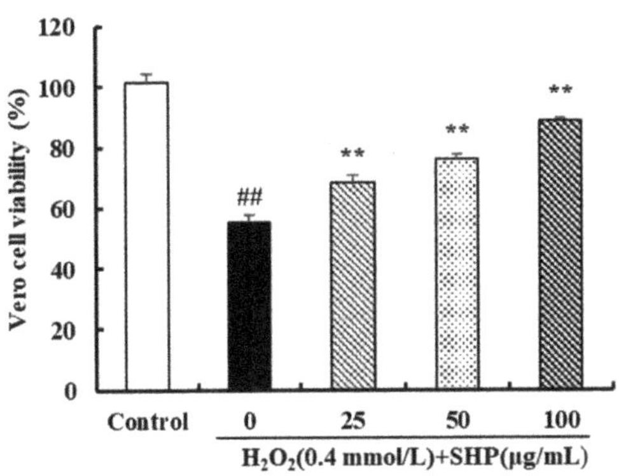

Figure 2. The effect of SHP on H_2O_2-induced oxidative damage in Vero cells. ## $p < 0.01$ vs. control group; ** $p < 0.01$ vs. H_2O_2 injury group.

3.4. The Scavenging Effect of SHP on ROS in Vero Cells

Excessive levels of ROS can disrupt redox homeostasis and damage cell macromolecular structures such as cell membranes, proteins, and lipids, ultimately leading to apoptosis or cell death. Thus, it is of significance to maintain healthy ROS levels within cells [42]. The effects of SHP on ROS levels in Vero cells are shown in Figure 3. Compared with the control group, the fluorescence intensity of H_2O_2-induced Vero cells was significantly increased by 158.06% ($p < 0.01$). However, when the concentration of SHP was increased to 100 μg mL^{-1}, the ROS level in Vero cells decreased to 116.01% ($p < 0.05$) (Figure 3). Meanwhile, the fluorescence microscopy images provide additional corroboration of this result (Figure 3). The aforementioned results demonstrate that SHP can effectively remove intracellular ROS to alleviate oxidative stress damage in Vero cells.

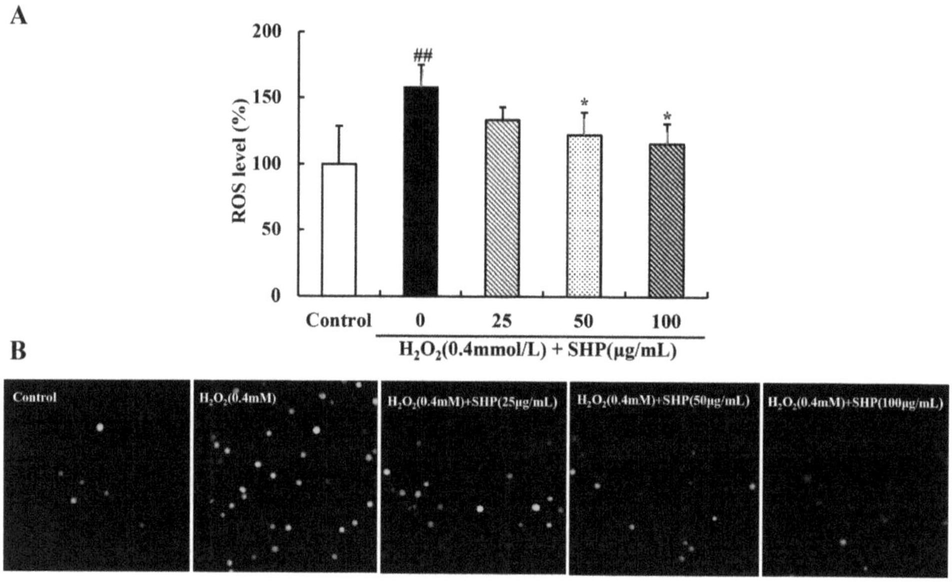

Figure 3. The effect of SHP on ROS generation in Vero cells induced by H_2O_2. (**A**) Results of the determination of fluorescence intensity; (**B**) fluorescence images. $^{\#\#}$ $p < 0.01$ vs. control group; * $p < 0.05$ vs. H_2O_2 injury group.

3.5. Effects of SHP on Intracellular MDA Content and SOD Activity

The SOD activity and MDA content in Vero cells are presented in Table 1. In contrast to the control group, the SOD activity of Vero cells was significantly decreased and the MDA content was significantly increased due to H_2O_2 induction ($p < 0.01$), indicating a high level of oxidative stress state. However, Vero cells treated with SHP (25, 50, and 100 μg mL^{-1}) showed a significant increase in SOD activity compared to the induction group, while MDA levels were significantly reduced and returned to normal levels ($p < 0.05$ or $p < 0.01$). It is proved that SHP can reduce H_2O_2-induced oxidative damage in Vero cells.

Table 1. The effect of SHP on SOD activity and MDA content in Vero cells.

Sample	(μg/mL)	Vero	
		SOD (U/mgprot)	MDA (nmol/mgprot)
Control	-	52.43 ± 1.31	7.17 ± 1.29
H_2O_2	-	22.9 ± 4.46 $^{\#\#}$	19.04 ± 2.53 $^{\#\#}$
	25	28.28 ± 5.77	14.09 ± 0.51 **
SHP	50	30.80 ± 6.04	11.09 ± 2.10 **
	100	40.70 ± 8.95 **	9.39 ± 0.92 **

Note: $^{\#\#}$ $p < 0.01$ (control group vs. H_2O_2 injury group); ** $p < 0.01$ (SHP vs. H_2O_2 injury group).

3.6. Effects of H_2O_2 on Survival Rate and Hatching Rate of Zebrafish

The survival rate and hatching rate of zebrafish are affected by the concentration of H_2O_2 (Figure 4). To establish an inflammation model of zebrafish, it is necessary to determine a safe dose of H_2O_2. A low concentration of H_2O_2 has little effect on the survival rate of zebrafish; however, the survival rate gradually decreases with the increase in time (Figure 4B). When the concentration of H_2O_2 exceeds 7 mmol L^{-1}, the survival rate of zebrafish is 55% at 96 hpf, approximately half of the mortality rate ($p < 0.01$ or $p < 0.001$) (Figure 4). All zebrafish treated with H_2O_2 (13 mmol L^{-1}) died at 96 hpf, and those with 15 mmol/L H_2O_2 all died starting at 72 hpf ($p < 0.01$ or $p < 0.001$) (Figure 4). The hatching rate of zebrafish also decreases with the increase in H_2O_2 concentration (Figure 4C). Thus, 7 mmol L^{-1} was chosen as the H_2O_2-induced concentration for further experiments.

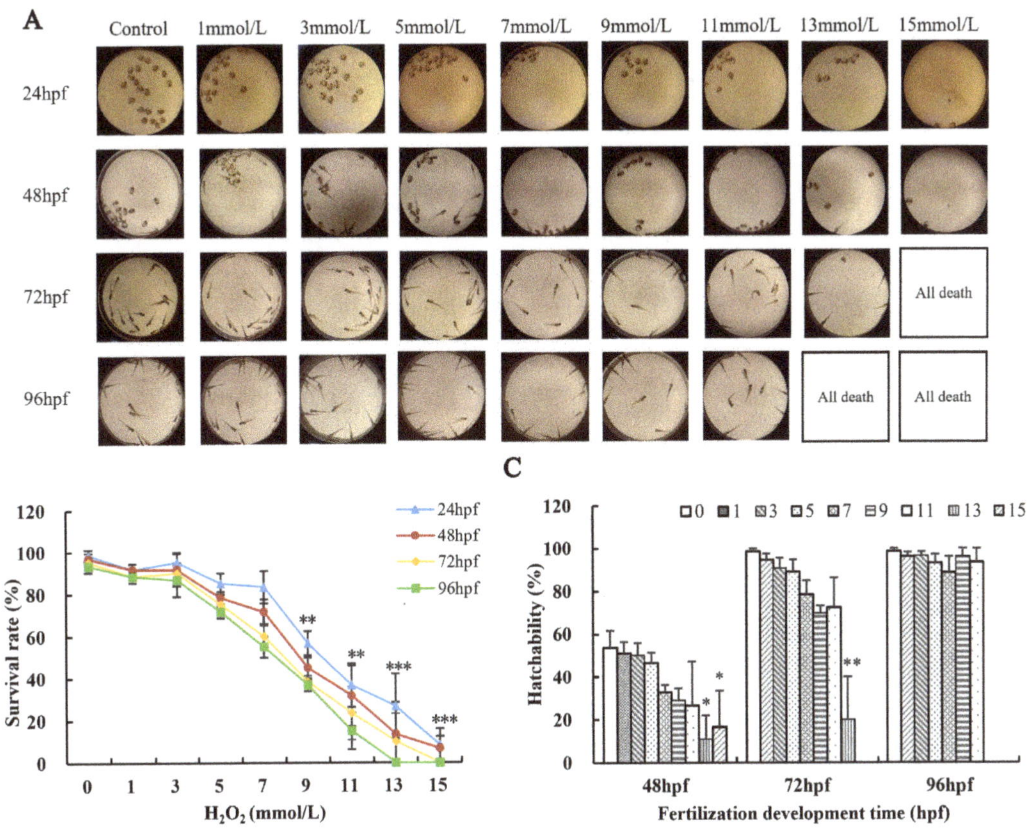

Figure 4. The effect of H_2O_2 on the survival and development of zebrafish. (**A**) Morphological growth and development diagram of zebrafish (×1.8 times); (**B**) survival rate of zebrafish; (**C**) hatching rate of zebrafish. Compared with the control group, * $p < 0.05$, ** $p < 0.01$, *** $p < 0.001$.

3.7. Effects of H_2O_2 on the Growth and Development of Zebrafish

High levels of H_2O_2 can exert an impact on the normal development of zebrafish, resulting in certain deformities (Figure 5). The body length of zebrafish was negatively correlated with an H_2O_2 concentration above 7 mmol L^{-1}, while the malformation rate was positively correlated with H_2O_2 ($p < 0.01$ or $p < 0.001$) (Figure 5). The morphology of zebrafish from fertilization to adult stage was observed under the microscope (Figure 5). It was found that the zebrafish in the control group showed normal body development and a uniform pigment distribution, whereas those treated with H_2O_2 showed various deformities (Figure 5C).

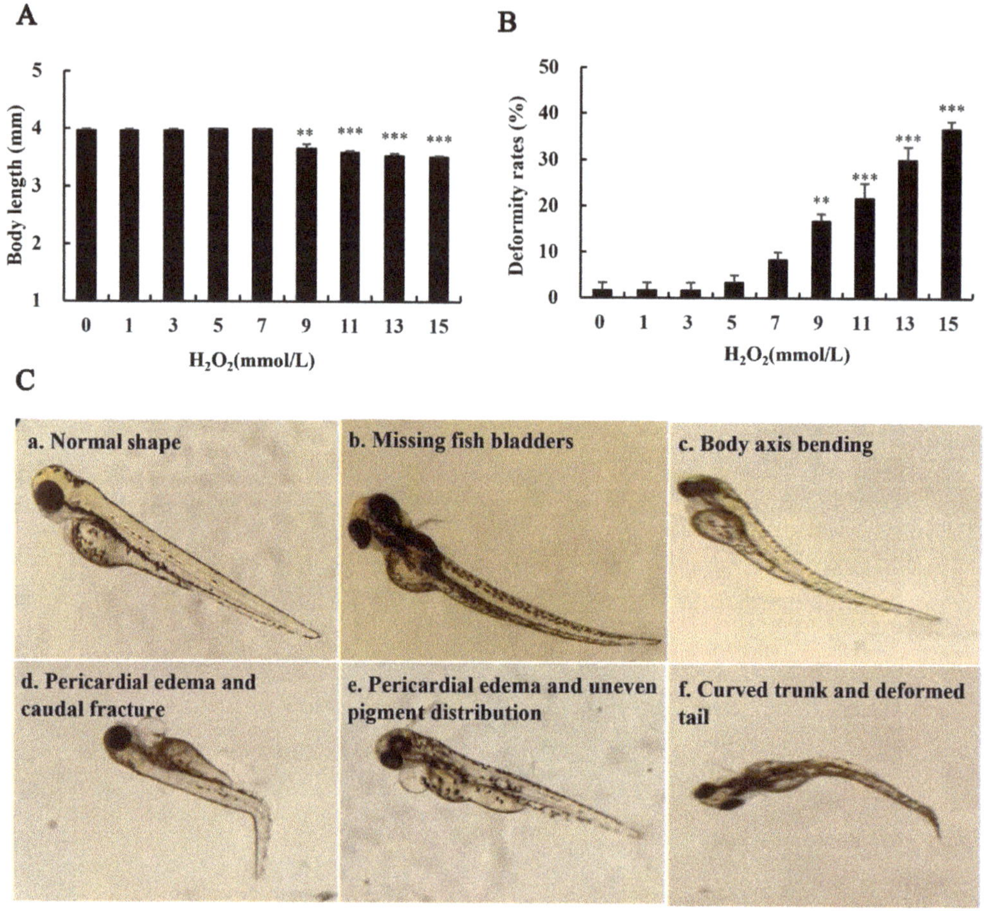

Figure 5. The effect of H_2O_2 on the morphological development of zebrafish (×4 times). (**A**) The body length of zebrafish; (**B**) deformity rates; (**C**) abnormal morphology of zebrafish (×4 times). Compared with the control group, ** $p < 0.01$, *** $p < 0.001$.

3.8. Effects of SHP on H_2O_2-Induced Zebrafish

We determined the survival rate of zebrafish embryos to assess the protective effect of SHP on H_2O_2-induced zebrafish embryos (Figure 6). The survival rate of zebrafish embryos decreased rapidly in the H_2O_2 (7 mmol L^{-1}) environment in contrast to the control group. However, upon the addition of SHP (25, 50, 100 μg mL^{-1}), the survival rate of zebrafish embryos improved with an ascending concentration of SHP ($p < 0.01$ or $p < 0.001$). These experimental findings imply that SHP has a protective effect on H_2O_2-induced zebrafish embryos. The influence of SHP on the H_2O_2-induced zebrafish heart rate was observed under the microscope (Figure 6B). The H_2O_2-induced heart rate in zebrafish increased significantly to 117.33% compared with the control group ($p < 0.01$). Nevertheless, after pre-treatment with SHP, the heart rate of the zebrafish decreased in a dose-dependent manner to 106.33% ($p < 0.01$). Additionally, the deformity of zebrafish was greatly improved under SHP treatment. The results showed that SHP possesses an outstanding antioxidant capacity.

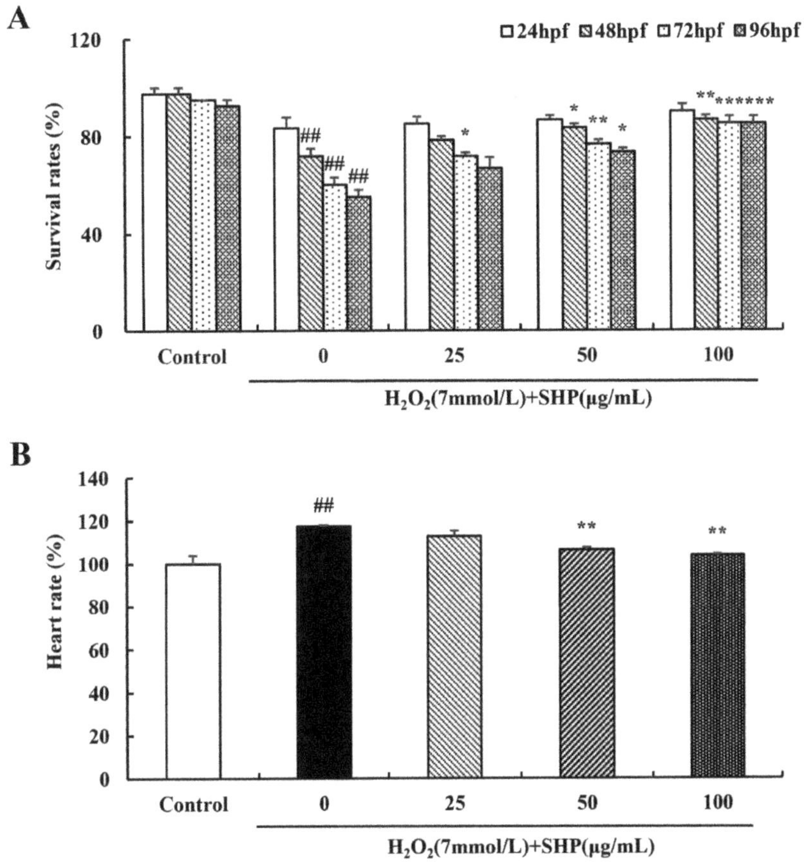

Figure 6. The effect of SHP on the survival and development of zebrafish embryos induced by H_2O_2 (**A**) The survival rate of zebrafish; (**B**) the heart rate of zebrafish. $^{\#\#}$ $p < 0.01$ vs. control group; * $p < 0.05$, ** $p < 0.01$, *** $p < 0.001$ vs. H_2O_2 model group.

3.9. Effects of SHP on the Activities of SOD and CAT and on the Levels of GSH-PX and MDA in Zebrafish

To further investigate the protective mechanism of SHP against oxidative stress in zebrafish embryos, we measured the activities of SOD and CAT, as well as the contents of GSH-PX and MDA in zebrafish at 96 hpf (Figure 7). The activities of SOD and CAT and the content of GSH-PX in zebrafish in the induction group (7 mmol L^{-1} H_2O_2) were significantly decreased compared with the control group. However, after pretreatment with different doses of SHP, the activities of SOD and CAT and the content of GSH-PX increased in a dose-dependent manner (Figure 7A–C). The effect of SHP on MDA content in zebrafish is shown in Figure 7D. The MDA content in zebrafish in the induction group showed a significant increase, reaching 147.14%, and the difference was statistically significant compared to the control group ($p < 0.05$). Nevertheless, after pretreatment with SHP, the MDA content in zebrafish decreased in a dose-dependent manner.

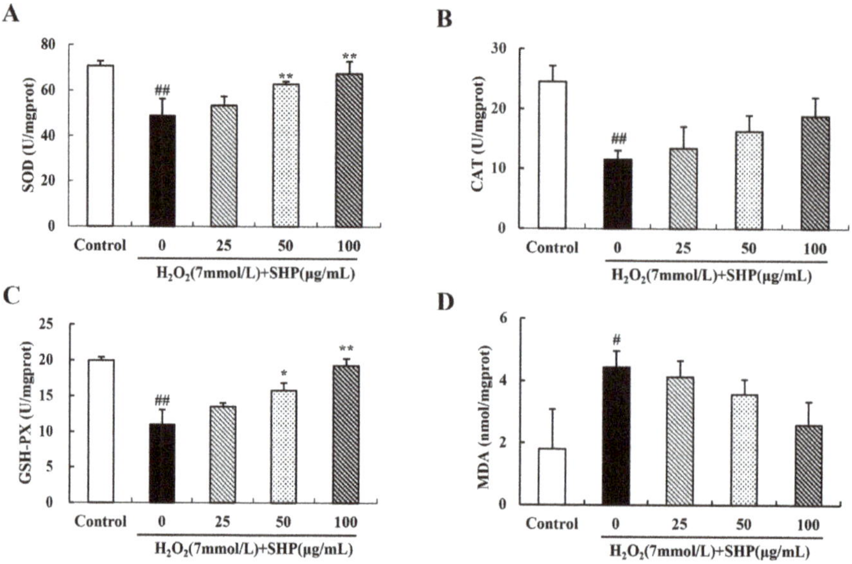

Figure 7. The effect of SHP on oxidative stress-related indexes of zebrafish larvae induced by H_2O_2 (**A**) SOD activity histogram; (**B**) CAT activity histogram; (**C**) histogram of GSH-PX content; (**D**) bar chart of MDA content. $^{\#}$ $p < 0.05$, $^{\#\#}$ $p < 0.01$ vs. control group; * $p < 0.05$, ** $p < 0.01$ vs. H_2O_2 model group.

3.10. The Role of SHP on H_2O_2-Induced ROS in Zebrafish

The DCFH-DA probe was used to determine the ROS content in zebrafish, thereby assessing the scavenging capacity of SHP for ROS [43]. The fluorescence of zebrafish in the control group was negligible, whereas that of zebrafish in the induction group was significantly enhanced. However, the fluorescence intensity gradually declined after SHP pretreatment (Figure 8A). Similar results were shown for the relative ROS level. Compared with the control group, the ROS level in the induction group was highly expressed (165.51%), but after SHP (25, 50, 100 μg mL^{-1}) pretreatment, the ROS level in the induction group decreased to 148.10%, 120.12%, 108.46% ($p < 0.05$ or $p < 0.01$) (Figure 8B).

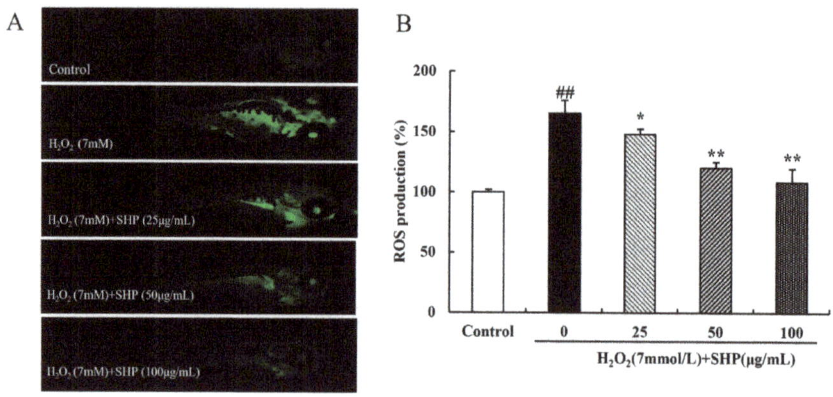

Figure 8. The effect of SHP on H_2O_2-induced ROS generation in zebrafish. (**A**) Fluorescence images of ROS expression in zebrafish; (**B**) relative ROS expression levels in zebrafish. $^{\#\#}$ $p < 0.01$ vs. control group; * $p < 0.05$, ** $p < 0.01$ vs. H_2O_2 injury group.

4. Discussion

Oxidative and antioxidative regulatory mechanisms exist in organisms. When the organism is stimulated by the external environment, if the balance between the two is disturbed, the organism will trigger oxidative stress, which destroys the macromolecular structure of cells and ultimately results in cell apoptosis or death [44]. Previous studies have reported that the activity or content of antioxidant defense factors under oxidative stress can indirectly indicate the degree of oxidative stress in the body [41,45]. Several previous studies have also evaluated the in vitro antioxidant activity of polysaccharides from *S. horneri*. For instance, Shao et al. [22] evaluated the capacity of polysaccharides derived from *S. horneri* to neutralize free radicals. Additionally, Wen et al. [23] further clarified that polysaccharides from *S. horneri* exhibit excellent antioxidant effects by regulating gene expression levels of key enzymes involved in oxidative stress in RAW264.7 cells. We not only selected Vero cells to evaluate the in vitro antioxidant capacity of polysaccharides from *S. horneri* (SHP) but also supplemented the in vivo antioxidant activity of SHP using zebrafish as the model organism for the first time.

In this research, Vero cells were exposed to diverse concentrations of H_2O_2. The concentration of 0.4 mmol L^{-1} induced by H_2O_2 was chosen to establish an oxidative stress model, which was then utilized to investigate the antioxidant effects of SHP in vitro. The findings indicated that SHP exerted no cytotoxic impact on Vero cells and enhanced cell viability. Additionally, the ROS levels in H_2O_2-induced Vero cells were significantly higher compared to the control group, while ROS levels decreased in a dose-dependent manner following SHP treatment.

Under normal conditions, cells possess an endogenous antioxidant system that shields them from oxidative harm by generating antioxidant enzymes such as SOD, CAT, and GSH-PX [46,47]. SOD constitutes the most critical line of defense within the antioxidant enzyme defense system. It is capable of eliminating ROS, particularly superoxide anion free radicals, from cells, thereby enhancing the antioxidant capacity of the organism [41,48]. Our findings demonstrated that SOD activity decreased significantly in Vero cells induced by H_2O_2 compared to the control group, suggesting that a large number of free radicals were produced by the cells at this time and the oxidative stress was intensified. Meanwhile, the SHP treatment augmented SOD activity in a dose-dependent manner, indicating that the extent of oxidative damage in cells was mitigated.

In biological systems, lipid peroxidation is considered a toxic phenomenon that leads to a wide range of pathological conditions. Moreover, H_2O_2 intensifies the status of lipid peroxidation in cells [49,50]. The MDA is regarded as the ultimate decomposition product of lipid peroxidation, and its content can mirror the level of oxidative damage in cells [39]. Hence, the MDA content was determined in this study, and it was discovered that SHP reduces MDA to prevent the accumulation of free radicals in Vero cells. In summary, the antioxidant potential of SHP in protecting Vero cells from H_2O_2-induced oxidative stress was confirmed.

Several studies have demonstrated that oxidative stress can reduce the survival of zebrafish and result in malformation [51–53]. Considering the survival rate, growth, and development of zebrafish, we chose 7 mmol L^{-1} of H_2O_2 for zebrafish induction. Furthermore, we discovered that SHP enhances the survival of zebrafish embryos in a dose-dependent manner and mitigates zebrafish damage caused by oxidative stress. Additionally, H_2O_2 induction significantly increased ROS in zebrafish, but SHP treatment exhibited a dose-dependent decrease in ROS levels in zebrafish, which was also confirmed by fluorescence images under microscopy. Moreover, the enzymatic activity of SOD, CAT, and GSH-PX was significantly increased in zebrafish, and the MDA content was effectively suppressed, thereby protecting zebrafish from H_2O_2-induced oxidative damage. In conclusion, firstly, SHP might lower ROS levels and eliminate oxidative free radicals by enhancing the enzymatic activity of the oxidative defense system, thereby reducing the oxidative stress response of the organism. Secondly, the sulfates and uronic acids present in polysaccharides also enhance hydrogen donation and thus exert antioxidation activities to convert ROS into inactive or stable compounds [42]. Thus, polysaccharides are indispensable contributors to maintaining the radical scavenging system.

5. Conclusions

SHP derived from *S. horneri* can effectively safeguard the vitality of Vero cells, enhance the activity of the SOD enzyme, lower the content of MDA, eliminate intracellular ROS, and mitigate the oxidative damage caused by H_2O_2. Meanwhile, SHP can enhance the activities of SOD, CAT, and GSH-PX enzymes in zebrafish, effectively suppressing the levels of MDA and ROS, thereby reducing the oxidative damage caused by H_2O_2 and exerting a protective effect on zebrafish. Utilizing zebrafish as a model organism, we present, for the first time, a detailed demonstration of the antioxidant effects of SHP in vivo. This study validates the potential of SHP as a natural antioxidant to augment the protective effects against oxidative stress and to be considered for applications in the biomedical, cosmetic, and functional food industries.

Author Contributions: S.W.: Data analysis. L.W.: Manuscript writing, Submission. J.Y.: Checking of the manuscript. R.X.: Conceptualization. R.J.: Funding acquisition, Supervision. P.H.: Funding acquisition. All authors have read and agreed to the published version of the manuscript.

Funding: This work was financed by the National Key Research and Development Program of China (2022YFC2601303), the National Key Research and Development Plan (2019YFC0312604), the Shanghai Agriculture Science and Technology Innovation Project (2017, No 1-13), the Shanghai Science and Technology Commission Innovation Project (Nos. 16391903500, and 17391902200), and the National Marine 863 Project (No. 2014AA093506).

Institutional Review Board Statement: The use of zebrafish adhered to the document of the Animal Ethics Committee of Shanghai Ocean University (SHOUDW-2016-003).

Informed Consent Statement: Not applicable.

Data Availability Statement: The datasets generated during and/or analyzed during the current study are available from the corresponding author on reasonable request.

Acknowledgments: The authors are grateful to the anonymous reviewers for their valuable comments and contributions to improve the manuscript.

Conflicts of Interest: There is no conflict of interests in this study.

References

1. Fernando, I.P.S.; Kim, M.; Son, K.T.; Jeong, Y.; Jeon, Y.J. Antioxidant activity of marine algal polyphenolic compounds: A mechanistic approach. *J. Med. Food* **2016**, *19*, 615–628. [CrossRef]
2. Wang, L.; Jo, M.J.; Katagiri, R.; Harata, K.; Ohta, M.; Ogawa, A.; Kamegai, M.; Ishida, Y.; Tanoue, S.; Kimura, S.; et al. Antioxidant effects of citrus pomace extracts processed by super-heated steam. *LWT Food Sci. Technol.* **2018**, *90*, 331–338. [CrossRef]
3. Kim, Y.S.; Hwang, J.W.; Sung, S.H.; Jeon, Y.J.; Jeong, J.H.; Jeon, B.T.; Moon, S.H.; Park, P.J. Antioxidant activity and protective effect of extract of Celosia cristata L. flower on tert-butyl hydroperoxide-induced oxidative hepatotoxicity. *Food Chem.* **2015**, *168*, 572–579. [CrossRef]
4. Begum, R.; Howlader, S.; Mamun-Or-Rashid, A.N.M.; Rafiquzzaman, S.M.; Ashraf, G.M.; Albadrani, G.M.; Sayed, A.A.; Peluso, I.; Abdel-Daim, M.M.; Uddin, M.S. Antioxidant and signal-modulating effects of brown seaweed-derived compounds against oxidative stress-associated pathology. *Oxidative Med. Cell. Longev.* **2021**, *2021*, 9974890. [CrossRef]
5. Sarniak, A.; Lipińska, J.; Tytman, K.; Lipińska, S. Endogenous mechanisms of reactive oxygen species (ROS) generation. *Adv. Hyg. Exp. Med.* **2016**, *70*, 1150–1165. [CrossRef]
6. Wojtovich, A.P.; Berry, B.J.; Galkin, A. Redox signaling through compartmentalization of reactive oxygen species: Implications for health and disease. *Antioxid. Redox Signal.* **2019**, *31*, 591–593. [CrossRef]
7. Rezayian, M.; Niknam, V.; Ebrahimzadeh, H. Oxidative damage and antioxidative system in algae. *Toxicol. Rep.* **2019**, *6*, 1309–1313. [CrossRef]
8. Wang, L.; Oh, J.Y.; Je, J.G.; Jayawardena, T.U.; Kim, Y.S.; Ko, J.Y.; Fu, X.T.; Jeon, Y.J. Protective effects of sulfated polysaccharides isolated from the enzymatic digest of Codium fragile against hydrogen peroxide-induced oxidative stress in in vitro and in vivo models. *Algal Res.* **2020**, *48*, 101891. [CrossRef]

9. Wang, L.; Oh, J.Y.; Yang, H.W.; Kim, H.S.; Jeon, Y.J. Protective effect of sulfated polysaccharides from a Celluclast-assisted extract of *Hizikia fusiforme* against ultraviolet B-induced photoaging in vitro in human keratinocytes and in vivo in zebrafish. *Mar. Life Sci. Technol.* **2019**, *1*, 104–111. [CrossRef]
10. Dimitrova-Shumkovska, J.; Krstanoski, L.; Veenman, L. Potential beneficial actions of fucoidan in brain and liver injury, disease, and intoxication-potential implication of Sirtuins. *Mar. Drugs* **2020**, *18*, 242. [CrossRef]
11. Saraswati; Giriwono, P.E.; Iskandriati, D.; Tan, C.P.; Andarwulan, N. *Sargassum* seaweed as a source of anti-inflammatory substances and the potential insight of the tropical species: A review. *Mar. Drugs* **2019**, *17*, 590. [CrossRef]
12. Riccio, G.; Lauritano, C. Microalgae with immunomodulatory activities. *Mar. Drugs* **2020**, *18*, 2. [CrossRef]
13. Marques, M.L.M.; Presa, F.B.; Viana, R.L.S.; Costa, M.; Amorim, M.O.R.; Bellan, D.L.; Alves, M.; Costa, L.S.; Trindade, E.S.; Rocha, H.A.O. Anti-thrombin, anti-adhesive, anti-migratory, and anti-proliferative activities of sulfated galactans from the tropical green seaweed, *Udotea flabellum*. *Mar. Drugs* **2019**, *17*, 5. [CrossRef] [PubMed]
14. Adrien, A.; Bonnet, A.; Dufour, D.; Baudouin, S.; Maugard, T.; Bridiau, N. Anticoagulant activity of sulfated *Ulvan* isolated from the green macroalga *Ulva rigida*. *Mar. Drugs* **2019**, *17*, 291. [CrossRef]
15. Jun, J.Y.; Jung, M.J.; Jeong, I.H.; Yamazaki, K.; Kawai, Y.; Kim, B.M. Antimicrobial and antibiofilm activities of sulfated polysaccharides from marine algae against dental plaque bacteria. *Mar. Drugs* **2018**, *16*, 301. [CrossRef]
16. Rasmussen, R.S.; Morrissey, M.T. Marine biotechnology for production of food ingredients. *Adv. Food Nutr. Res.* **2007**, *52*, 237–292. [CrossRef]
17. Wang, L.; Oh, J.Y.; Kim, H.S.; Lee, W.; Cui, Y.R.; Lee, H.G.; Kim, Y.T.; Ko, J.Y.; Jeon, Y.J. Protective effect of polysaccharides from Celluclast-assisted extract of *Hizikia fusiforme* against hydrogen peroxide-induced oxidative stress in vitro in Vero cells and in vivo in zebrafish. *Int. J. Biol. Macromol.* **2018**, *112*, 483–489. [CrossRef]
18. Nichols, J.A.; Katiyar, S.K. Skin photoprotection by natural polyphenols: Anti-inflammatory, antioxidant and DNA repair mechanisms. *Arch. Dermatol. Res.* **2010**, *302*, 71–83. [CrossRef]
19. Kim, E.A.; Lee, S.H.; Ko, C.I.; Cha, S.H.; Kang, M.C.; Kang, S.M.; Ko, S.C.; Lee, W.W.; Ko, J.Y.; Lee, J.H.; et al. Protective effect of fucoidan against AAPH-induced oxidative stress in zebrafish model. *Carbohydr. Polym.* **2014**, *102*, 185–191. [CrossRef]
20. Jayawardena, T.U.; Wang, L.; Sanjeewa, K.K.A.; Kang, S.I.; Lee, J.S.; Jeon, Y.J. Antioxidant potential of sulfated polysaccharides from *Padina boryana*: Protective effect against oxidative stress in an in vitro and in vivo zebrafish model. *Mar. Drugs* **2020**, *18*, 212. [CrossRef]
21. Shao, P.; Chen, X.; Sun, P. Improvement of antioxidant and moisture-preserving activities of *Sargassum horneri* polysaccharide enzymatic hydrolyzates. *Int. J. Biol. Macromol.* **2015**, *74*, 420–427. [CrossRef] [PubMed]
22. Shao, P.; Chen, X.; Sun, P. Chemical characterization, antioxidant and antitumor activity of sulfated polysaccharide from *Sargassum horneri*. *Carbohydr. Polym.* **2014**, *105*, 260–269. [CrossRef] [PubMed]
23. Wen, Z.S.; Liu, L.J.; OuYang, X.K.; Qu, Y.L.; Chen, Y.; Ding, G.F. Protective effect of polysaccharides from *Sargassum horneri* against oxidative stress in RAW264.7 cells. *Int. J. Biol. Macromol.* **2014**, *68*, 98–106. [CrossRef]
24. Jesumani, V.; Du, H.; Pei, P.; Zheng, C.; Cheong, K.L.; Huang, N. Unravelling property of polysaccharides from *Sargassum* sp. as an anti-wrinkle and skin whitening property. *Int. J. Biol. Macromol.* **2019**, *140*, 216–224. [CrossRef]
25. Wen, Z.S.; Xiang, X.W.; Jin, H.X.; Guo, X.Y.; Liu, L.J.; Huang, Y.N.; OuYang, X.K.; Qu, Y.L. Composition and anti-inflammatory effect of polysaccharides from *Sargassum horneri* in RAW264.7 macrophages. *Int. J. Biol. Macromol.* **2016**, *88*, 403–413. [CrossRef]
26. Sanjeewa, K.K.A.; Fernando, I.P.S.; Kim, S.Y.; Kim, H.S.; Ahn, G.; Jee, Y.; Jeon, Y.J. In vitro and in vivo anti-inflammatory activities of high molecular weight sulfated polysaccharide; containing fucose separated from *Sargassum horneri*: Short communication. *Int. J. Biol. Macromol.* **2018**, *107*, 803–807. [CrossRef] [PubMed]
27. Sanjeewa, K.K.A.; Jayawardena, T.U.; Kim, S.Y.; Kim, H.S.; Ahn, G.; Kim, J.; Jeon, Y.J. Fucoidan isolated from invasive *Sargassum horneri* inhibit LPS-induced inflammation via blocking NF-κB and MAPK pathways. *Algal Res.* **2019**, *41*, 101561. [CrossRef]
28. Sanjeewa, K.K.A.; Jayawardena, T.U.; Kim, S.Y.; Lee, H.G.; Je, J.G.; Jee, Y.; Jeon, Y.J. *Sargassum horneri* (Turner) inhibit urban particulate matter-induced inflammation in MH-S lung macrophages via blocking TLRs mediated NF-kappa B and MAPK activation. *J. Ethnopharmacol.* **2020**, *249*, 112363. [CrossRef] [PubMed]
29. Shao, P.; Liu, J.; Chen, X.; Fang, Z.; Sun, P. Structural features and antitumor activity of a purified polysaccharide extracted from *Sargassum horneri*. *Int. J. Biol. Macromol.* **2015**, *73*, 124–130. [CrossRef]
30. Lee, P.T.; Quan Tran, H.T.; Huang, H.T.; Nan, F.H.; Lee, M.C. *Sargassum horneri* extracts stimulate innate immunity, enhance growth performance, and upregulate immune genes in the white shrimp *Litopenaeus vannamei*. *Fish Shellfish. Immunol.* **2020**, *102*, 276–285. [CrossRef] [PubMed]
31. Dias, M.K.H.M.; Madusanka, D.M.D.; Han, E.J.; Kim, H.S.; Jeon, Y.J.; Jee, Y.; Kim, K.N.; Lee, K.; Fernando, I.P.S.; Ahn, G. *Sargassum horneri* (Turner) C. Agardh ethanol extract attenuates fine dust-induced inflammatory responses and impaired skin barrier functions in HaCaT keratinocytes. *J. Ethnopharmacol.* **2021**, *273*, 114003. [CrossRef]
32. Cruz-Rivera, E.; Flores-Diaz, M.; Hawkins, A. A fish kill coincident with dense *Sargassum* accumulation in a tropical bay. *Bull. Mar. Sci.* **2015**, *91*, 455–456. [CrossRef]
33. Smetacek, V.; Zingone, A. Green and golden seaweed tides on the rise. *Nature* **2013**, *504*, 84–88. [CrossRef]
34. Schell, J.M.; Goodwin, D.S.; Siuda, A.N.S. Recent *Sargassum* inundation events in the Caribbean shipboard observations reveal dominance of a previously rare form. *Oceanography* **2015**, *28*, 8–10. [CrossRef]

35. Sfriso, A.; Facca, C. Annual growth and environmental relationships of the invasive species *Sargassum muticum* and *Undaria pinnatifida* in the lagoon of Venice. *Estuar. Coast. Shelf Sci.* **2013**, *129*, 162–172. [CrossRef]
36. Wei, S.Y.; Cai, C.N.; He, P.M.; Jia, R. Protective Effect of Polysaccharide from *Sargassum horneri* on H_2O_2-induced Oxidative Stress in HaCaT Cells. *J. Trop. Subtrop. Bot.* **2023**, *31*, 232–240. [CrossRef]
37. Wang, L.; Kim, H.S.; Oh, J.Y.; Je, J.G.; Jeon, Y.J.; Ryu, B. Protective effect of diphlorethohydroxycarmalol isolated from *Ishige okamurae* against UVB-induced damage in vitro in human dermal fibroblasts and in vivo in zebrafish. *Food Chem. Toxicol.* **2020**, *136*, 110963. [CrossRef]
38. Pallavi, S.; Bhushan, J.A.; Shanker, D.R.; Mohammad, P. Reactive oxygen species, oxidative damage, and antioxidative defense mechanism in plants under stressful conditions. *J. Bot.* **2012**, *2012*, 217037. [CrossRef]
39. Zhou, Y.H.; Yu, J.P.; Liu, Y.F.; Teng, X.J.; Ming, M.; Lv, P.; An, P.; Liu, S.Q.; Yu, H.G. Effects of Ginkgo biloba extract on inflammatory mediators (SOD, MDA, TNF-alpha, NF-kappa Bp65, IL-6) in TNBS-induced colitis in rats. *Mediat. Inflamm.* **2006**, *2006*, 92642. [CrossRef] [PubMed]
40. Kim, S.Y.; Kim, E.A.; Kim, Y.S.; Yu, S.K.; Choi, C.; Lee, J.S.; Kim, Y.T.; Nah, J.W.; Jeon, Y.J. Protective effects of polysaccharides from *Psidium guajava* leaves against oxidative stresses. *Int. J. Biol. Macromol.* **2016**, *91*, 804–811. [CrossRef]
41. Ighodaro, O.M.; Akinloye, O.A. First line defence antioxidants-superoxide dismutase (SOD), catalase (CAT) and glutathione peroxidase (GPX): Their fundamental role in the entire antioxidant defence grid. *Alex. J. Med.* **2018**, *54*, 287–293. [CrossRef]
42. Zeng, F.S.; Yao, Y.F.; Wang, L.F.; Li, W.J. Polysaccharides as antioxidants and prooxidants in managing the double-edged sword of reactive oxygen species. *Biomed. Pharmacother.* **2023**, *159*, 114221. [CrossRef]
43. Kang, S.M.; Heo, S.J.; Kim, K.N.; Lee, S.H.; Jeon, Y.J. Isolation and identification of new compound, 2,7″-phloroglucinol-6,6′-bieckol from brown algae, *Ecklonia cava* and its antioxidant effect. *J. Funct. Foods* **2012**, *4*, 158–166. [CrossRef]
44. Hajam, Y.A.; Rani, R.; Ganie, S.Y.; Sheikh, T.A.; Javaid, D.; Qadri, S.S.; Pramodh, S.; Alsulimani, A.; Alkhanani, M.F.; Harakeh, S.; et al. Oxidative stress in human pathology and aging: Molecular Mechanisms and Perspectives. *Cells* **2022**, *11*, 552. [CrossRef]
45. Stara, A.; Kristan, J.; Zuskova, E.; Velisek, J. Effect of chronic exposure to prometryne on oxidative stress and antioxidant response in common carp (*Cyprinus carpio* L.). *Pestic. Biochem. Physiol.* **2013**, *105*, 18–23. [CrossRef]
46. Vijay Sankar, N.P.; Jagtap, A.S.; Baghel, R.S.; Imchen, T.; Manohar, C.S. Chapter 28—Elucidation of the antioxidant potential of marine macroalgal biomolecules for healthcare applications: Current status and future prospects. In *Marine Antioxidants*; Kim, S.-K., Shin, K.-H., Venkatesan, J., Eds.; Academic Press: Cambridge, MA, USA, 2023; pp. 365–377. [CrossRef]
47. Miao, L.; St Clair, D.K. Regulation of superoxide dismutase genes: Implications in disease. *Free Radic. Biol. Med.* **2009**, *47*, 344–356. [CrossRef]
48. Imen, M.Y.B. Free radical metabolism in human erythrocytes. *Clin. Chim. Acta* **2008**, *390*, 1–11. [CrossRef]
49. Yokozawa, T.; Liu, Z.W.; Chen, C.P. Protective effects of Glycyrrhizae radix extract and its compounds in a renal hypoxia (ischemia)-reoxygenation (reperfusion) model. *Phytomedicine* **2000**, *6*, 439–445. [CrossRef]
50. Lee, J.H.; Ko, J.Y.; Oh, J.Y.; Kim, E.A.; Kim, C.Y.; Jeon, Y.J. Evaluation of phlorofucofuroeckol-A isolated from *Ecklonia cava* (Phaeophyta) on anti-lipid peroxidation in vitro and in vivo. *Algae* **2015**, *30*, 313–323. [CrossRef]
51. Chowdhury, S.; Saikia, S.K. Use of zebrafish as a model organism to study oxidative stress: A review. *Zebrafish* **2022**, *19*, 165–176. [CrossRef]
52. Gyimah, E.; Zhu, X.; Zhang, Z.; Guo, M.; Xu, H.; Mensah, J.K.; Dong, X.; Zhang, Z.; Gyimah, G.N.W. Oxidative stress and apoptosis in bisphenol AF-induced neurotoxicity in zebrafish embryos. *Environ. Toxicol. Chem.* **2022**, *41*, 2273–2284. [CrossRef] [PubMed]
53. Hahn, M.E.; Timme-Laragy, A.R.; Karchner, S.I.; Stegeman, J.J. Nrf2 and Nrf2-related proteins in development and developmental toxicity: Insights from studies in zebrafish (*Danio rerio*). *Free Radic. Biol. Med.* **2015**, *88*, 275–289. [CrossRef] [PubMed]

biology

MDPI

Article

Phytoplankton Diversity, Spatial Patterns, and Photosynthetic Characteristics Under Environmental Gradients and Anthropogenic Influence in the Pearl River Estuary

Jing Xia [1,†], Haojie Hu [1,†], Xiu Gao [1], Jinjun Kan [2], Yonghui Gao [3,4,*] and Ji Li [3,4,*]

1 School of Oceanography, Shanghai Jiao Tong University, Shanghai 200030, China; xiajingsherry@sjtu.edu.cn (J.X.); huhaojie@sjtu.edu.cn (H.H.); shayegao@sjtu.edu.cn (X.G.)
2 Stroud Water Research Center, 970 Spencer Rd., Avondale, PA 19311, USA; jkan@stroudcenter.org
3 Key Laboratory of Polar Ecosystem and Climate Change, Ministry of Education, School of Oceanography, Shanghai Jiao Tong University, Shanghai 200030, China
4 Key Laboratory for Polar Science, Polar Research Institute of China, Ministry of Natural Resources, Shanghai 200136, China
* Correspondence: ygao80@sjtu.edu.cn (Y.G.); liji81@sjtu.edu.cn (J.L.)
† These authors contributed equally to this work.

Simple Summary: The Pearl River Estuary (PRE) in China, a highly urbanized coastal area, presents a unique opportunity to study the effects of environmental changes on the phytoplankton community. In September 2018, a field study was conducted to examine how the photosynthetic status and spatial distribution of these organisms varied from freshwater to oceanic waters. Dinophyta and Haptophyta were prevalent in seawater, while Chlorophyta and Cryptophyta dominated from freshwater to estuarine water. Bacillariophyta were found across all regions. Phytoplankton in the mixing zone showed signs of stress due to fluctuating environmental conditions, whereas those in freshwater and oceanic areas appeared more photosynthetically active. Human activities have in-creased nutrient levels in the estuary, leading to higher chlorophyll concentrations and more diverse phytoplankton communities upstream. Understanding these patterns helps us assess the health of coastal ecosystems, which is crucial for managing the impacts of climate change and human development on marine environments.

Abstract: The Pearl River Estuary (PRE) is one of the world's most urbanized subtropical coastal systems. It presents a typical environmental gradient suitable for studying estuarine phytoplankton communities' dynamics and photosynthetic physiology. In September 2018, the maximum photochemical quantum yield (Fv/Fm) of phytoplankton in different salinity habitats of PRE (oceanic, estuarine, and freshwater zones) was studied, revealing a complex correlation with the environment. Fv/Fm of phytoplankton ranged from 0.16 to 0.45, with taxa in the upper Lingdingyang found to be more stressed. Community composition and structure were analyzed using 18S rRNA, accompanied by a pigment analysis utilized as a supplementary method. Nonmetric multidimensional scaling analysis indicated differences in the phytoplankton spatial distribution along the estuarine gradients. Specificity-occupancy plots identified different specialist taxa for each salinity habitat. Dinophyta and Haptophyta were the predominant taxa in oceanic areas, while Chlorophyta and Cryptophyta dominated freshwater. Bacillariophyta prevailed across all salinity gradients. Canonical correlation analysis and Mantel tests revealed that temperature, salinity, and elevated nutrient levels (i.e., NO_3^--N, PO_4^{3-}-P, and SiO_3^{2-}-Si) associated with anthropogenic activities significantly influenced the heterogeneity of community structure. The spatial distribution of phytoplankton, along with in situ photosynthetic characteristics, serves as a foundational basis to access estuarine primary productivity, as well as community function and ecosystem health.

Keywords: estuary; phytoplankton; 18S rRNA; environmental factors; photosynthesis

1. Introduction

Estuaries serve as transitional zones connecting freshwater and marine environments [1–4], where complex hydrodynamics create various physical and chemical habitats, thus shaping the diversity of freshwater and saltwater organisms. A typical feature of estuaries is the salinity gradient formed by the mixing of two distinct water masses, which strongly influences community composition and diversity [1–3,5–7]. The physiochemical properties in estuaries, such as light availability, salinity, and nutrient availability, dynamically regulate the photosynthetic carbon fixation, biomass, and community composition of phytoplankton, resulting in spatial–temporal heterogeneity in phytoplankton distribution [5,8–11]. In addition, site-specific disturbance, climatology, and nutrient enrichment account for large-scale patterns of phytoplankton variability [12]. In recent decades, as estuaries accumulate both natural and excessive anthropogenic inputs of nutrients, there has been a trend of eutrophication, leading to an increase in phytoplankton biomass and the occurrence of harmful algal blooms [13,14]. As the primary producers in aquatic ecosystems, phytoplankton play a pivotal role not only in shaping the food webs but also in regulating the biogeochemical cycles. Indeed, a comprehensive understanding of the relationship between environmental conditions and phytoplankton composition is critical for the carbon cycles in the estuaries [15].

Phytoplankton photosynthesis accounts for most aquatic primary production, regulating phytoplankton growth, population dynamics, community composition, and succession [16,17]. Estuarine environmental gradients affect phytoplankton photosynthesis and metabolism, leading to variations in carbon assimilation and primary production [10,18–20]. Active variable chlorophyll fluorescence is a highly sensitive and rapid measurement of phytoplankton photosynthetic activity, indicating the impact of environmental stress, such as nutrient limitation and light inhibition [19,21,22]. The maximal photochemical efficiency (Fv/Fm) is a well-recognized indicator for assessing photosynthetic performance, directly reflecting the efficiency of photons that are converted into electrons by phytoplankton [16,19,23]. The Fv/Fm ratios typically show substantial variability across different marine environments, with increases from stressful to favorable conditions [16,17,19,24]. Specifically, Fv/Fm values can reach up to 0.65 in nutrient-rich areas and decrease below 0.3 in oligotrophic regions [24,25]. Overall, chlorophyll fluorescence contributes significantly to aquatic ecosystem studies, providing significant insights into the photochemical mechanisms of phytoplankton [10,16,17].

The Pearl River is China's third longest river with the second largest discharge, forming complex branches over its watershed and entering the South China Sea through eight outlets [26,27]. The Pearl River Estuary (PRE) is influenced by both the river discharge and the South China coastal current, contributing to its complex estuarine dynamics [26]. The Pearl River Delta urban agglomeration is one of the three major city clusters in China, with a high concentration of industries and an urbanization rate of 85.3% in 2018, resulting in a significant increase in urban land use [28,29]. Over the past 20 years, its population has increased by 1.8 times, reaching 78.01 million in 2020 [30]. These changes have contributed to the large nutrient loading into the PRE [31,32]. The average dissolved inorganic nitrogen (DIN) increased from 2.05×10^{-5} t/a in 1985–1995 to 3.91×10^{-5} t/a in 2014–2021 [32]. Dissolved inorganic phosphate (DIP) has also gradually increased over the past 20 years [31]. Consequently, eutrophication has caused frequent algal blooms [33]. The hydrology, physical transport, and biogeochemical processes of this region have been extensively investigated [27,34,35], along with records and descriptions of the spatial and temporal distribution of phytoplankton [7,8,36]. However, there is a lack of studies on the spatial patterns of the photosynthetic physiology of phytoplankton in this estuary and its connections with the environment and community. Understanding the spatial patterns of phytoplankton community structure in the PRE, along with the distribution characteristics of photosynthetic physiology related to primary productivity under environmental gradients and human impacts, is crucial for comprehending the responses of estuarine and coastal ecosystems to global climate change. The integration of active chlorophyll fluorescence with multiple analytical methods can facilitate a comprehensive understanding of phytoplankton physiology and community composition [37–39]. The application of 18S ribosomal RNA genes analysis provides a novel approach for gaining insights into the composition of eukaryotic unicellular phytoplankton communities [39,40]. Additionally, chemotaxonomy based on photosynthetic pigments is often used to characterize phytoplankton biomass and taxonomic groups [41–43]. This study employs high-throughput sequencing complemented by photosynthetic pigment analysis to access the spatial patterns of phytoplankton in the PRE. Concurrently, Fv/Fm is used as an indicator of the photosynthetic characteristics, providing a comprehensive investigation of the health status of the phytoplankton community under natural and anthropogenic stressors. This study aims to investigate the changes in the population structure and function of primary producers in a typical subtropical eutrophic estuary, providing baseline data for the response of local or similar highly urbanized estuarine ecosystems to climate change and offering factual evidence for coastal management policies.

2. Methods and Materials

2.1. Study Sites and Sample Collection

A comprehensive biological survey was conducted in the PRE on R.V. Haiyang-Dizhi 10 during 27–28 September 2018 (Figure 1). Water samples were collected using an underway system that consisted of a pump mounted at a depth of approximately 4 m. Based on phytoplankton biomass, 150–250 mL of water samples was filtered through each GF/F filter (25 mm, Whatman, UK) using a low-pressure vacuum pump. At least 3 filter samples were collected at each station, and the filters were stored in sterile bags and preserved at −80 °C in the dark until subsequent high-performance liquid chromatography (HPLC) and high-throughput sequencing. The filtrate water was stored in clean polyethylene bottles at −20 °C for nutrient analysis.

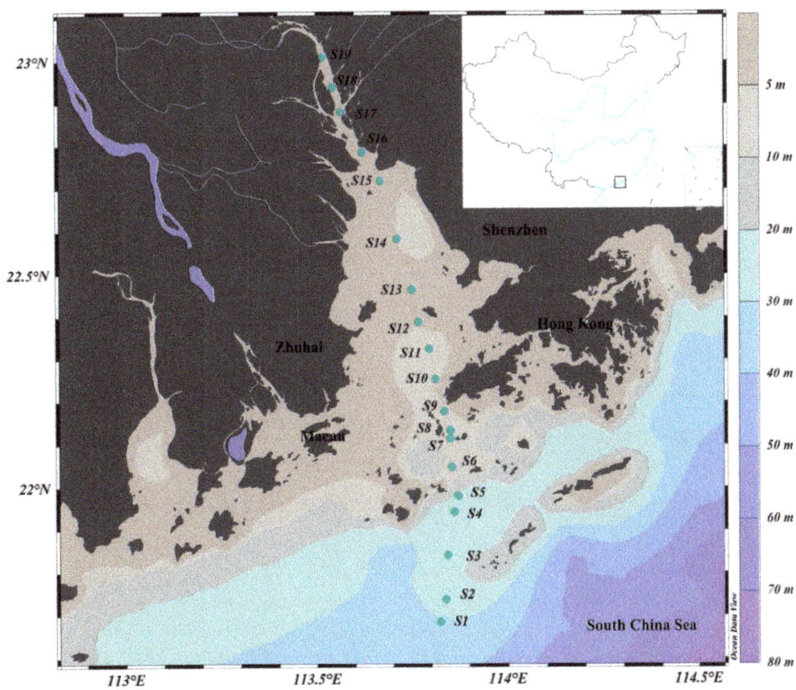

Figure 1. Sampling stations in the Pearl River Estuary. Color gradients represent ocean bathymetry.

2.2. Determination of Environmental Parameters

The water properties of temperature and salinity were measured using a thermosalinograph (SBE45, Sea-Bird Inc., Bellevue, WA, USA). Nitrate (NO^{3-}-N) and silicate ($SiO_3{}^{2-}$-Si) were determined using a SEAL Analytical AQ400 autoanalyzer (Seal Analytical, Mequon, WI, USA). Ammonium ($NH_4{}^{+}$-N) was measured by hypobromite oxidation spectrophotometry at 540 nm, and phosphate ($PO_4{}^{3-}$-P) was determined by molybdenum blue spectrophotometry at 882 nm [44]. The composition of photosynthetic pigments was analyzed by high-performance liquid chromatography (Dionex UltiMate 3000, Thermo Fisher Scientific, Waltham, MA, USA), according to the method of Van Heukelem and Thomas [45].

Stations were divided according to salinity: oceanic water (salinity $\geq$ 30; S1~S4), estuarine water (salinity 5~30; S5~S13 in the Lingdingyang area), and freshwater (salinity < 5; S14~S19).

2.3. Measurement of In Situ Photosynthetic Activity

Water samples were stored in brown bottles immediately after collection for 15 min dark adaption. Subsequently, the maximum photochemical quantum yield (Fv/Fm) of photosystem II (PS II) was measured using a Phyto-PAM II (Heinz Walz, Effeltrich, Germany). The minimal fluorescence level (F_0) was first measured under minimum light (~0 μmol/(s·m^2), and then a saturating pulse (SP) was applied to obtain the maximal fluorescence level (Fm). The Fv/Fm ratio was calculated using Equation (1) [46]:

$$Fv/Fm = \frac{Fm - F_0}{Fm} \tag{1}$$

2.4. DNA Extraction, PCR Amplification, and Sequencing

The collected filters were used for high-throughput sequencing. Total genomic DNA was extracted using the CTAB method according to the manufacturer's instructions, and stored at −20 °C for further analysis. The quality and concentration of the extracted DNA were assessed using a NanoDrop NC2000 spectrophotometer (Thermo Fisher Scientific, Waltham, MA, USA) and agarose gel electrophoresis. Polymerase chain reaction (PCR) was subsequently performed on the DNA samples.

The 18S rRNA genes V4 region was amplified through polymerase chain reaction (PCR) with forward primer 547F (5′-CCAGCASCYGCGGTAATTCC-3′) and reverse primer V4R (5′-ACTTTCGTTCTTGATYRA-3′) [47]. The 25 μL PCR components contained 5 μL of buffer (5×), 0.25 μL of FastPfu DNA Polymerase (5U/μL), 2 μL (2.5 mM) of dNTPs, 1 μL (10 μM) of each forward and reverse primer, 1 μL of DNA template, and 14.75 μL of ddH_2O. The PCR amplification protocol featured an initial denaturation at 98 °C for 3 min, followed by 33 cycles at 98 °C for 30 s, 46 °C for 30 s, and 72 °C for 45 s, with a final extension at 72 °C for 5 min. PCR amplicons were purified with Vazyme VAHTSTM DNA Clean Beads (Vazyme, Nanjing, China) and quantified using the Quant-iT PicoGreen dsDNA Assay Kit (Invitrogen, Carlsbad, CA, USA). Amplicons were pooled in equal amounts. DNA samples were processed by Shanghai Personal Biotechnology Co., Ltd. (Shanghai, China), and paired-end 2 × 250 bp sequencing was performed using the Illlumina NovaSeq platform (San Diego, CA, USA) with the NovaSeq 6000 SP Reagent Kit (500 cycles).

2.5. Sequencing Data Processing

The obtained 18S rRNA sequencing data were processed using Quantitative Insights Into Microbial Ecology (QIIME2, version 2019.4) [48]. The demux plugin was utilized for demultiplexing the sequencing libraries, and the cutadapt plugin was employed to remove primers [49]. Sequence quality filtering, denoising, merging, and chimera removal were conducted using DADA2 [50]. The resulting Amplicon Sequence Variants (ASVs) table had singleton ASVs removed. Taxonomic classification of the ASVs was performed against the National Center for Biotechnology Information (NCBI) database (https://www.ncbi.

nlm.nih.gov/, accessed in August 2019) using the classify-sklearn Naive Bayes taxonomy classifier within the feature-classifier plugin [51]. ASVs represented by protozoa, fungi, and macroalgae were removed before rarefaction. The raw sequencing data were uploaded to NCBI, with the accession number PRJNA1101256.

2.6. Statistical Analysis and Visualization

MATLAB (2023b, Natick, MA, USA) and Ocean Data View (version 5.6.7, https://odv.awi.de, accessed on 25 October 2023) [52] were used for map visualization. Statistics and visualization were performed in R Studio (version 4.2.2). The "rstatix" and "FSA" packages were used to perform Kruskal–Wallis tests and Dunn's multiple comparisons test on environmental factors.

Alpha diversity indices (Chao1, Simpson, Shannon, Pielou's evenness, Faith's PD, and observed species richness) were calculated using GenesCloud tools (https://www.genescloud.cn, accessed on 28 February 2024) and subjected to Kruskal–Wallis tests and Dunn's multiple comparisons test. The "vegan" package (version 2.6-4) was employed to plot rarefaction curves to assess sequencing saturation. The relative abundances at the phylum and genus levels of phytoplankton were calculated using the "plyr" and "reshape2" packages, and visualized using the "ggplot2" and "circlize" packages [53]. Differences in phytoplankton community structure among salinity habitats were evaluated using non-metric multidimensional scaling (NMDS) based on Bray–Curtis distance with the environmental fitting test analysis (envfit) method, followed by an analysis of similarities (ANOSIM). Specificity–occupancy (SPEC–OCCU) plots were drawn following the method described by Gweon et al. [54] to identify potential key species of phytoplankton. To compare the correlation between environmental factors, major phytoplankton phyla, and photosynthetic physiological parameters along the salinity gradient, the "psych" package was used for Spearman correlation analysis. To evaluate the relationship between overall community structure and environmental factors, detrended correspondence analysis (DCA) was performed on species data transformed via Hellinger transformation, and canonical correlation analysis (CCA) was employed due to the length of the first axis being greater than 4. During CCA analysis, the explained variance was calculated after correcting for R^2, and all constrained axes were tested for significance using permutation tests. After that, Monte Carlo permutation tests were used to assess the significance of environmental factors on community impact. Mantel tests were performed using the "ggcor" package.

CHEMTAX [55] was used to calculate the relative abundances of phytoplankton at the phylum level based on HPLC analysis, using a matrix adapted from Wang, et al. [53] in their study of the South China Sea in R studio.

3. Results

3.1. Physical and Chemical Property Along the PRE

Investigations on environmental factors were conducted along the PRE (Figure 2). Results revealed significant differences ($p < 0.05$) in the distributions of temperature, salinity, and nutrients from the lower to the upper estuary, displaying a distinct gradient trend (Figure 2). Concentrations of PO_4^{3-}-P and SiO_3^{2-}-Si, associated with temperatures, decreased significantly from the upper estuary to the lower estuary ($p < 0.05$). Significant differences in environmental factors, except for NH_4^+-N, were observed between oceanic and freshwater zones ($p < 0.05$). NO_3^--N and NH_4^+-N concentrations were generally higher in the upper estuary, and the nitrogen levels in estuarine water were significantly higher than those in oceanic water ($p < 0.05$).

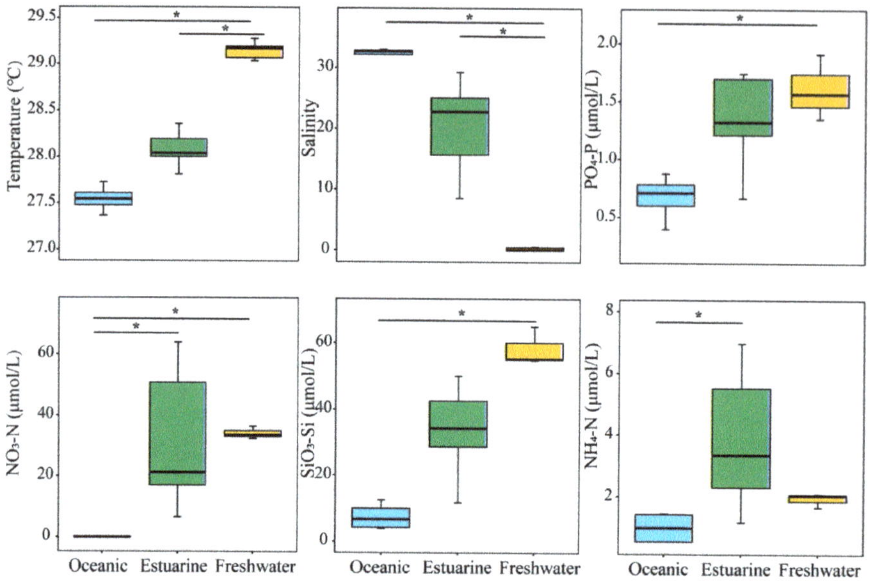

Figure 2. Variation in environmental factors across different regions of the Pearl River Estuary based on salinity levels. "*" denotes statistically significant differences at $p < 0.05$.

3.2. Distribution of Phytoplankton Photosynthetic Parameters

The spatial distribution of the Fv/Fm of the phytoplankton community in the PRE (Figure 3) exhibits a "U-shaped" trend along the salinity gradient. The Fv/Fm values ranged from 0.16 to 0.45, with higher photosynthetic activity in the oceanic zone (average 0.38) and the upper estuary (average 0.35) compared to the tidal mixing water of the lower estuary (average 0.27) (Figure 3).

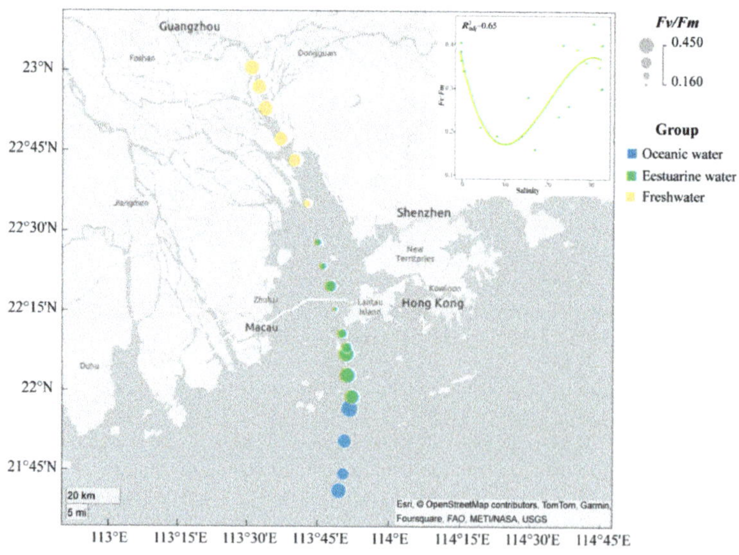

Figure 3. Variation in maximum photochemical quantum yield (Fv/Fm) of phytoplankton in the Pearl River Estuary. A smooth-fit curve showing Fv/Fm as a function of salinity.

3.3. Spatial Variation in the Taxonomic Composition of the Phytoplankton Community

Rarefaction curves for different samples (Figure S1) showed a plateau at a reads number of 20,000, indicating that the sequencing depth was representative of the majority of species. The 18S rRNA sequencing results exhibited the composition of phytoplankton at the phylum level, which varied regionally across the PRE (Figure 4), including Dinophyta (average relative abundance 37.06%), Chlorophyta (19.61%), Cryptophyta (17.60%), Bacillariophyta (12.14%), and Haptophyta (2.65%). Dinophyta (an average relative abundance of 76.36%) was the predominant group in oceanic areas (Figure 4B), followed by Bacillariophyta (11.81%) and Haptophyta (6.92%). The abundance of Dinophyta decreased with reducing salinity, from 35.88% in estuarine areas to 12.42% in freshwater areas. Similarly, Haptophyta followed this trend (Figure 4). Conversely, the relative abundance of Cryptophyta and Chlorophyta significantly increased in the upper estuary, (Figure 4), becoming the dominant group in estuarine and freshwater areas. The relative abundance of Bacillariophyta showed minor variations across different salinity habitats.

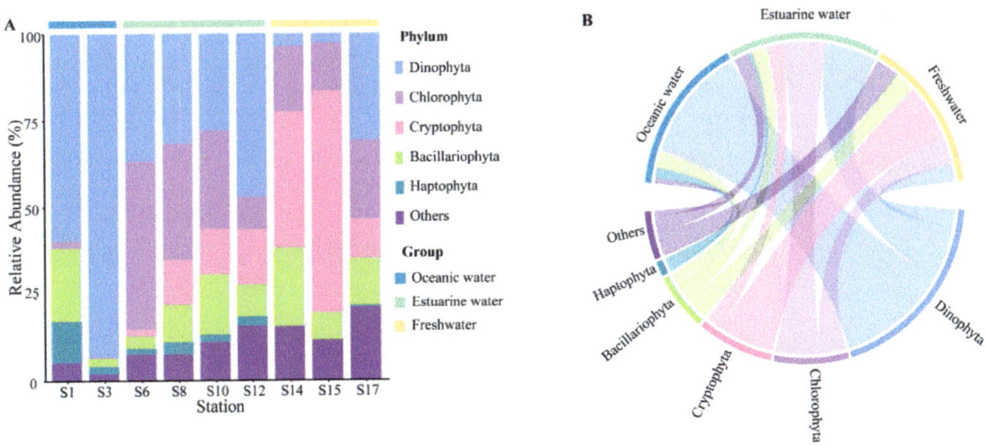

Figure 4. Phytoplankton community composition at the phylum level in the Pearl River Estuary based on 18S rDNA sequencing. (**A**) Relative abundance. (**B**) Chord diagram illustrating species–sample relationships.

At the genus level, the representative taxonomy of each salinity habitat exhibited significant differences (Table S1). *Gyrodinium*, *Chaetoceros*, and *Chrysochromulina* are the primary representative groups in oceanic areas. *Chlorella*, *Cyclotella*, and *Ostreococcus* are abundant in mixed areas. In freshwater areas, *Cryptomonas*, *Cyclotella*, and *Chlamydomonas* were the dominant species, with *Cryptomonas* particularly accounting for 25.81%.

The community composition, derived from pigment analysis, in the PRE (Figure 5), showed that Cyanobacteria were only present at stations S1, S3, S8, and S14 (Figure 5A), with higher abundances noted in high salinity areas (Figure 5B). Both methods showed that Dinophyta and Haptophyta predominantly inhabited polyhaline and euhaline zones, and Bacillariophyta were widely distributed along the salinity gradient, while Cryptophyta tended to inhabit mesohaline and oligohaline zones.

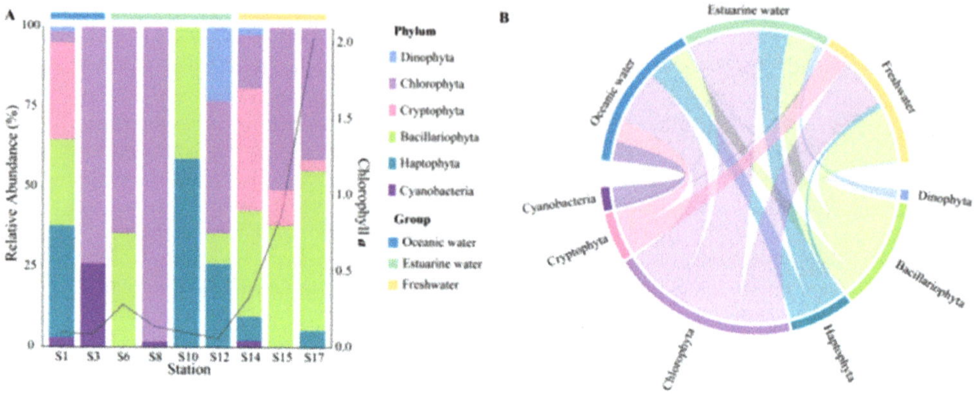

Figure 5. Phytoplankton community composition at the phylum level in the Pearl River Estuary based on pigment analysis. (**A**) Relative abundance. (**B**) Chord diagram illustrating species–sample relationships.

3.4. Spatial Patterns of Phytoplankton Community Diversity

Table 1 displays the alpha diversity indices of different salinity habitats, including Chao1, Simpson, Shannon, Pielou's evenness, Faith's PD, and observed species richness. The freshwater-oligohaline zone had the highest average values for all diversity indices, without significant differences among the three habitats ($p > 0.05$).

Table 1. Alpha diversity indices of phytoplankton based on 18S rRNA in the Pearl River Estuary.

Group	Chao1	Faith's PD	Observed Species	Pielou's Evenness	Shannon	Simpson
Oceanic water	670.766 ± 242.410	66.662 ± 40.004	661.150 ± 229.598	0.717 ± 0.0252	6.690 ± 0.602	0.956 ± 0.005
Estuarine water	703.047 ± 170.311	104.138 ±32.528	644.200 ± 133.957	0.6733 ± 0.065	6.280 ± 0.803	0.945 ± 0.045
Freshwater	1129.983 ± 87.660	122.425 ± 15.439	987.100 ± 87.167	0.757 ± 0.035	7.530 ± 0.426	0.977 ± 0.008

NMDS based on Bray–Curtis distance revealed that the stress value was less than 0.05, and samples from different salinity habitats clustered well together, clearly separating into three distinct clusters with good interpretability (Figure 6A). Results from ANOSIM also indicated significant differences in phytoplankton community composition between oceanic, estuarine, and freshwater zones, with clear distinctions ($p < 0.05$) (Figure 6A). To reveal the environmental gradient and the relationship between the photosynthetic physiological activity of phytoplankton (Fv/Fm) and community structure, the envfit method was performed (Figure 6A). Nutrient levels were negatively correlated with salinity, which significantly affected phytoplankton communities in freshwater areas. Fv/Fm was negatively correlated with the estuarine population, indicating that the photosynthetic activity of the phytoplankton community in the mixed zone was lower, which may be related to phytoplankton cellular stress caused by the drastic change in salinity.

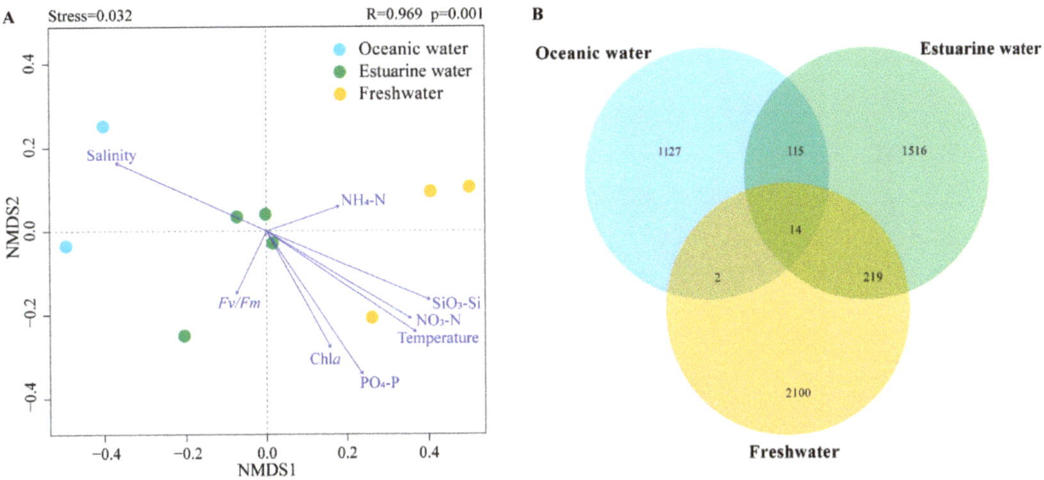

Figure 6. (**A**) NMDS (Non-metric multidimensional scaling) with envfit method of phytoplankton communities across different salinities; R and *p* values were calculated by an analysis of similarities (ANOSIM). (**B**) Venn diagram illustrating phytoplankton distribution across different salinity levels.

Venn diagrams showing the overlap of phytoplankton taxonomic groups among different salinity habitats are presented in Figure 6B. The freshwater zone had the highest number of ASVs, totaling 2335, while the oceanic zone had the fewest. There were 14 ASVs common across the three salinity zones. The freshwater and estuarine zones shared more ASVs, totaling 233; however, the overlap with the oceanic zone was the least, with only 16 shared ASVs.

3.5. Differences in Phytoplankton Community Composition across Different Salinity Habitats

The SPEC–OCCU plots illustrate the distribution and relative abundance of the top 500 most abundant ASVs across various habitats in the PRE, encompassing taxa such as Dinophyta, Chlorophyta, Cryptophyta, Bacillariophyta, Haptophyta, and other Ochrophyta (Figure 7). ASVs in each habitat showed a degree of variation in the occupancy (x-axis), with fewer species found in all stations (Figure 7A). Specialists, defined by both specificity and occupancy rates of $\geq$0.7, were identified in each habitat (Figure 7B). The number of specialist ASVs varied significantly across the salinity gradient, with 37 in oceanic areas, 108 in estuarine areas, and 117 in freshwater areas, representing diverse taxonomic groups. ASVs representing Dinophyta, Bacillariophyta, and Chlorophyta were present as specialists in all salinity habitats, though their proportions varied (Figure 7B). Among specialists, Dinophyta tended to dominate in higher salinity environments and were less prevalent in low salinity areas. Haptophyta accounted for 10.8% of the specialists in oceanic areas and 6.5% in estuarine areas but were not specialists in freshwater areas. In estuarine and freshwater areas, Chlorophyta and Cryptophyta constituted the majority of specialists, but their specificity was low in oceanic areas.

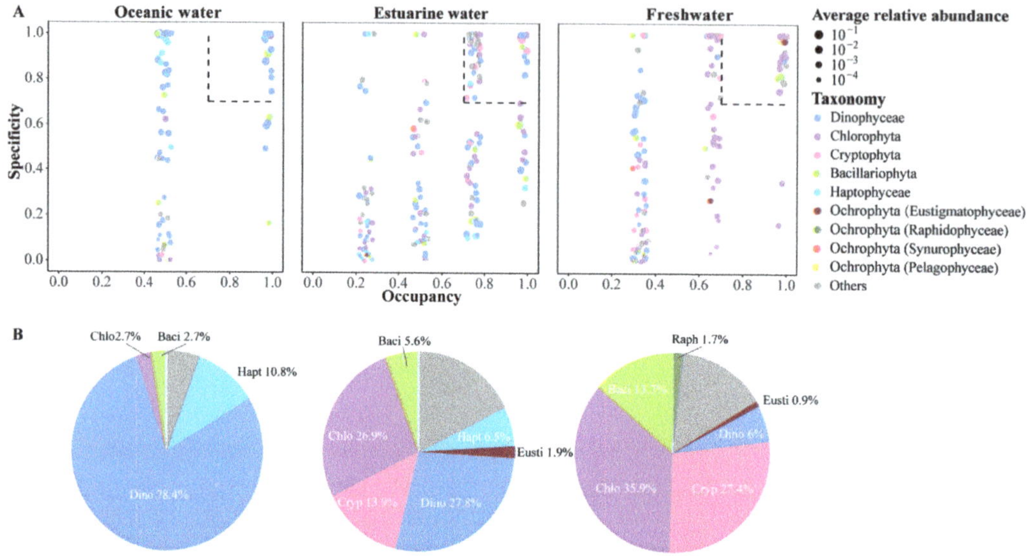

Figure 7. (**A**) SPEC–OCCU plots presenting the distribution of the top 500 ASVs across different salinity habitats in the Pearl River Estuary. Occupancy (x-axis) indicates the distribution of an ASV across all samples in salinity habitats, while specificity (y-axis) indicates whether it also present in other habitats. The dotted line includes phytoplankton with occupancy and specificity $\geq$0.7. Scatter colors and sizes represent different phylum (class) and relative abundances, respectively. (**B**) Pie charts depict the percentage of specialist ASVs in each salinity habitat.

3.6. Correlations between Phytoplankton Community, Photosynthetic Physiology, and Environmental Factors

Figure 8A shows the Spearman correlations between the major phytoplankton groups, diversity indices, and environmental factors in the PRE. Among the main groups, Bacillariophyta and Chlorophyta showed no significant correlations with environmental factors ($p > 0.05$). Cryptophyta was significantly positively correlated with all nutrients except NH_4^+-N but was negatively correlated with salinity ($p < 0.05$). Dinophyta was significantly positively correlated with salinity but was negatively correlated with SiO_3^{2-}-Si and temperature ($p < 0.05$). Diversity indices were significantly influenced by temperature and salinity as well as nutrient levels. Overall, the diversity indices were positively correlated with nutrient levels and temperature, but negatively correlated with salinity.

Furthermore, CCA was used to explore the overall regulation of environmental factors (excluding photosynthetic activity) on the community structure of phytoplankton in the PRE (Figure 8B; Table S2), which is consistent with NMDS and envfit. Compared to NMDS with envfit, CCA provides the explanation rate of the variation in the community structure. The eigenvalues of axes CCA 1 and CCA 2 together explained 17.7% of the variation in the phytoplankton community structure. According to permutation test results, environmental factors significantly influenced the variation in phytoplankton community structure ($p < 0.05$). Temperature, salinity, NO_3^--N, PO_4^{3-}-P, and SiO_3^{2-}-Si had significant impacts on the phytoplankton community structure ($R^2 = 0.7381$~0.9261, $p < 0.05$). Phytoplankton in oceanic areas were positively correlated with salinity but negatively correlated with temperature, NO_3^--N, PO_4^{3-}-P, and SiO_3^{2-}-Si. The correlations for the phytoplankton in freshwater were the exact opposite. Nutrient levels significantly influenced phytoplankton in the upper estuary with high Chl *a* content.

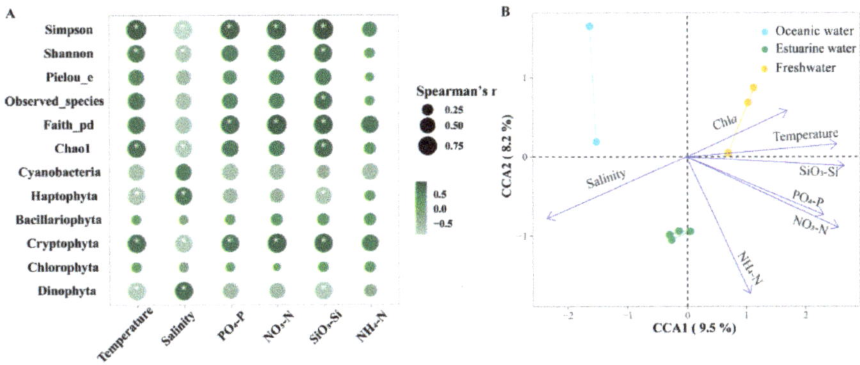

Figure 8. (**A**) Spearman correlation analysis between environmental factors and the relative abundance of major phytoplankton taxa and alpha biodiversity in the Pearl River Estuary. Cyanobacteria richness was determined based on pigment analysis, while the richness of other algae was determined based on 18S rRNA. Bubble size indicates the absolute value of the Spearman rank correlation coefficient. "*" denotes statistically significant differences at $p < 0.05$. (**B**) Canonical correlation analysis (CCA) of environmental factors and phytoplankton communities.

Mantel tests provide correlations of phytoplankton communities across the estuary with environmental changes and photosynthetic responses of *Fv/Fm* (Figure 9, Table S3). Under broad gradients in estuaries, phytoplankton communities are strongly shaped by temperature, salinity, and nutrient levels ($p < 0.05$). Meanwhile, there is a certain correlation between environmental factors, which will directly or indirectly affect the changes in phytoplankton communities.

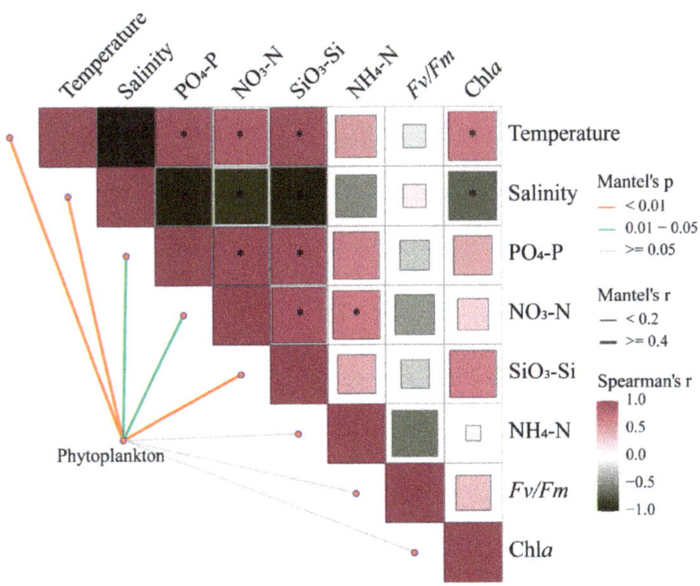

Figure 9. Pairwise comparisons between 18S rRNA-based phytoplankton communities and environmental factors in the Pearl River Estuary by Mantel tests. The upper triangle is the Spearman correlation between the environmental factor, Chl *a*, and *Fv/Fm*. The "*" denotes statistically significant differences at $p < 0.05$, and the color represents Spearman's r. The line width represents Mantel's r between each factor and the phytoplankton community, and the color represents Mantel's *p-value*.

4. Discussion

4.1. Community Composition of Phytoplankton in the Pearl River Estuary during Autumn

Combining molecular biology and chemotaxonomy provides detailed insights into the plankton community composition [39,56,57]. The PRE is an ideal experimental area, presenting the typical spatial and temporal differentiation of phytoplankton, and facilitating comparisons between different methods [39]. Both approaches in this study indicate that Dinophyta, Chlorophyta, Cryptophyta, Bacillariophyta, and Haptophyta are the dominant phytoplankton phyla in the estuary, consistent with previous research [26,39,58,59]. The taxonomy of phytoplankton at the genus level is also similar to Ding, et al. [58]. According to 18S rRNA sequencing results, in late summer and early fall, Dinophyta appeared at all stations, particularly dominating in the high and medium salinity areas (Figure 4). Similar observations of increased Dinophyta abundance during autumn, accompanied by massive blooms of *Cochlodinium geminatum*, were reported by Dong, et al. [33] and Ke, et al. [60]. Guo, et al. [61] also observed dinoflagellate blooms in the PRE in summer. Higher temperatures and salinity during this season were one of the main reasons for the enrichment of dinoflagellate, which is consistent with previous observations [33,60]. Additionally, both methods detected significant presences of Bacillariophyta at all stations (Figures 4 and 5), which are important freshwater and brackish species in the estuary [36]. Due to the suitable water temperature and turbidity, diatoms prevail in both wet and dry seasons in the PRE [26]. The hydrological dynamics of the estuary, strongly influenced by seasonal variations, drive temporal shifts in phytoplankton composition [7,8,11,26,36].

Generally, the abundance of 18S rRNA genes was well consistent with pigment content in phytoplankton [39,56,62], representing the major algae along the PRE (Figures 4B and 5B). Relative to CHEMTAX results, the 18S rRNA results showed a higher average relative abundance of Dinophyta, while Haptophyta appeared to be underestimated (Figures 4 and 5), similar to observations from a 2020 cruise in the PRE by Xu, et al. [39]. The relative abundance of Chlorophyta was also higher in the pigment analysis. The underestimation of Haptophyta in the 18S dataset could be due to a single base mismatch on the 3′ end of the V4R primer in Prymnesiales (Haptophyta), leading to possible underestimation [59]. Developing primers with better specificity could help avoid the widespread underestimation of this significant marine group [63]. Furthermore, the variability in 18S rRNA gene copies among different eukaryotic species can lead to discrepancies between PCR-based and pigment-based methods [64,65], which could also explain the discrepancy in the content of other algae in the two methods in this study. Compared to other algae, Dinophyta, possessing considerably more rRNA gene copies, is more readily detected in sequencing studies [64]. Dinophyta's unique nutritional strategies might also contribute to such inconsistencies. Most dinophytes are not strictly autotrophic and can achieve mixotrophy through predatory strategies or even incorporating phototrophy through the acquisition of chloroplasts from other algae [66,67]. This phenomenon can result in the potential underreporting of mixotrophic Dinophyta in chemotaxonomic analyses based on pigment contents. In summary, combining multiple approaches helps to comprehensively understand the composition of marine phytoplankton.

4.2. Spatial Heterogeneity of Photosynthetic Characteristics in Phytoplankton of the Pearl River Estuary

Fv/Fm, recognized as a sensitive in situ indicator of stressors for phytoplankton [16,17,19], effectively reflects the status of photosynthetic physiology of phytoplankton along the environmental gradient of the PRE (Figures 3 and 6A). Li, et al. [19] categorized estuarine phytoplankton based on Fv/Fm values into stress (low, <0.3), transitional (moderate, 0.3–0.5), and blooming conditions (high, >0.5). In this study, phytoplankton in the upper estuary (dominant by Dinophyta, Bacillariophyta, and Haptophyta) and lower estuary with high nutrient levels (dominant by Cryptophyta, Chlorophyta, and Bacillariophyta) appeared healthier and more active, whereas communities in the midstream were more stressed (Figures 3 and 6A).

The salinity gradient of estuaries is well-known as one of the most significant and fundamental causes of phytoplankton stress, resulting in slowed growth, reduced photosynthesis, and increased respiration. These factors contribute to the spatial succession of estuarine phytoplankton communities [68,69]. Our results demonstrate a complex correlation between photosynthetic activity in phytoplankton of the PRE and variations in salinity (Figure 3). Different species exhibit varying adaptations to salinity [68,69]. After river water flows into the upper Lingdingyang, osmotic stress caused by the shift inhibits stenohaline species originating from freshwater environments [68], leading to decreased algal photosynthetic activity [10]. As the environment transitions to higher salinity marine conditions, the community becomes dominated by euryhaline species such as Dinophyta and Bacillariophyta [70].

Light availability is always a primary factor limiting the photosynthetic physiology of phytoplankton. Within brackish water environments, in contrast to freshwater and seawater, phytoplankton cells are often more sensitive to changes in ultraviolet (UV) radiation, resulting in significant inhibition of photosynthesis [10]. Reduced light availability due to high turbidity in estuaries also contributes to the limitation of phytoplankton physiology and growth [19,26], despite higher nutrient levels in the middle sections of the estuary (Figure 2). The Estuarine Turbidity Maxima (ETM) zone, observed in many estuaries worldwide [71], exerts a substantial influence on the distribution of marine microbes [72,73]. Shi, et al. [27] found that the ETM zone in the PRE is located on the western side of the middle channel of the PRE during the dry season, with high bacterial abundance but low phytoplankton abundance. Therefore, the decline in photosynthetic activity within the mid-channel (an area characterized by moderate salinity) of the PRE may be caused by the cumulative influence of turbidity-induced alterations in light availability and dramatic changes in salinity. In contrast, nutrient levels were also higher upstream, but lower turbidity resulted in higher Fv/Fm in the phytoplankton community.

4.3. Environmental Factors Shaping Spatial Distribution of Phytoplankton

The significant environmental gradient in the PRE (Figure 2) leads to heterogeneity within the phytoplankton community, thereby instigating shifts in both structure and abundance (Figures 4–7). Salinity strongly governs the spatial succession of the phytoplankton community in the estuary [7,11,58,74], dividing it into three distinct groups: the estuarine community, the mixing waters community, and the coastal community [36]. SPEC–OCCU plots provide detailed insights into species specialization (Figure 7). Along the salinity gradient, unique species primarily belonging to Chlorophyta and Cryptophyta, adapted to oligohaline and mesohaline conditions, transition into species like Dinophyta and Haptophyta, which thrive in polyhaline and euhaline conditions. Meanwhile, Bacillariophyta species dominate across all salinity habitats (Figures 4 and 5). This succession pattern is similar to that observed in other estuaries [10,70]. Generally, freshwater species are confined to the upper reaches of the river, with only a few extending into saline areas, whereas marine species can extend into oligohaline and even freshwater regions, although with reduced diversity [6].

Nutrient levels regulate the composition, abundance, diversity, and primary productivity of estuarine phytoplankton [8,19,34,75]. Due to strong terrestrial influences, the upper and middle estuaries are nutrient-rich, leading to significant increases in Chl *a* (Figure 5). As the river widens downstream, nutrients are quickly diluted, resulting in decreased Chl *a* content. CCA analysis and NMDS with envfit indicate that the low-salinity community is positively correlated with nutrient concentrations (Figure 8B), highlighting the stimulatory effect of nutrients. Elevated nutrient levels also significantly enhance diversity (Figures 6A and 8A). Studies have shown that SiO_3^{2-}-Si concentration significantly affects the growth of phytoplankton in the PRE [11,36], and this study likewise confirms the positive role of silicate on the major phytoplankton groups and diversity (Figure 8A). Additionally, the enrichment of phytoplankton in the estuary is controlled by residence time/turnover rate, closely related to seasonality and flow rates [76]. Compared to the

Mississippi River and Yangtze River, the residence time of PRE is relatively short, ranging from a couple of weeks in the dry season to a few days in the wet season [77,78], resulting in complex temporal and spatial dynamics of phytoplankton. Higher nutrient levels and the stability of the water column during autumn contribute to phytoplankton enrichment in the middle to upper parts of the estuary as well as higher photosynthetic activity.

With the fast development in the Pearl River Delta area, research on a 40-year PRE data set suggested that there was a significant decrease in the number of phytoplankton species [79], accompanied by an increase in harmful algal bloom events [80]. Although *Chaetoceros* sp. is still one of the dominant diatoms, *Cyclotella* adapted and bloomed in the estuary prompted by the anthropogenic dissolved organic phosphorus (DOP) [81]. The low light availability and high nutrients have favored the blooms of pico-level brackish Chlorophyta (e.g., *Ostreococcus tauri*) [58], while the continuous supply of DIN with an enrichment of DOP supported the recurrent harmful algal blooms of *Gyrodinium* in the lower estuary [82]. In the upper river, mixotrophic *Chlorella* and *Chlamydomonas* also bloom [83,84]. All of these species are present in our sample at the genus level. The ongoing anthropogenic influences on the PRE have shaped the dynamics of the phytoplankton community [34,75,85], and assessing the function and health of coastal ecosystems aids in developing long-term management strategies.

5. Conclusions

The 18S rRNA and pigment analysis showed that environmental gradients in the PRE play a dominant role in shaping the differentiation of phytoplankton communities into oceanic, estuarine, and freshwater communities. The relative abundance of Dinophyta and Haptophyta decreased from the coast to the upper estuary, while Chlorophyta and Cryptophyta became increasingly dominant. Bacillariophyta was identified as an important species present in both saline and freshwater environments. The in situ photosynthetic characteristic of phytoplankton along the PRE indicated complex correlations with environmental factors. Due to a wide variety of environmental gradients and the adaptability of phytoplankton, communities in the mixed zone were observed to be under obvious stress, while those in freshwater and oceanic regions appeared relatively active and healthy. The combination of environmental factors revealed that, in addition to the temperature and salinity gradients of the PRE, the increase in nutrient levels related to anthropogenic influence also had significant effects on the concentration of Chl *a* and the phytoplankton community in the upper estuary. Focusing on the photosynthetic characteristics and community structure of coastal phytoplankton contributes to understanding the functioning and health of coastal ecosystems.

Supplementary Materials: The following supporting information can be downloaded at: https://www.mdpi.com/article/10.3390/biology13070550/s1, Figure S1: Rarefaction curve of phytoplankton sequencing; Table S1: Top 20 phytoplankton taxonomy at the genus level in the Pearl River Estuary based on 18S rDNA sequencing; Table S2: Permutation test results of environmental factors for the CCA; Table S3: Pairwise comparisons between 18S rRNA-based phytoplankton communities and environmental factors in the Pearl River Estuary by Mantel tests.

Author Contributions: Conceptualization: J.X., J.K. and J.L.; methodology: J.X., H.H., X.G. and J.K.; formal Analysis: J.X., H.H. and X.G.; investigation: J.X., H.H., X.G. and J.L.; resources: J.L. and Y.G.; writing—original draft preparation, J.X., H.H. and X.G.; writing—review and editing, J.X., X.G., J.K. and J.L.; visualization: J.X.; supervision: J.K.; funding acquisition, Y.G. and J.L. All authors have read and agreed to the published version of the manuscript.

Funding: This work was supported by the National Key Research and Development Plan (Grant No. 2023YFC3107403) and the Oceanic Interdisciplinary Program of Shanghai Jiao Tong University (Grant No. WH410260402/035).

Institutional Review Board Statement: Not applicable.

Informed Consent Statement: Not applicable.

Data Availability Statement: Data will be made available on request. The raw sequencing data were uploaded to NCBI (https://www.ncbi.nlm.nih.gov/, accessed on 20 April 2024), with the accession number PRJNA1101256.

Conflicts of Interest: The authors declare no conflicts of interest.

References

1. Cloern, J.E.; Jassby, A.D.; Schraga, T.S.; Nejad, E.; Martin, C. Ecosystem variability along the estuarine salinity gradient: Examples from long-term study of San Francisco Bay. *Limnol. Oceanogr.* **2017**, *62*, S272–S291. [CrossRef]
2. Elliott, M.; Whitfield, A.K. Challenging paradigms in estuarine ecology and management. *Estuar. Coast. Shelf Sci.* **2011**, *94*, 306–314. [CrossRef]
3. Muylaert, K.; Sabbe, K.; Vyverman, W. Changes in phytoplankton diversity and community composition along the salinity gradient of the Schelde estuary (Belgium/The Netherlands). *Estuar. Coast. Shelf Sci.* **2009**, *82*, 335–340. [CrossRef]
4. Quinlan, E.L.; Phlips, E.J. Phytoplankton assemblages across the marine to low-salinity transition zone in a blackwater dominated estuary. *J. Plankton Res.* **2007**, *29*, 401–416. [CrossRef]
5. Herlemann, D.P.R.; Labrenz, M.; Jürgens, K.; Bertilsson, S.; Waniek, J.J.; Andersson, A.F. Transitions in bacterial communities along the 2000 km salinity gradient of the Baltic Sea. *ISME J.* **2011**, *5*, 1571–1579. [CrossRef] [PubMed]
6. Whitfield, A.K.; Elliott, M.; Basset, A.; Blaber, S.J.M.; West, R.J. Paradigms in estuarine ecology—A review of the Remane diagram with a suggested revised model for estuaries. *Estuar. Coast. Shelf Sci.* **2012**, *97*, 78–90. [CrossRef]
7. Zhang, X.; Zhang, J.P.; Huang, X.P.; Huang, L.M. Phytoplankton assemblage structure shaped by key environmental variables in the Pearl River Estuary, South China. *J. Ocean Univ.* **2014**, *13*, 73–82. [CrossRef]
8. Li, G.; Lin, Q.; Lin, J.; Song, X.; Tan, Y.; Huang, L. Environmental gradients regulate the spatial variations of phytoplankton biomass and community structure in surface water of the Pearl River estuary. *Acta Ecol. Sin.* **2014**, *34*, 129–133. [CrossRef]
9. Wang, H.L.; Chen, F.; Zhang, C.L.; Wang, M.; Kan, J.J. Estuarine gradients dictate spatiotemporal variations of microbiome networks in the Chesapeake Bay. *Environ. Microbiome* **2021**, *16*, 22. [CrossRef]
10. Li, G.; Gao, K.S.; Yuan, D.X.; Zheng, Y.; Yang, G.Y. Relationship of photosynthetic carbon fixation with environmental changes in the Jiulong River estuary of the South China Sea, with special reference to the effects of solar UV radiation. *Mar. Pollut. Bull.* **2011**, *62*, 1852–1858. [CrossRef]
11. Xu, S.N.; Liu, Y.; Fan, J.T.; Xiao, Y.Y.; Qi, Z.H.; Lakshmikandan, M. Impact of salinity variation and silicate distribution on phytoplankton community composition in Pearl River estuary, China. *Ecohydrol. Hydrobiol.* **2022**, *22*, 466–475. [CrossRef]
12. Cloern, J.E.; Jassby, A.D. Patterns and Scales of Phytoplankton Variability in Estuarine-Coastal Ecosystems. *Estuaries Coasts* **2010**, *33*, 230–241. [CrossRef]
13. Anderson, D.M.; Glibert, P.M.; Burkholder, J.M. Harmful algal blooms and eutrophication: Nutrient sources, composition, and consequences. *Estuaries* **2002**, *25*, 704–726. [CrossRef]
14. Gilbert, P.M. Eutrophication, harmful algae and biodiversity—Challenging paradigms in a world of complex nutrient changes. *Mar. Pollut. Bull.* **2017**, *124*, 591–606. [CrossRef] [PubMed]
15. Chen, D.W.; Shi, Z.; Li, R.H.; Li, X.F.; Cheng, Y.Y.; Xu, J. Hydrodynamics drives shifts in phytoplankton community composition and carbon-to-chlorophyll *a* ratio in the northern South China Sea. *Front. Mar. Sci.* **2023**, *10*, 1293354. [CrossRef]
16. Li, J.L.; Sun, X.X.; Zheng, S. In situ study on photosynthetic characteristics of phytoplankton in the Yellow Sea and East China Sea in summer 2013. *J. Mar. Syst.* **2016**, *160*, 94–106. [CrossRef]
17. Wang, F.; Guo, S.J.; Liang, J.H.; Sun, X.X. In situ phytoplankton photosynthetic characteristics and their controlling factors in the eastern Indian Ocean. *Mar. Pollut. Bull.* **2024**, *198*, 115869. [CrossRef]
18. Cloern, J.E. The relative importance of light and nutrient limitation of phytoplankton growth: A simple index of coastal ecosystem sensitivity to nutrient enrichment. *Aquat. Ecol.* **1999**, *33*, 3–16. [CrossRef]
19. Li, J.; Gao, Y.H.; Bao, Y.L.; Gao, X.; Glibert, P.M. Summer phytoplankton photosynthetic characteristics in the Changjiang River Estuary and the adjacent East China Sea. *Front. Mar. Sci.* **2023**, *10*, 1111557. [CrossRef]
20. Zhang, L.K.; Li, G.; Xiang, C.H.; Huang, Y.D.; Fu, X.M.; Zheng, C.Y.; Wang, Z.; Ouyang, Z.Y.; Song, X.Y. Plankton Metabolism in Coastal Waters of the Guangdong-Hong Kong-Macao Greater Bay: Regional Variance and Driving Factors. *Front. Mar. Sci.* **2022**, *9*, 844970. [CrossRef]
21. Bergmann, T.; Richardson, T.L.; Paerl, H.W.; Pinckney, J.L.; Schofield, O. Synergy of light and nutrients on the photosynthetic efficiency of phytoplankton populations from the Neuse River Estuary, North Carolina. *J. Plankton Res.* **2002**, *24*, 923–933. [CrossRef]
22. Xia, J.; Bao, Y.L.; Gao, Y.H.; Li, J. The effects of temperature and sulfamethoxazole on the growth and photosynthetic characteristics of *Phaeodactylum tricornutum*. *Mar. Pollut. Bull.* **2024**, *200*, 116122. [CrossRef] [PubMed]
23. Hanelt, D. 9-Photosynthesis assessed by chlorophyll fluorescence. In *Bioassays*; Häder, D.-P., Erzinger, G.S., Eds.; Elsevier: Amsterdam, The Netherlands, 2018; pp. 169–198.
24. Falkowski, P.G.; Kolber, Z. Variations in chlorophyll fluorescence yields in phytoplankton in the world oceans. *Aust. J. Plant Physiol.* **1995**, *22*, 341–355. [CrossRef]

25. Bao, Y.; Li, J. Photosynthetic Characteristics of Phytoplankton in the Surface Water of Changjiang Estuary and Its Adjacent Sea Area in Summer. *Adv. Mar. Sci.* **2023**, *41*, 87–99.
26. Shen, P.P.; Li, G.; Huang, L.M.; Zhang, J.L.; Tan, Y.H. Spatio-temporal variability of phytoplankton assemblages in the Pearl River estuary, with special reference to the influence of turbidity and temperature. *Cont. Shelf Res.* **2011**, *31*, 1672–1681. [CrossRef]
27. Shi, Z.; Xu, J.; Huang, X.P.; Zhang, X.; Jiang, Z.J.; Ye, F.; Liang, X.M. Relationship between nutrients and plankton biomass in the turbidity maximum zone of the Pearl River Estuary. *J. Environ. Sci.* **2017**, *57*, 72–84. [CrossRef] [PubMed]
28. Cheshmehzangi, A.; Tang, T. Pearl River Delta City Cluster: From Dual-Core Structure Economic Development Strategies to Regional Economic Plans. In *China's City Cluster Development in the Race to Carbon Neutrality*; Springer Nature: Singapore, 2022; pp. 63–75.
29. Chen, S.L.; Zhu, Z.H.; Liu, X.T.; Yang, L. Variation in Vegetation and Its Driving Force in the Pearl River Delta Region of China. *Int. J. Environ. Res. Public Health* **2022**, *19*, 10343. [CrossRef]
30. Shen, Y.J. Analysis of Urban Expansion in the Pearl River Delta Urban Agglomeration Based on Multi-Source Time-Series Remote Sensing Data. Master's Thesis, Central South University, Changsha, China, 2023.
31. Tao, W.; Niu, L.X.; Dong, Y.H.; Fu, T.; Lou, Q.S. Nutrient Pollution and Its Dynamic Source-Sink Pattern in the Pearl River Estuary (South China). *Front. Mar. Sci.* **2021**, *8*, 713907. [CrossRef]
32. Zhang, S.Y.; Zhang, H. Anthropogenic impact on long-term riverine COD_{Mn}, BOD, and nutrient flux variation in the Pearl River Delta. *Sci. Total Environ.* **2023**, *859*, 160197. [CrossRef]
33. Dong, Y.L.; Cui, L.; Cao, R.B.; Cen, J.Y.; Zou, J.; Zhou, X.Y.; Lu, S.H. Ecological characteristics and teratogenic retinal determination of *Cochlodinium geminatum* blooms in Pearl River Estuary, South China. *Ecotox. Environ. Safe.* **2020**, *191*, 110226. [CrossRef]
34. Harrison, P.J.; Yin, K.D.; Lee, J.H.W.; Gan, J.P.; Liu, H.B. Physical-biological coupling in the Pearl River Estuary. *Cont. Shelf Res.* **2008**, *28*, 1405–1415. [CrossRef]
35. Hu, J.T.; Li, S.Y.; Geng, B.X. Modeling the mass flux budgets of water and suspended sediments for the river network and estuary in the Pearl River Delta, China. *J. Mar. Syst.* **2011**, *88*, 252–266. [CrossRef]
36. Huang, L.M.; Jian, W.J.; Song, X.Y.; Huang, X.P.; Liu, S.; Qian, P.Y.; Yin, K.D.; Wu, M. Species diversity and distribution for phytoplankton of the Pearl River estuary during rainy and dry seasons. *Mar. Pollut. Bull.* **2004**, *49*, 588–596. [CrossRef] [PubMed]
37. Charvet, S.; Vincent, W.F.; Lovejoy, C. Chrysophytes and other protists in High Arctic lakes: Molecular gene surveys, pigment signatures and microscopy. *Polar Biol.* **2012**, *35*, 733–748. [CrossRef]
38. Xiao, X.; Sogge, H.; Lagesen, K.; Tooming-Klunderud, A.; Jakobsen, K.S.; Rohrlack, T. Use of high throughput sequencing and light microscopy show contrasting results in a study of phytoplankton occurrence in a freshwater environment. *PLoS ONE* **2014**, *9*, e106510. [CrossRef]
39. Xu, S.M.; Li, G.H.; He, C.; Huang, Y.; Yu, D.; Deng, H.W.; Tong, Z.Y.; Wang, Y.C.; Dupuy, C.; Huang, B.Q.; et al. Diversity, community structure, and quantity of eukaryotic phytoplankton revealed using 18S rRNA and plastid 16S rRNA genes and pigment markers: A case study of the Pearl River Estuary. *Mar. Life Sci. Technol.* **2023**, *5*, 415–430. [CrossRef] [PubMed]
40. Yang, J.; Lv, J.P.; Liu, Q.; Nan, F.R.; Li, B.; Xie, S.L.; Feng, J. Seasonal and spatial patterns of eukaryotic phytoplankton communities in an urban river based on marker gene. *Sci. Rep.* **2021**, *11*, 23147. [CrossRef]
41. Mackey, D.J.; Higgins, H.W.; Mackey, M.D.; Holdsworth, D. Algal class abundances in the western equatorial Pacific: Estimation from HPLC measurements of chloroplast pigments using CHEMTAX. *Deep-Sea Res. Part I-Oceanogr. Res. Pap.* **1998**, *45*, 1441–1468. [CrossRef]
42. Swan, C.M.; Vogt, M.; Gruber, N.; Laufkoetter, C. A global seasonal surface ocean climatology of phytoplankton types based on CHEMTAX analysis of HPLC pigments. *Deep-Sea Res. Part I-Oceanogr. Res. Pap.* **2016**, *109*, 137–156. [CrossRef]
43. Wright, S.W.; van den Enden, R.L.; Pearce, I.; Davidson, A.T.; Scott, F.J.; Westwood, K.J. Phytoplankton community structure and stocks in the Southern Ocean (30–80°E) determined by CHEMTAX analysis of HPLC pigment signatures. *Deep-Sea Res. Part II-Top. Stud. Oceanogr.* **2010**, *57*, 758–778. [CrossRef]
44. *GB 17378.4-2007*; The Specification for Marine Monitoring-Part 4: Seawater Analysis. National Marine Environmental Monitoring Center: Dalian, China, 2007.
45. Van Heukelem, L.; Thomas, C.S. Computer-assisted high-performance liquid chromatography method development with applications to the isolation and analysis of phytoplankton pigments. *J. Chromatogr. A* **2001**, *910*, 31–49. [CrossRef] [PubMed]
46. Butler, W.L.; Kitajima, M. Fluorescence quenching in photosystem-II of chloroplasts. *Biochim. Biophys. Acta* **1975**, *376*, 116–125. [CrossRef] [PubMed]
47. Stoeck, T.; Bass, D.; Nebel, M.; Christen, R.; Jones, M.D.M.; Breiner, H.W.; Richards, T.A. Multiple marker parallel tag environmental DNA sequencing reveals a highly complex eukaryotic community in marine anoxic water. *Mol. Ecol.* **2010**, *19*, 21–31. [CrossRef] [PubMed]
48. Bolyen, E.; Rideout, J.R.; Dillon, M.R.; Bokulich, N.; Abnet, C.C.; Al-Ghalith, G.A.; Alexander, H.; Alm, E.J.; Arumugam, M.; Asnicar, F.; et al. Reproducible, interactive, scalable and extensible microbiome data science using QIIME 2. *Nat. Biotechnol.* **2019**, *37*, 852–857. [CrossRef] [PubMed]
49. Martin, M. Cutadapt removes adapter sequences from high-throughput sequencing reads. *EMBnet. J.* **2011**, *17*, 10–12. [CrossRef]
50. Callahan, B.J.; McMurdie, P.J.; Rosen, M.J.; Han, A.W.; Johnson, A.J.A.; Holmes, S.P. DADA2: High-resolution sample inference from Illumina amplicon data. *Nat. Methods* **2016**, *13*, 581–583. [CrossRef] [PubMed]

51. Bokulich, N.A.; Kaehler, B.D.; Rideout, J.R.; Dillon, M.; Bolyen, E.; Knight, R.; Huttley, G.A.; Caporaso, J.G. Optimizing taxonomic classification of marker-gene amplicon sequences with QIIME 2′s q2-feature-classifier plugin. *Microbiome* **2018**, *6*, 90. [CrossRef] [PubMed]
52. Schlitzer, R. Ocean Data View. Available online: https://odv.awi.de/ (accessed on 25 October 2023).
53. Gu, Z.G.; Gu, L.; Eils, R.; Schlesner, M.; Brors, B. *circlize* implements and enhances circular visualization in R. *Bioinformatics* **2014**, *30*, 2811–2812. [CrossRef]
54. Gweon, H.S.; Bowes, M.J.; Moorhouse, H.L.; Oliver, A.E.; Bailey, M.J.; Acreman, M.C.; Read, D.S. Contrasting community assembly processes structure lotic bacteria metacommunities along the river continuum. *Environ. Microbiol.* **2021**, *23*, 484–498. [CrossRef]
55. Mackey, M.D.; Mackey, D.J.; Higgins, H.W.; Wright, S.W. CHEMTAX—A program for estimating class abundances from chemical markers: Application to HPLC measurements of phytoplankton. *Mar. Ecol.-Prog. Ser.* **1996**, *144*, 265–283. [CrossRef]
56. Lin, Y.J.; Gifford, S.; Ducklow, H.; Schofield, O.; Cassar, N. Towards quantitative microbiome community profiling using internal standards. *Appl. Environ. Microbiol.* **2019**, *85*, e02634-18. [CrossRef]
57. Piquet, A.M.T.; van de Poll, W.H.; Visser, R.J.W.; Wiencke, C.; Bolhuis, H.; Buma, A.G.J. Springtime phytoplankton dynamics in Arctic Krossfjorden and Kongsfjorden (Spitsbergen) as a function of glacier proximity. *Biogeosciences* **2014**, *11*, 2263–2279. [CrossRef]
58. Ding, X.; Liu, J.X.; Liu, W.W.; Dai, S.; Ke, Z.X.; Guo, J.; Lai, Y.J.; Tan, Y.H. Phytoplankton Communities Miniaturization Driven by Extreme Weather in Subtropical Estuary under Climate Changes. *Water Res.* **2023**, *245*, 120588. [CrossRef] [PubMed]
59. Yeh, Y.C.; McNichol, J.; Needham, D.M.; Fichot, E.B.; Berdjeb, L.; Fuhrman, J.A. Comprehensive single-PCR 16S and 18S rRNA community analysis validated with mock communities, and estimation of sequencing bias against 18S. *Environ. Microbiol.* **2021**, *23*, 3240–3250. [CrossRef] [PubMed]
60. Ke, Z.X.; Huang, L.M.; Tan, Y.H.; Song, X.Y. A dinoflagellate *Cochlodinium geminatum* bloom in the Zhujiang (Pearl) River estuary in autumn 2009. *Chin. J. Oceanol. Limnol.* **2012**, *30*, 371–378. [CrossRef]
61. Guo, Y.P.; Lin, S.J.; Huang, L.M.; Chen, Y.Q.; Hu, S.M.; Liu, S.; Tan, Y.H.; Huang, X.P.; Qiu, D.J. Dissipation of a *Polykrikos geminatum* Bloom after Wind Events in Pearl River Estuary. *Water* **2022**, *14*, 2313. [CrossRef]
62. Gong, W.D.; Hall, N.; Paerl, H.; Marchetti, A. Phytoplankton composition in a eutrophic estuary: Comparison of multiple taxonomic approaches and influence of environmental factors. *Environ. Microbiol.* **2020**, *22*, 4718–4731. [CrossRef] [PubMed]
63. Liu, H.; Probert, I.; Uitz, J.; Claustre, H.; Aris-Brosou, S.; Frada, M.; Not, F.; de Vargas, C. Extreme diversity in noncalcifying haptophytes explains a major pigment paradox in open oceans. *Proc. Natl. Acad. Sci. USA* **2009**, *106*, 12803–12808. [CrossRef] [PubMed]
64. Potvin, M.; Lovejoy, C. PCR-based diversity estimates of artificial and environmental 18s rRNA gene libraries. *J. Eukaryot. Microbiol.* **2009**, *56*, 174–181. [CrossRef]
65. Zhu, F.; Massana, R.; Not, F.; Marie, D.; Vaulot, D. Mapping of picoeucaryotes in marine ecosystems with quantitative PCR of the 18S rRNA gene. *FEMS Microbiol. Ecol.* **2005**, *52*, 79–92. [CrossRef]
66. Hansen, P.J.; Nielsen, L.T.; Johnson, M.; Berge, T.; Flynn, K.J. Acquired phototrophy in *Mesodinium* and *Dinophysis*—A review of cellular organization, prey selectivity, nutrient uptake and bioenergetics. *Harmful Algae* **2013**, *28*, 126–139. [CrossRef]
67. Mitra, A.; Flynn, K.J.; Tillmann, U.; Raven, J.A.; Caron, D.; Stoecker, D.K.; Not, F.; Hansen, P.J.; Hallegraeff, G.; Sanders, R.; et al. Defining planktonic protist functional groups on mechanisms for energy and nutrient acquisition: Incorporation of diverse mixotrophic strategies. *Protist* **2016**, *167*, 106–120. [PubMed]
68. D'Ors, A.; Bartolomé, M.C.; Sánchez-Fortún, S. Repercussions of salinity changes and osmotic stress in marine phytoplankton species. *Estuar. Coast. Shelf Sci.* **2016**, *175*, 169–175. [CrossRef]
69. Li, G. Fast acclimation of phytoplankton assemblies to acute salinity stress in the Jiulong River Estuary. *Acta Oceanol. Sin.* **2019**, *38*, 78–85. [CrossRef]
70. Nche-Fambo, F.A.; Scharler, U.M.; Tirok, K. Resilience of estuarine phytoplankton and their temporal variability along salinity gradients during drought and hypersalinity. *Estuar. Coast. Shelf Sci.* **2015**, *158*, 40–52. [CrossRef]
71. Burchard, H.; Schuttelaars, H.M.; Ralston, D.K. Sediment Trapping in Estuaries. *Annu. Rev. Mar. Sci.* **2018**, *10*, 371–395. [CrossRef]
72. Goosen, N.K.; Kromkamp, J.; Peene, J.; van Rijswik, P.; van Breugel, P. Bacterial and phytoplankton production in the maximum turbidity zone of three European estuaries: The Elbe, Westerschelde and Gironde. *J. Mar. Syst.* **1999**, *22*, 151–171. [CrossRef]
73. Huang, X.; Huang, L. Progress in researches on dynamical processes of phytoplankton ecology in maximum turbidity zone of estuary. *Acta Ecol. Sin.* **2002**, *22*, 1527–1533.
74. Wu, J.N.; Zhu, Z.; Waniek, J.J.; Niu, M.Y.; Wang, Y.T.; Zhang, Z.R.; Zhou, M.; Zhang, R.F. The biogeography and co-occurrence network patterns of bacteria and microeukaryotes in the estuarine and coastal waters. *Mar. Environ. Res.* **2023**, *184*, 105873. [CrossRef]
75. Niu, L.X.; Luo, X.X.; Hu, S.; Liu, F.; Cai, H.Y.; Ren, L.; Ou, S.Y.; Zeng, D.N.; Yang, Q.S. Check impact of anthropogenic forcing on the environmental controls of phytoplankton dynamics between 1974 and 2017 in the Pearl River estuary, China. *Ecol. Indic.* **2020**, *116*, 106484. [CrossRef]
76. Lu, Z.M.; Gan, J.P. Controls of seasonal variability of phytoplankton blooms in the Pearl River Estuary. *Deep-Sea Res. Part II-Top. Stud. Oceanogr.* **2015**, *117*, 86–96. [CrossRef]

77. Rabouille, C.; Conley, D.J.; Dai, M.H.; Cai, W.J.; Chen, C.T.A.; Lansard, B.; Green, R.; Yin, K.; Harrison, P.J.; Dagg, M.; et al. Comparison of hypoxia among four river-dominated ocean margins: The Changjiang (Yangtze), Mississippi, Pearl, and Rhone rivers. *Cont. Shelf Res.* **2008**, *28*, 1527–1537. [CrossRef]
78. Sun, J.; Lin, B.L.; Li, K.M.; Jiang, G.Q. A modelling study of residence time and exposure time in the Pearl River Estuary, China. *J. Hydro-Environ. Res.* **2014**, *8*, 281–291. [CrossRef]
79. Chen, W.L.; Guo, F.; Huang, W.J.; Wang, J.G.; Zhang, M.; Wu, Q. Advances in phytoplankton population ecology in the Pearl river estuary. *Front. Environ. Sci.* **2023**, *11*, 1084888. [CrossRef]
80. Gu, H.; Wu, Y.; Lu, S.; Lu, D.; Tang, Y.Z.; Qi, Y. Emerging harmful algal bloom species over the last four decades in China. *Harmful Algae* **2022**, *111*, 102059. [CrossRef]
81. Li, K.; Huang, L.; Zhang, J.; Yin, J.; Luo, L. Characteristics of phytoplankton community in the Pearl River Estuary during saline water intrusion period. *J. Trop. Oceanogr.* **2010**, *29*, 62–68.
82. Wang, Z.H.; Guo, X.; Qu, L.J.; Lin, L.C. Effects of nitrogen and phosphorus on the growth of *Levanderina fissa*: How it blooms in Pearl River Estuary. *J. Ocean Univ.* **2017**, *16*, 114–120. [CrossRef]
83. Ren, H.; Tian, T.; Yang, Y.; Wang, Q. Spatial and temporal distribution of phytoplankton community and its relationship with environment factors in Nansha's Rivers, Pearl River estuary. *Acta Ecol. Sin.* **2017**, *37*, 7729–7740.
84. Gou, T.; Xu, Z.; Li, J.; Ma, Q.; Wang, L.; Zhao, X.; Liang, R.; Guo, J. Phytoplankton community structure and water quality assessment of Hejiang River, a branch of Xijiang River, Pearl River drainage basin. *Hupo Kexue* **2015**, *27*, 412–420.
85. Qiu, D.J.; Huang, L.M.; Zhang, J.L.; Lin, S.J. Phytoplankton dynamics in and near the highly eutrophic Pearl River Estuary, South China Sea. *Cont. Shelf Res.* **2010**, *30*, 177–186. [CrossRef]

biology

MDPI

Article

Cloning and Functional Analysis of a Zeaxanthin Epoxidase Gene in *Ulva prolifera*

Hongyan He [1,†], Xiuwen Yang [1,†], Aurang Zeb [1], Jiasi Liu [1], Huiyue Gu [1], Jieru Yang [1], Wenyu Xiang [2] and Songdong Shen [1,*]

[1] School of Biology & Basic Medical Sciences, Soochow University, Suzhou 215101, China
[2] Suzhou Industrial Park Environmental Law Enforcement Brigade, Suzhou 215021, China; xwy@sipac.gov.cn
* Correspondence: shensongdong@suda.edu.cn; Tel.: +86-0512-65880276
† These authors contributed equally to this work.

Simple Summary: The xanthophyll cycle plays an important role in plants' responses to abiotic stress. Zeaxanthin epoxidase (*ZEP*) contributes to the xanthophyll cycle and the biosynthesis of abscisic acid (ABA) and carotenoids. The biological function of *ZEP* has been extensively studied in higher plants but has not been reported in *Ulva prolifera*. In this study, the relative content of xanthophylls was determined in *U. prolifera*. We also cloned and analyzed a *ZEP* gene from *U. prolifera*, namely *UpZEP*. The expression and biological function of the *UpZEP* gene were examined. The results of this study provide a theoretical basis for understanding the high-salt resistance of *U. prolifera*.

Abstract: The xanthophyll cycle is a photoprotective mechanism in plants and algae, which protects the photosynthetic system from excess light damage under abiotic stress. Zeaxanthin is considered to play a pivotal role in this process. In this study, the relative content of xanthophylls was determined using HPLC-MS/MS in *Ulva prolifera* exposed to different salinities. The results showed that high-salt stress significantly increased the relative content of xanthophylls and led to the accumulation of zeaxanthin. It was speculated that the accumulated zeaxanthin may contribute to the response of *U. prolifera* to high-salt stress. Zeaxanthin epoxidase (*ZEP*) is a key enzyme in the xanthophyll cycle and is also involved in the synthesis of abscisic acid and carotenoids. In order to explore the biological function of *ZEP*, a *ZEP* gene was cloned and identified from *U. prolifera*. The CDS of *UpZEP* is 1122 bp and encodes 373 amino acids. Phylogenetic analysis showed that *UpZEP* clusters within a clade of green algae. The results of qRT-PCR showed that high-salt stress induced the expression of *UpZEP*. In addition, heterologous overexpression of the *UpZEP* gene in yeast and *Chlamydomonas reinhardtii* improved the salt tolerance of transgenic organisms. In conclusion, the *UpZEP* gene may be involved in the response of *U. prolifera* to high-salt stress and can improve the high-salt tolerance of transgenic organisms.

Keywords: *Ulva prolifera*; zeaxanthin epoxidase; high-salt stress; overexpression

1. Introduction

Green tides are a marine ecological anomaly that usually occur in eutrophicated estuaries, inner bays and urban dense coastal zones [1]. Under certain environmental conditions, green seaweed rapidly grows and propagates, accumulating in piles and thus causing green tides [2]. The first large-scale outbreak of green tides occurred in France, followed by the United States, Japan and China [3]. In recent years, green tides have increasingly become a global environmental concern because of the rising severity, geographical scale and frequency of macroalgal blooms [4]. Since 2007, green tides have been breaking out along the Yellow Sea coast every year. Green tides cover a wide area and have a large influence, seriously threatening the ecological safety of China [3,5]. The common species causing green tide disasters in China are *Ulva*, *Cladophora* and *Chaetomorpha*, with *Ulva prolifera* being the predominant species [6,7].

U. prolifera grows in nearshore ponds, estuaries and intertidal zones, where the environment is very complex and susceptible to abiotic stresses, such as salt stress, light stress and temperature stress [8]. To cope with abiotic stress, *U. prolifera* has evolved numerous defense mechanisms [9,10]. Salt stress is one of the abiotic stresses that restricts the growth and development of *U. prolifera* [11,12]. In general, both too high and too low salinity have adverse effects on photosynthesis, respiration and other physiological activities of *U. prolifera* [11]. In order to understand how *U. prolifera* responds to salt stress, biochemical and physiological response mechanisms have been extensively investigated [12,13]. Under salt stress, the photosynthetic performance of *U. prolifera* decreases, and the expression levels of several stress-resistance protein-related genes, such as heat shock protein and tubulin, are significantly up-regulated. These play an important role in sustaining life activities [12]. *U. prolifera* can also cope with high-salt stress by enhancing non-photochemical quenching (NPQ) [13]. Moreover, *U. prolifera* cells accumulate carbohydrates, such as glucose and fructan, under salt stress, with their main functions being osmotic protection and scavenging reactive oxide species (ROS) [14]. Interestingly, *U. prolifera* has salinity-dependent morphological variations; it has more branches under low salinity conditions and longer branches under high salt conditions [15].

In general, plants capture and use light energy efficiently. However, under environmental stress, plants can absorb excess light energy, which can lead to the destruction of the photosystem [16]. To avoid destroying the photosystems, plants have evolved a variety of photoprotection strategies, among which the xanthophyll cycle is considered an important photoprotection mechanism [17]. The xanthophyll cycle involves three components: violaxanthin, antheraxanthin and zeaxanthin. When plants receive excess light energy, zeaxanthin is produced through two de-epoxidation reactions under the action of violaxanthin de-epoxidase (VDE). Conversely, when light energy is insufficient, violaxanthin is produced by two epoxidation reactions under the action of zeaxanthin epoxidase, with antheraxanthin serving as the intermediate product in both the de-epoxidation and epoxidation reactions [18]. The xanthophylls are uniformly distributed in the thylakoid membrane, where they bind to the light-harvesting complex (LHC), which changes its conformation, leading to the dissipation of excess light energy [19]. Zeaxanthin, a xanthophyll with strong antioxidant activity, is widely used in feed, cosmetics, food and medical treatment [20]. Moreover, zeaxanthin can eliminate ROS generated by photosynthesis, such as triplet oxygen, singlet oxygen and superoxide anion, thereby safeguarding the integrity of the photosystem [21].

Zeaxanthin epoxidase (*ZEP*) is a bi-functional monooxygenase that epoxidizes zeaxanthin to antheraxanthin and then further to violaxanthin. This epoxidation reaction not only contributes to the xanthophyll cycle but also to the biosynthesis of abscisic acid (ABA) and carotenoids [22,23]. Several reports have suggested that *ZEP* plays an important role in plant growth, development and abiotic stress resistance [24]. Under abiotic stress, plants can regulate the content of carotenoids and other secondary metabolites by regulating the expression of the *ZEP* gene, thereby removing excessive ROS caused by abiotic stress. The expression of *ZEP* in tobacco and tomato roots significantly increased under drought stress [25,26]. The expression of *ZEP* in *Arabidopsis thaliana* roots was up-regulated under drought stress, and the content of ABA in roots and leaves also increased [27]. In addition, overexpression of a *ZEP* gene from *Medicago sativa* in tobacco significantly improved its salt tolerance [28]. The high expression of *ZEP* genes involved in ABA synthesis in *Brassica napus* increased drought tolerance [29]. These studies demonstrate that higher plants respond to abiotic stress by up-regulating *ZEP* gene expression.

The biological function of *ZEP* has been extensively studied in higher plants but has not been reported in *U. prolifera*. Here, the HPLC-MS/MS technique was employed to determine the relative contents of lutein, violaxanthin, antheraxanthin and zeaxanthin in *U. prolifera* cultivated under normal and high-salt conditions. Then, a *ZEP* gene was cloned and analyzed from *U. prolifera*, and the expression of the *UpZEP* gene was analyzed using qRT-PCR. In addition, to detect the function of the *UpZEP* gene, it was heterologously overexpressed in *Chlamydomonas reinhardtii* and *Saccharomyces cerevisiae*, which provided scientific insight into the high-salt resistance of *U. prolifera*.

2. Materials and Methods

2.1. Alga Material and Treatment

U. prolifera was collected from the coast of Qingdao, China (36°48′39.75″ N; 121°38′10.88″ E). Pure lines were obtained through genetic and morphological identification and were cultured in our laboratory. These lines were then cultured in distilled seawater supplemented with Von Stosch's Enriched (VSE) medium under conditions of 20 ± 1 °C, a light intensity of 120 μmol photons $m^{-2} \cdot s^{-1}$ and a photoperiod of 12 L:12 D [30]. After 15 d of cultivation, the thalli with healthy growth and similar lengths were selected and placed in a constant temperature light incubator for high-salt stress (50‰) treatment at different time points, with other culture conditions remaining unchanged.

The *C. reinhardtii* wild-type (WT) strain was provided by Dr. Yingjuan Wang (College of Life Science, Northwest University, Xi'an, China) and was cultivated in Tris-acetate-phosphate (TAP) medium under the following conditions: a light intensity of 50–60 μmol photons $m^{-2} \cdot s^{-1}$, a photoperiod of 12 L:12 D and 21 ± 1 °C. High-salt treatment was performed in 120 mM sodium chloride at different time points. The cell growth rate was assessed by measuring OD_{750}.

2.2. High-Performance Liquid Chromatography-Tandem Mass Spectrometry (HPLC-MS/MS) Analysis

The frozen *U. prolifera* samples were pulverized in liquid nitrogen. About 0.5 g of the powder was aliquoted into a tube and 5 mL of acetone was added. After ultrasonication in a water bath for 15 min, the sample was centrifuged. The upper phase was collected, and the precipitate was repeatedly extracted until it was colorless. The resulting sample solution was dried with nitrogen gas. The dry extracts were dissolved in 25 mL of methanol and then filtered through a 0.22 μm polyvinylidene fluoride (PVDF) membrane filter. All the samples were analyzed using HPLC-MS/MS.

We employed an Agilent 1290 high performance liquid chromatograph in tandem (Agilent, Santa Clara, CA, USA) with an AB Qtrap6500 mass spectrometer (Allen-Bradley, Milwaukee, WI, USA)to quantify lutein, zeaxanthin, violaxanthin and antheraxanthin contents. The samples were injected into a CORTECS UPLC C18+ column (1.6 μm, 2.1 mm × 75 mm) and maintained at 35 °C. Mobile phase A consisted of 0.1% (v/v) formic acid in water and mobile phase B consisted of methanol. The column was eluted at a flow rate of 450 μL/min, with the gradient elution profile starting at 10% A and 90% B for 3 min. The proportion of mobile phase B was then increased to 100%, and mobile phase A was decreased to 0%. Finally, the initial mobile phase proportions were restored within 4 min to equilibrate the column for a further 2 min. The samples were then analyzed using atmospheric pressure chemical ionization (APCI) in the positive mode under the following conditions: ion source temperature TEM (transmission electron microscopy), 380 °C; curtain gas, 40 Psi; nebulizer gas (Gas1), 50 Psi; auxiliary heating gas (Gas2), 55 Psi; and collision gas CAD (collision activated dissociation), medium. The parameters for the MS analysis of xanthophylls are listed in Table S1. The relative contents of xanthophylls were quantified by integrating and comparing peak areas with those of the matrix-matched calibration curve.

2.3. RNA Isolation and qRT-PCR

The total RNA was extracted from *U. prolifera* and *C. reinhardtii* using TRIzol reagents (Thermo Fisher, Waltham, MA, USA), followed by reverse-transcription into cDNA using the Hifair™ II 1st Strand cDNA Synthesis Kit (Yeasen, Shanghai, China). Then, the levels of transcripts were detected using the Hieff® qPCR SYBR Green Master Mix (Low Rox Plus) (Yeasen, Shanghai, China). The primers used for qRT-PCR are listed in Table S2. The *18Sr*

RNA and *Tubulin* were used as the internal reference genes for *U. prolifera* and *C. reinhardtii*, respectively. All the samples were analyzed in triplicate, and the $2^{-\Delta\Delta CT}$ method was used to determine the transcript levels.

2.4. Full-Length of the UpZEP Gene Cloning and Bioinformatics Analysis

The sequence of the *ZEP* gene was obtained from the *U. prolifera* genome [31]. The full-length sequences and coding DNA sequences (CDS) of the *UpZEP* gene were amplified using the genomic DNA and cDNA from *U. prolifera* as templates, respectively. The primers were designed using Primer Premier 5.0 and are listed in Table S2. The physicochemical properties of the *UpZEP* protein were analyzed using the Prot Param program [32]. To analyze the phylogenetic relationship, *ZEP* protein sequences from representative Plants, Chlorophyta, Bacillariophyta, Rhodophyta and Cyanobacteria were downloaded from NCBI. Multiple sequence alignment was performed using ClustalW2.1, and the phylogenetic tree was constructed using the neighbors-joining method with MEGA6 (Bootstrap value = 1000). The Newick tree was exported from MEGA6 and improved using the Evolview website (https://evolgenius.info//evolview-v2 (accessed on 26 July 2024)).

2.5. C. reinhardtii Transformation and Screening of Positive Clones

We used the pChlamy-3 vector, an expression vector containing the Hsp70A-Rbc S2 fusion promoter, to explore the function of the *UpZEP* gene. The FAD (flavin adenine dinucleotide) domain coding sequences of the *UpZEP* gene were amplified to construct the recombinant vector pChlamy-3_*UpZEP*. Then, the pChlamy-3_*UpZEP* recombinant expression vector was transformed into *C. reinhardtii* using the "glass bead transformation method", and monoclonal algal colonies were selected on hygromycin-B-containing medium (50 mg mL^{-1}). The hygromycin-B-resistant and wild-type strains of *C. reinhardtii* were randomly selected, genomic DNA was extracted and the *UpZEP* gene sequences were amplified using gene-specific primers (Table S2).

2.6. Functional Characterization via Heterologous Expressions of the UpZEP Gene in Yeasts

The FAD domain coding sequences of the *UpZEP* gene were recombined into the pYES2 vector. The recombinant plasmid pYES2-*UpZEP* was transformed into the *S. cerevisiae* strain INVSc1 (Coolaber, Beijing, China), and yeast cells harboring pYES2-*UpZEP* were selected on SD-Ura medium (Coolaber, Beijing, China). Single colonies were picked and transferred into liquid SD-Ura medium, then grown overnight at 200 rpm in a shaker at 30 °C. An aliquot of 1 mL of the suspension was centrifuged, resuspended in 100 µL of ddH_2O and added to SG-Ura medium (Coolaber, Beijing, China) containing different concentrations of NaCl to observe growth.

2.7. Measurement of Chlorophyll Contents in C. reinhardtii

The samples were dissolved in 95% ethanol and crushed using an ultrasonic homogenizer. Then, they were placed in a dark environment at 4 °C for 24 h, and the absorbance values at OD665 and OD649 were measured. The contents of chlorophyll were calculated using the following equations:

$$\text{Chl. } a = 13.95 \times \text{OD665} - 6.88 \times \text{OD649}$$
$$\text{Chl. } b = 24.96 \times \text{OD649} - 7.32 \times \text{OD665}$$
$$\text{Total chlorophyll} = \text{Chl. } a + \text{Chl. } b$$

2.8. Statistical Analyses

All statistical analyses were conducted using Tukey's test with GraphPad Prism software. Statistical significance was indicated by *p*-values of $p < 0.05$, $p < 0.01$, $p < 0.001$ and $p < 0.0001$, marked as *, **, *** and ****, respectively. Three biological replicates were performed for all experimental treatments. The results are expressed as the mean ± standard deviation (SD).

3. Results

3.1. Determination of Relative Content of Xanthophylls in *U. prolifera*

The relative contents of xanthophylls in *U. prolifera* grown under different salinity conditions were tested using HPLC-MS/MS technology. The results indicated the presence of several kinds of xanthophylls in *U. prolifera*, such as lutein (L), zeaxanthin (Z), antheraxanthin (A) and violaxanthin (V), with lutein and zeaxanthin being the main components. Compared to the control group (25‰), the relative contents of lutein, zeaxanthin and violaxanthin increased significantly ($p < 0.05$) under high-salt stress (50‰), whereas the content of antheraxanthin remained unchanged (Figure 1A–D). Under high-salt stress (50‰), the content of de-epoxidized xanthophylls increased significantly ($p < 0.05$) compared to the control group (25‰), resulting in the xanthophyll cycle being in a de-epoxidation state ([Z/(V + A + Z)]; [(Z + 0.5A)/(V + A + Z)]) (Figure 1E,F).

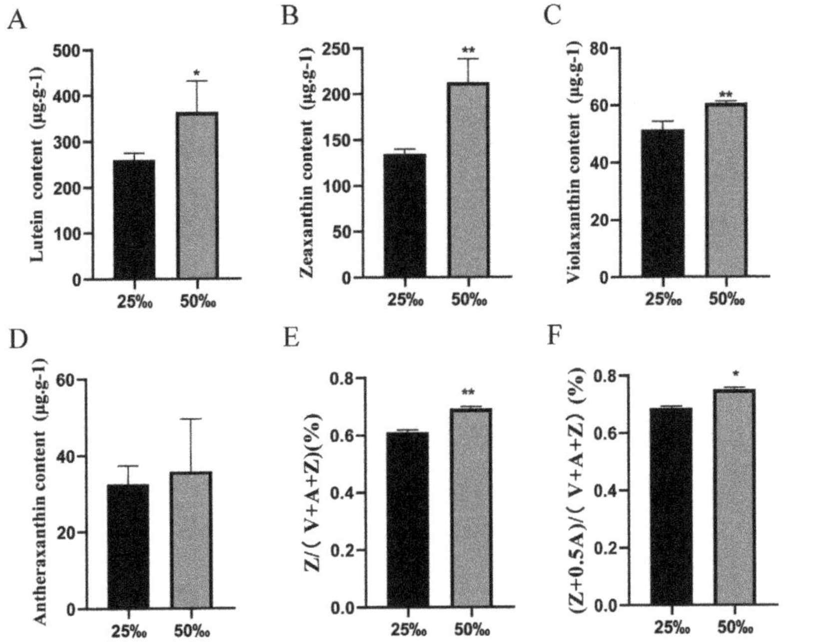

Figure 1. Relative content of xanthophylls in *U. prolifera*. (**A**) Lutein content; (**B**) Zeaxanthin content; (**C**) Violaxanthin content; (**D**) Antheraxanthin content; (**E**) De-epoxidation state, Z/(V + A + Z); (**F**) De-epoxidation state, (Z + 0.5A)/(V + A + Z). Zeaxanthin, Z; Violaxanthin, V; Antheraxanthin, A. 25‰, control group; 50‰, high-salt stress. Values are means ± SD (n = 3; compared with 25‰, * $p < 0.05$, ** $p < 0.01$, Student's *t*-test). All experiments were repeated at least three times with similar results.

3.2. Full-Length Cloning and Characterization of the *UpZEP* Gene

We amplified the full-length sequence and CDS of the *U. prolifera* gene using PCR (Figure S1). After a BLAST search, the gene was identified as the *ZEP* gene and named *UpZEP*. The full-length sequence of the *UpZEP* gene is 1609 bp, containing three exons and two introns, with a CDS of 1122 bp, encoding 373 amino acids. The ProtParam tool was used to analyze the properties of the protein, revealing that the molecular mass of the *UpZEP* protein is about 41.37 kDa, and the theoretical isoelectric point is 5.68.

The *UpZEP* protein sequences were aligned with *ZEP* proteins from *A. thaliana*, *Oryza sativa*, *Dunaliella tertiolecta*, *Haematococcus lacustris* and *C. reinhardtii*. As shown in Figure 2, the amino acid sequence of *UpZEP* exhibits high homology with their *ZEPs*. Furthermore, there are two highly conserved regions, which were predicted to be the FAD domain and the FHA (fork head-associated) domain, respectively. This suggests that *UpZEP* likely performs the same role as other *ZEP*s, catalyzing epoxidation using NADPH and FAD as cofactors [33].

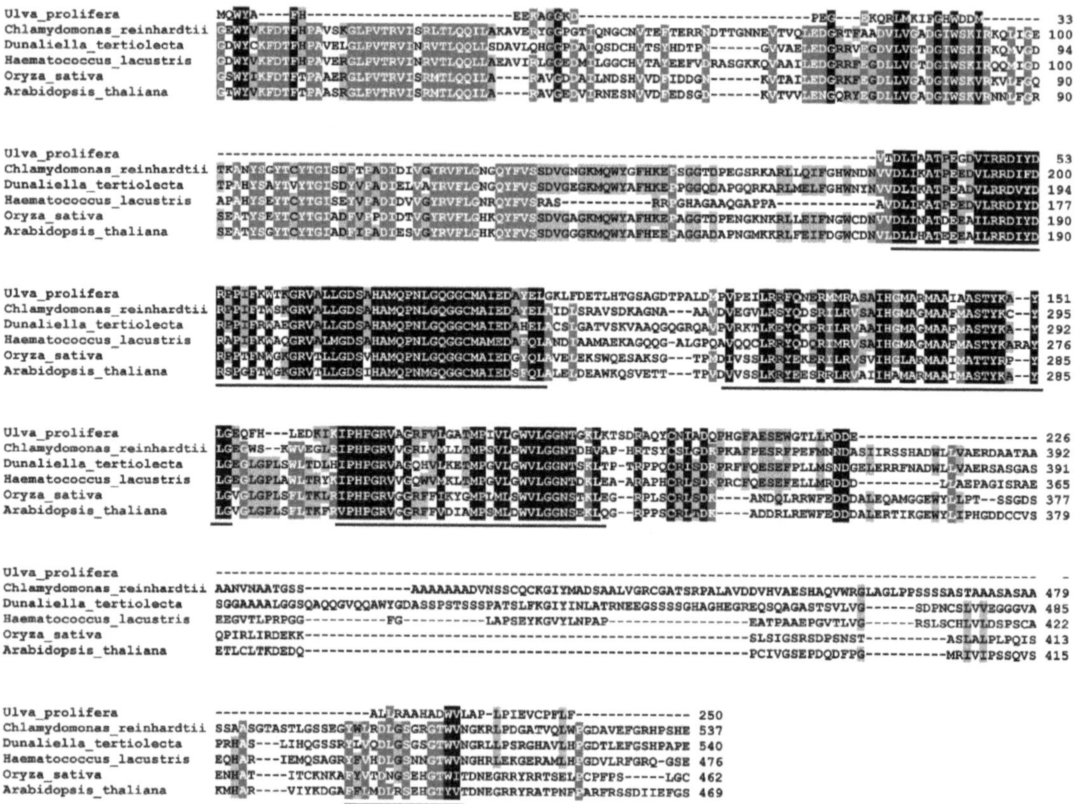

Figure 2. Alignment of the amino acid sequence of *UpZEP* with other *ZEP*s. The residues that are identical in more than 80% of all the sequences are shaded in gray. The conserved residues are shaded in black. The FAD domain is indicated with a black underline, and the FHA domain is indicated with a red underline. The species and corresponding GenBank accession number are as follows: *Chlamydomonas reinhardtii* (AAO48940.1); *Dunaliella tertiolecta* (AZK89898.2); *Haematococcus lacustris* (GFH20616.1); *Oryza sativa* (BAB39765.1); *Arabidopsis thaliana* (AAG38877.1).

In order to further investigate the evolutionary relationship of *UpZEP*, MEGA 6 software was used to construct a phylogenetic tree of *ZEP*s from several Plants, Chlorophyta, Rhodophyta, Bacillariophyta and Cyanobacteria using the neighbor-joining method. The results showed that *UpZEP* clustered with other Chlorophyta *ZEP*s in a single branch, with *UpZEP* being most closely related to the *ZEP* of *D. tertiolecta* (Figure 3). *UpZEP* and other Chlorophyta *ZEP*s were evolutionarily closer to the *ZEP*s of land plants. These results suggest that the function of *UpZEP* may be similar to the *ZEP*s of Chlorophyta and land plants.

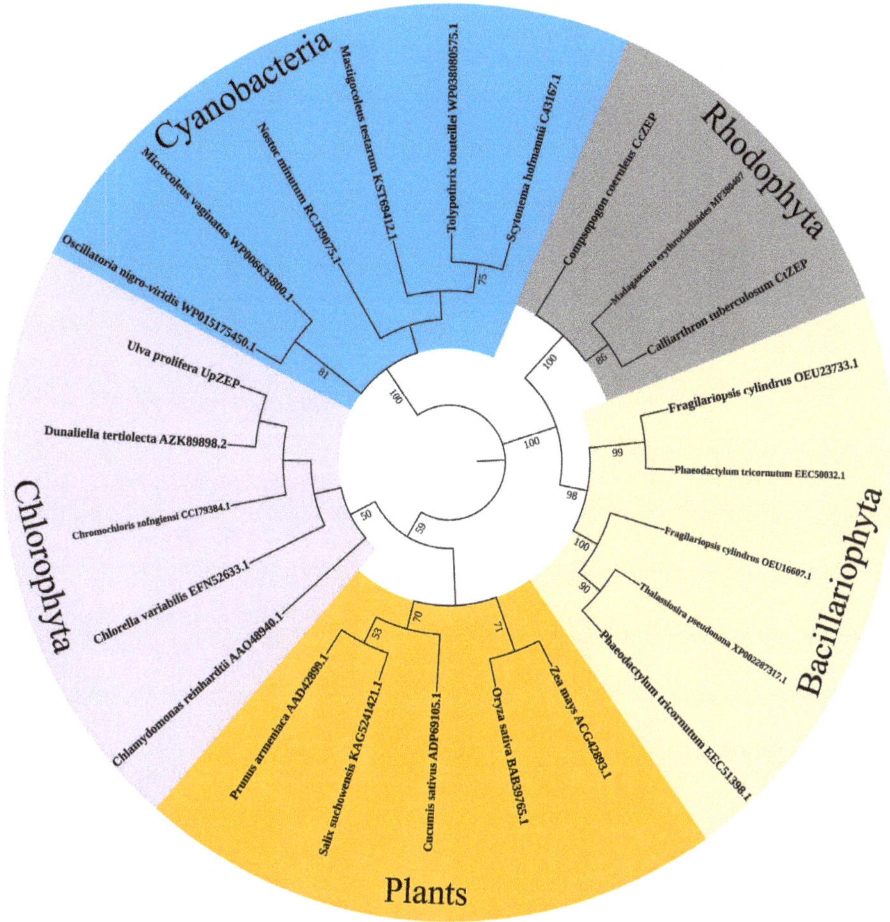

Figure 3. Phylogenetic trees of *ZEPs* from several Plants, Chlorophyta, Rhodophyta, Bacillariophyta and Cyanobacteria. The neighbor-joining method was used to reconstruct the phylogenetic trees using the MEGA6 software (Bootstrap value = 1000). *UpZEP* refers to *U. prolifera*. The GenBank IDR is shown right after each species name.

3.3. Expression Levels of the UpZEP Gene under High-Salt Stress

In order to investigate the expression patterns of the *UpZEP* gene under high-salt stress (50‰), the expression of the *UpZEP* gene was analyzed at 12 h, 24 h and 48 h under different salinity conditions. The internal reference gene used was *18Sr RNA*. Compared to the control group (25‰), high-salt stress (50‰) significantly induced the expression of the *UpZEP* gene ($p < 0.01$). The gene expression reached maximum at 24 h, then decreased at 48 h but remained significantly higher than in the control group (25‰) ($p < 0.01$) (Figure 4).

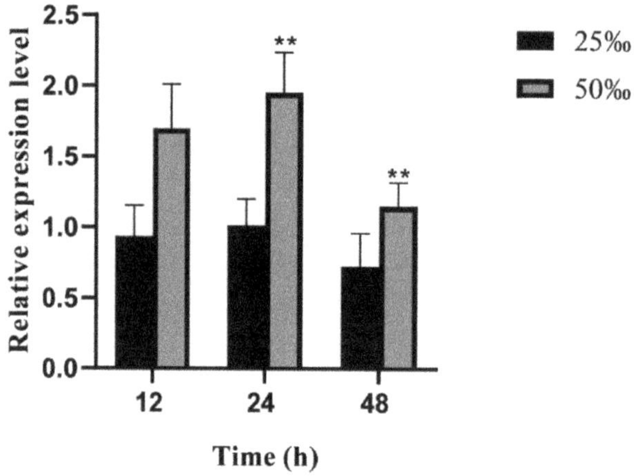

Figure 4. Expression levels of the *UpZEP* gene under different salinity conditions at 12 h, 24 h and 48 h. The 25‰ condition represents the control group and the 50‰ condition represents high-salt stress. The values are expressed as means ± SD (n = 3; compared with 25‰, ** $p < 0.01$, Student's t-test).

3.4. Functional Characterization via Heterologous Expressions of the UpZEP Gene in Yeast

To investigate whether heterologous overexpression of the *UpZEP* gene affects high-salt tolerance in yeast, the recombinant vector pYES2-*UpZEP* was transformed into *S. cerevisiae* INVSc1 strains. Yeasts transformed with pYES2-*UpZEP* (experimental group) and those with empty pYES2 (control group) were respectively cultured on SG-Ura medium with different salinities. The results showed that both strains could grow on SG-Ura medium containing 0 M, 0.5 M and 1 M NaCl. On the SG-Ura medium with 0 M NaCl, there was no difference in the biomass between the two groups. On the SG-Ura medium with 0.5 M and 1 M NaCl, the biomass of the strains in the experimental group increased significantly compared to the control group; however, on the SG-Ura medium with 1.3 M NaCl, the control strains showed no growth, whereas the experimental strains were able to grow (Figure 5).

3.5. Validation of C. reinhardtii Transformation and UpZEP Gene Expression Levels under High-Salt Stress

To further explore the biological function of the *UpZEP* gene in vivo, we generated a transgenic *C. reinhardtii* strain with heterologous overexpression of the *UpZEP* gene. Then, the phenotype, physiological parameters and gene expression patterns of transgenic *C. reinhardtii* were analyzed and compared to the wild-type. Firstly, the overexpression vector pChlamy-3-*UpZEP* was constructed and transformed into *C. reinhardtii* using the "glass bead transformation method". Then, hygromycin-B-resistant monoclonal algal colonies were obtained and screened using PCR amplification with special primers (Table S2). The selected monoclonal colonies produced target bands of the expected size, whereas the wild-type colonies did not produce bands, suggesting that *UpZEP* was successfully integrated into the genome of *C. reinhardtii* (Figure 6A).

To explore the expression patterns of the *UpZEP* gene in the transgenic *C. reinhardtii* strain, algae cultured under normal conditions were used as the control group, and strains cultured under 120 mM NaCl conditions were used as the experimental group. The expression levels of the *UpZEP* gene at 24 h, 48 h and 72 h in the two groups were detected using a qRT-PCR assay. *Tubulin* was used as the internal reference gene. The results showed that the transgenic *C. reinhardtii* exhibited a short adaptation period under high-salt stress. At 24 h, there was no significant difference in *UpZEP* gene expression between the two groups. However, compared to the control group, the *UpZEP* transcription level in the experimental group was significantly increased ($p < 0.01$) and showed an upward trend at 48 h and 72 h, indicating that the *UpZEP* gene was induced by high-salt stress (Figure 6B).

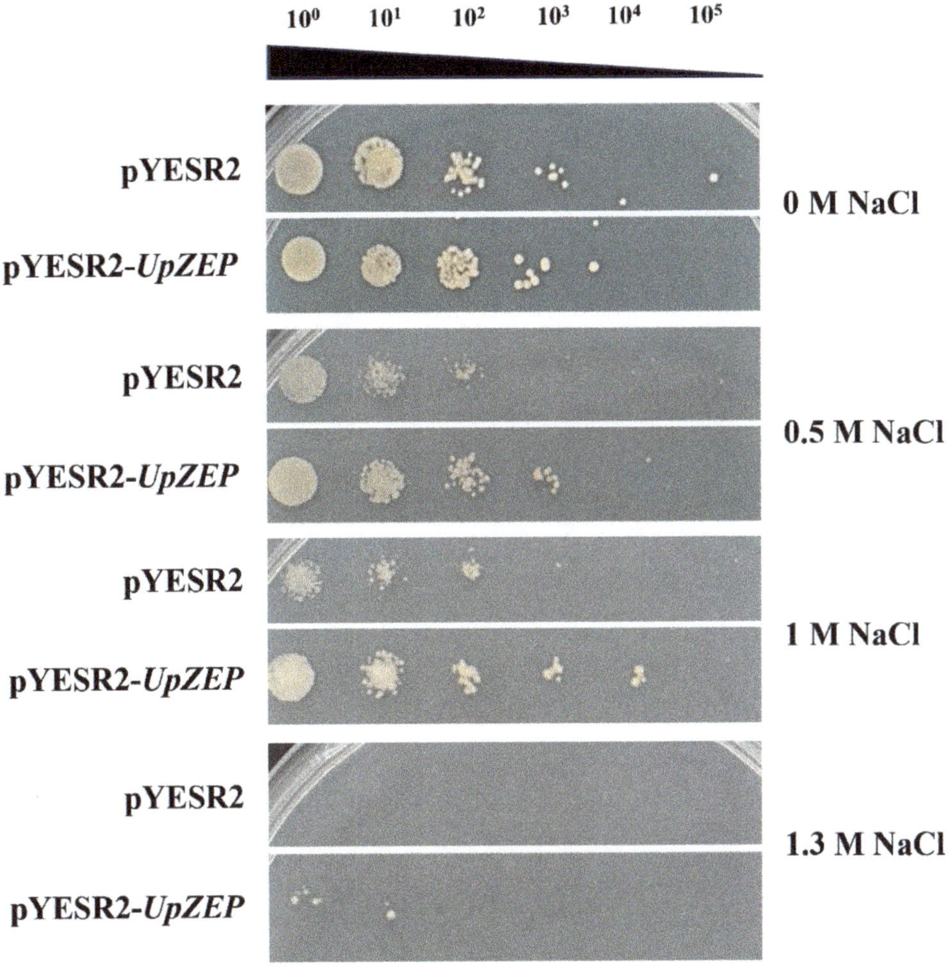

Figure 5. Verification experiment of yeast phenotypes. Yeast strains transformed with empty vector pYESR2 were used as the control group. Yeast strains transformed with recombinant vector pYESR2-*UpZEP* were used as the experimental group. Transformants of pYESR2 and pYESR2-*UpZEP* were grown in SG-Ura medium supplemented with different concentrations of NaCl.

3.6. Analysis of High-Salt Stress Tolerance in Transgenic C. reinhardtii

In order to explore the high-salt stress resistance of *C. reinhardtii* containing the *UpZEP* gene, wild-type and transgenic *C. reinhardtii* strains were cultured under 120 mM NaCl for 6 d. The color of transgenic *C. reinhardtii* changed from light green to dark green, whereas the wild-type did not change color. The transgenic *C. reinhardtii* appeared greener than the wild-type (Figure 7A). Additionally, biomass was measured using OD_{750} light absorption values (Figure 7B). The results showed that the biomass of transgenic *C. reinhardtii* was higher than that of the wild-type after 1 d of exposure to 120 mM NaCl. The corresponding phenotypic observations and biomass measurements were consistent, indicating that the *UpZEP* gene enhanced the salt tolerance of *C. reinhardtii*.

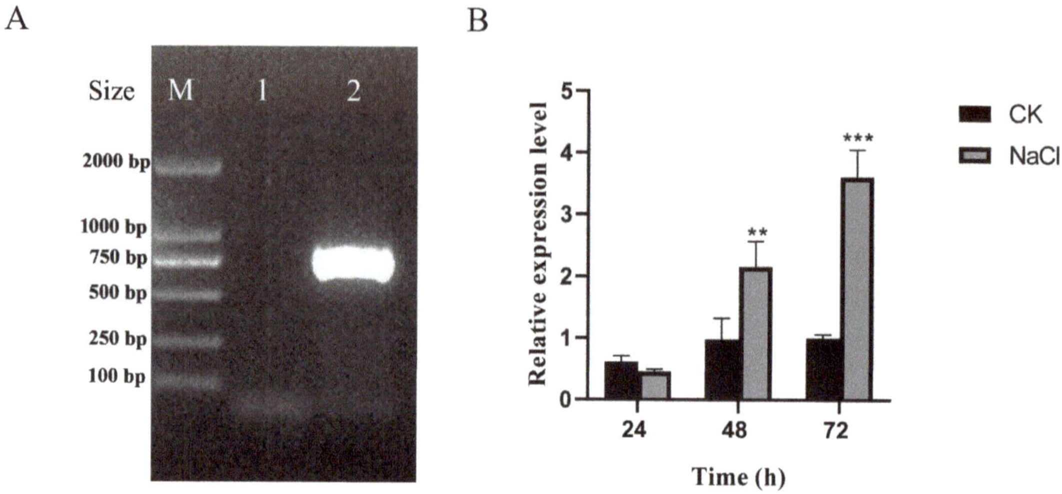

Figure 6. Validation of transgenic *C. reinhardtii* and the expression pattern of the *UpZEP* gene under high-salt stress. (**A**) Validation of transgenic *UpZEP*-positive clones using gel electrophoresis: 1, wild-type *C. reinhardtii*; 2, transgenic *UpZEP*-positive *C. reinhardtii*. (**B**) Expression levels of the *UpZEP* gene under high-salt stress. CK, control group; NaCl, 120 mM. The values are expressed as means ± SD (n = 3; compared with CK, ** $p < 0.01$, *** $p < 0.001$, Student's *t*-test).

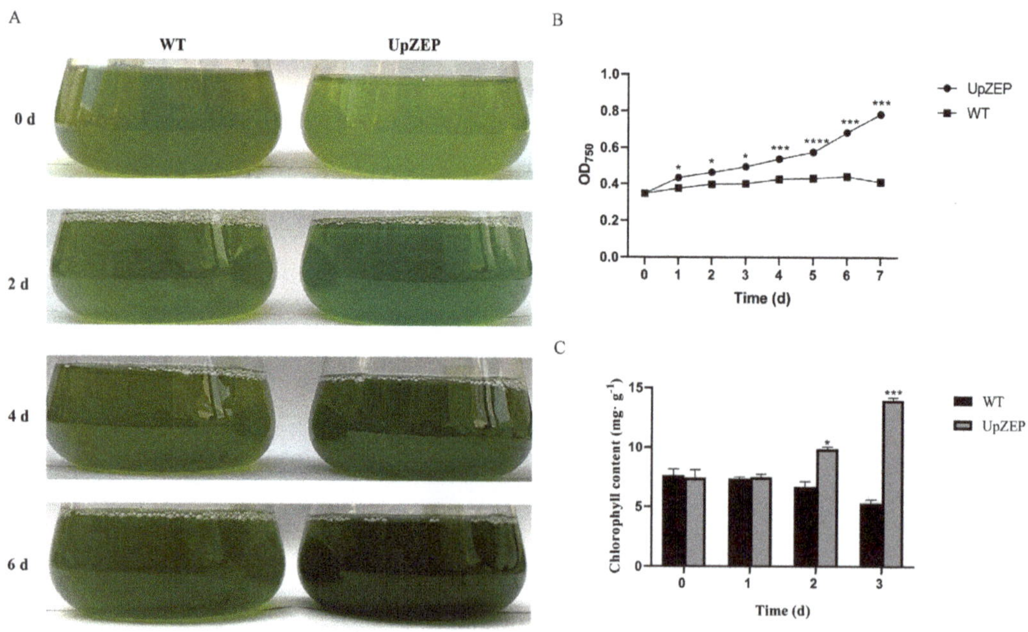

Figure 7. The biomass and chlorophyll content in *C. reinhardtii* under high-salt stress. Transgenic and wild-type (WT) *C. reinhardtii* phenotypes (**A**), biomass measurements (**B**) and chlorophyll content (**C**). WT, control group. The values are expressed as means ± SD (n = 3; compared with WT, * $p < 0.05$, *** $p < 0.001$, **** $p < 0.0001$, Student's *t*-test).

Simultaneously, the wild-type and transgenic *C. reinhardtii* strains were cultured under 120 mM NaCl for 0, 1, 2 and 3 d, and the chlorophyll content was measured. The transgenic *C. reinhardtii* had a short adaptation period under high-salt stress. At 1 d, there was no significant difference in chlorophyll content between the transgenic and wild-type strains; however, the chlorophyll content of transgenic *C. reinhardtii* was significantly higher compared to the wild-type strains at 2 and 3 d (Figure 7C).

4. Discussion

Under abiotic stress, the decreased growth rate of plants and algae is often associated with reduced photosynthesis [34]. This decline in photosynthesis in plants and algae will lead to an increase in excess light energy. If the excess excitation energy is not dissipated in time, this will lead to the accumulation of ROS, resulting in oxidative damage [16]. To avoid or minimize this photoinhibition damage, plants have evolved a variety of protective mechanisms, with carotenoids playing an important role in plant photoprotection [17]. Carotenoids dissipate excess light energy through NPQ, which is mediated by zeaxanthin [13,17]. In plants and algae, carotenoids are synthesized and accumulated in the thylakoid membrane, especially within the LHC [18]. Carotenoids are classified into carotenes and xanthophylls, and α-and β-carotene can be converted into xanthophylls. Seaweeds, a special class of marine plants, each have their own unique carotenoid composition [35]. In this study, we analyzed the carotenoids of *U. prolifera*, a dominant green tide species in China. The results showed that the carotenoids of *U. prolifera* are mainly lutein and zeaxanthin, and salt stress induced an increase in the contents of lutein, zeaxanthin and violaxanthin. In our previous study, we also found that low-light and high-salt stress promoted carotenoid synthesis [36]. In addition, carotenoid synthesis is induced by copper stress in *Ulva compressa* [37]. Collectively, these findings suggest that abiotic stress commonly promotes the synthesis and accumulation of carotenoids in *Ulva*. Under abiotic stresses, such as light stress, salinity stress and nitrogen deprivation, the carotenoid content also increased in *D. salina*, which is thought to be a strategy for resisting abiotic stress [38]. In this study, the carotenoid content increased under high-salt stress, indicating that their important role in the salt tolerance of *U. prolifera*.

Lutein is considered to be a xanthophyll involved in NPQ and acts as a direct quencher of chlorophyll *a* [38]. However, zeaxanthin is an allosteric modulator of non-photochemical energy dissipation [39]. In *U. pertusa*, accumulated zeaxanthin protects the thylakoid membrane and enhances thermal quenching during desiccation [40]. Slow zeaxanthin accumulation in *U. prolifera* under high-light stress is considered atypical for NPQ [41]. In this study, high-salt stress significantly increased the relative content of xanthophylls, leading to the accumulation of zeaxanthin, which may have a positive effect on the high-salt stress response of *U. prolifera*. The xanthophyll cycle significantly contributes to photoprotection in plants [19]. Violaxanthin and antheraxanthin, the epoxidation products of zeaxanthin, are the key precursors of carotenoids and also participate in the xanthophyll cycle. The catalytic function of *ZEP* has diversified through plant evolution [42]. The *ZEP* gene exists widely in photosynthetic eukaryotes. At present, the *ZEP* gene has been identified in several plants and algae. The genome of the diatom *Phaeodactylum tricornutum* harbors three *ZEP* genes, but only one *ZEP* gene is involved in NPQ [43]. *Nannochloropsis oceanica* contains two *ZEP* genes [44], and one *ZEP* gene has been identified in *D. tertiolecta* [33]. In this study, the *UpZEP* gene was cloned from *U. prolifera*. The amino acid sequence of *UpZEP* exhibited high homology with other *ZEP*s, and the *UpZEP* protein contained FAD and FHA domains, indicating that *UpZEP* is a typical *ZEP* protein. Phylogenetic analysis showed that the *UpZEP* clustered into a clade with *ZEP*s from green algae, and most closed to *D. tertiolecta*. Although red algae has no xanthophyll cycle, it has a *ZEP* gene that is not the same as that in terrestrial plants and other algae. This *ZEP* introduces an epoxide group into zeaxanthin and produces antheraxanthin instead of violaxanthin, indicating that the *ZEP* gene of red algae is relatively primitive. At the same time, this also indicates that *ZEP* first only catalyzed the transformation of zeaxanthin into antheraxanthin during the evolutionary process, and later acquired the ability to catalyze antheraxanthin into violaxanthin [42].

The *ZEP* gene is involved in the synthesis of carotenoids and ABA and plays a vital role in plant responses to abiotic stress [22,23]. In *D. tertiolecta*, with increasing salinity, the expression of the *ZEP* gene increased gradually [45]. In this study, *UpZEP* gene expression was up-regulated under high-salt stress, which is consistent with that of *D. tertiolecta* under salt stress, indicating that *UpZEP* gene expression was induced by salt stress in *U. prolifera*. However, down-regulation of the *ZEP* gene is thought to be associated with chilling sensitivities in rice [46]. Hoang et al. [47] proposed that the reversible down-regulation of *ZEP* gene expression plays an indispensable role in plants, and this reversible down-regulation mechanism is a common mechanism in plants under various light conditions at room temperature. Considering the existence of these opposite expression patterns, the role of the *UpZEP* gene *U. prolifera* under salt stress warrants further in-depth research.

Cao et al. reported that overexpression of the *ZEP* gene from *Medicago sativa* enhanced the tolerance of transgenic tobacco to low-light stress [28]. Overexpression of the alfalfa *ZEP* gene conferred drought and salt tolerance in transgenic tobacco [29]. Park et al. [48] found that overexpression of the *AtZEP* gene made transgenic *A. thaliana* plants grow more vigorously under salt and drought stress. In this study, the *UpZEP* gene was transformed into yeast cells and was overexpressed. The results showed that *UpZEP* transgenic yeast exhibited growth on SG-Ura medium containing 1.3 M NaCl, whereas the empty pYES2 vector transgenic yeast strain did not, indicating that overexpression of the *UpZEP* gene enhances salt tolerance in yeast. Meanwhile, the *UpZEP* gene was heterogeneously overexpressed in *C. reinhardtii*. Transcription level analysis showed that the *UpZEP* gene expression level was significantly up-regulated under high-salt stress, indicating that the gene was induced by salt stress. Under continuous salt stress, transgenic *C. reinhardtii* had a higher survival rate and chlorophyll content than the wild-type. These results suggest that the *UpZEP* gene enhances plant salt tolerance. As a euryhaline marine organism, the discovery of the *Ulva* salt-tolerant gene will be useful for breeding salt-tolerant crops in the future. For example, functional expression of an animal type-Na+-ATPase gene from the red seaweed *Porphyra yezoensis* increases salinity tolerance in rice plants [49].

5. Conclusions

The current study analyzed the relative content of xanthophylls in *U. prolifera* under different salinity conditions. We found that high-salt stress significantly increased the relative content of xanthophylls. A *ZEP* gene was cloned and identified from *U. prolifera*, and the *UpZEP* gene expression was significantly induced by high-salt stress. Moreover, heterologous overexpression of the *UpZEP* gene in yeast and *C. reinhardtii* increased their tolerance to salt. These results provide a new perspective for studying the resistance mechanisms of *U. prolifera* to high-salt stress.

Supplementary Materials: The following supporting information can be downloaded at: https://www.mdpi.com/article/10.3390/biology13090695/s1. Table S1 Mass spectrometry analysis parameters. Table S2 Primers and sequences used in this experiment. Figure S1 Gel electrophoresis of the cloned product of the *UpZEP* gene. (A) PCR products of full-length sequence of *UpZEP* gene; (B) PCR products of CDS of *UpZEP* gene.

Author Contributions: Methodology, investigation, data curation and writing—original draft, H.H.; methodology, formal analysis and writing—review and editing, X.Y.; methodology, A.Z.; data curation, J.L.; methodology, H.G.; software, J.Y.; resources, W.X.; conceptualization, formal analysis and funding acquisition, S.S. All authors have read and agreed to the published version of the manuscript.

Funding: This work was supported by the National Nature Science Foundation of China (grant number 42276100) and the project was funded by the Priority Academic Program Development of Jiangsu Higher Education Institutions (PAPD).

Institutional Review Board Statement: Not applicable.

Informed Consent Statement: Not applicable.

Data Availability Statement: The datasets generated and/or analyzed during this study are available from the corresponding author on reasonable request.

Conflicts of Interest: The authors declare that they have no known competing financial interests or personal relationships that could have appeared to influence the work reported in this paper. There were no conflicts of interest in this study.

References

1. Smetacek, V.; Zingone, A. Green and golden seaweed tides on the rise. *Nature* **2013**, *504*, 84–88. [CrossRef]
2. Cui, J.; Zhang, J.; Huo, Y.; Zhou, L.; Wu, Q.; Chen, L.; Yu, K.; He, P. Adaptability of free-floating green tide algae in the Yellow Sea to variable temperature and light intensity. *Mar. Pollut. Bull.* **2015**, *101*, 660–666. [CrossRef]
3. Ye, N.; Zhang, X.; Mao, Y.; Liang, C.-W.; Xu, D.; Zou, J.; Zhuang, Z.-M.; Wang, Q.-Y. 'Green tides' are overwhelming the coastline of our blue planet: Taking the world's largest example. *Ecol. Res.* **2011**, *26*, 477–485. [CrossRef]
4. Wang, Z.; Xiao, J.; Fan, S.; Li, Y.; Liu, X.; Liu, D. Who made the world's largest green tide in China?—An integrated study on the initiation and early development of the green tide in Yellow Sea. *Limnol. Oceanogr.* **2015**, *60*, 1105–1117. [CrossRef]
5. Zhang, Y.; He, P.; Li, H.; Li, G.; Liu, J.; Jiao, F.; Zhang, J.; Huo, Y.; Shi, X.; Su, R.; et al. *Ulva prolifera* green-tide outbreaks and their environmental impact in the Yellow Sea, China. *Natl. Sci. Rev.* **2019**, *6*, 825–838. [CrossRef]
6. Zhang, J.; Huo, Y.; Yu, K.; Chen, Q.; He, Q.; Han, W.; Chen, L.; Cao, J.; Shi, D.; He, P. Growth characteristics and reproductive capability of green tide algae in Rudong coast, China. *J. Appl. Phycol.* **2013**, *25*, 795–803. [CrossRef]
7. Zhao, J.; Jiang, P.; Liu, Z.; Wei, W.; Lin, H.; Li, F.; Wang, J.; Qin, S. The yellow sea green tides were dominated by one species, *Ulva (Enteromorpha) prolifera*, from 2007 to 2011. *Chin. Sci. Bull.* **2012**, *58*, 2298–2302. [CrossRef]
8. Xiao, J.; Zhang, X.; Gao, C.; Jiang, M.; Li, R.; Wang, Z.; Li, Y.; Fan, S.; Zhang, X. Effect of temperature, salinity and irradiance on growth and photosynthesis of *Ulva prolifera*. *Acta Oceanol. Sin.* **2016**, *35*, 114–121. [CrossRef]
9. Zhao, X.; Zhong, Y.; Zhang, H.; Qu, T.; Jiang, Y.; Tang, X.; Wang, Y. Cooperation between photosynthetic and antioxidant systems: An important factor in the adaptation of *Ulva prolifera* to abiotic factors on the sea surface. *Front. Plant Sci.* **2019**, *10*, 648. [CrossRef]
10. Wu, H.; Gao, G.; Zhong, Z.; Li, X.; Xu, J. Physiological acclimation of the green tidal alga *Ulva prolifera* to a fast-changing environment. *Mar. Environ. Res.* **2018**, *137*, 1–7. [CrossRef]
11. Zheng, M.; Lin, J.; Zhou, S.; Zhong, J.; Li, Y.; Xu, N. Salinity mediates the effects of nitrogen enrichment on the growth, photosynthesis, and biochemical composition of *Ulva prolifera*. *Environ. Sci. Pollut. Res.* **2019**, *26*, 19982–19990. [CrossRef]
12. Zhuo, J.; Wang, H.; Du, Y.; Shi, M.; Huan, L.; Wang, G. Transcriptomic analysis of *Ulva prolifera* in response to salt stress. *Water* **2022**, *15*, 63. [CrossRef]
13. Zheng, Z.; Gao, S.; Wang, G.; Cock, M. High salt stress in the upper part of floating mats of *Ulva prolifera*, a species that causes green tides, enhances non-photochemical quenching. *J. Phycol.* **2019**, *55*, 1041–1049. [CrossRef]
14. Luo, M.; Liu, F. Salinity-Induced oxidative stress and regulation of antioxidant defense system in the marine macroalga *Ulva prolifera*. *J. Exp. Mar. Biol. Ecol.* **2011**, *409*, 223–228. [CrossRef]
15. Gao, G.; Zhong, Z.; Zhou, X.; Xu, J. Changes in morphological plasticity of *Ulva prolifera* under different environmental conditions: A laboratory experiment. *Harmful Algae* **2016**, *59*, 51–58. [CrossRef]
16. Demmig-Adams, B.; Adams, W. Xanthophyll cycle and light stress in nature: Uniform response to excess direct sunlight among higher plant species. *Planta* **1996**, *198*, 460–470. [CrossRef]
17. Fernández-Marín, B.; Roach, T.; Verhoeven, A.; García-Plazaola, J.I. Shedding light on the dark side of xanthophyll cycles. *New Phytol.* **2021**, *230*, 1336–1344. [CrossRef]
18. Demmig-Adams, B.; Adams, W. The role of xanthophyll cycle carotenoids in the protection of photosynthesis. *Trends Plant Sci.* **1996**, *1*, 21–26. [CrossRef]
19. Jahns, P.; Holzwarth, A.R. The role of the xanthophyll cycle and of lutein in photoprotection of photosystem II. *Biochim. Biophys. Acta Bioenerg.* **2012**, *1817*, 182–193. [CrossRef] [PubMed]
20. Croce, R.; Cinque, G.; Holzwarth, A.; Bassi, R. The Soret absorption properties of carotenoids and chlorophylls in antenna complexes of higher plants. *Photosynth. Res.* **2000**, *64*, 221–231. [CrossRef]
21. Welc, R.; Luchowski, R.; Kluczyk, D.; Zubik-Duda, M.; Grudzinski, W.; Maksim, M.; Reszczynska, E.; Sowinski, K.; Mazur, R.; Nosalewicz, A.; et al. Mechanisms shaping the synergism of zeaxanthin and PsbS in photoprotective energy dissipation in the photosynthetic apparatus of plants. *Plant J.* **2021**, *107*, 418–433. [CrossRef]
22. Hieber, A.; Bugos, R.; Yamamoto, H. Plant lipocalins: Violaxanthin de-epoxidase and zeaxanthin epoxidase. *Biochim. Biophys. Acta Protein Struct. Mol. Enzymol.* **2000**, *1482*, 84–91. [CrossRef]

23. Jia, K.-P.; Mi, J.; Ali, S.; Ohyanagi, H.; Moreno, J.C.; Ablazov, A.; Balakrishna, A.; Berqdar, L.; Fiore, A.; Diretto, G.; et al. An alternative, zeaxanthin epoxidase-independent abscisic acid biosynthetic pathway in plants. *Mol. Plant* **2022**, *15*, 151–166. [CrossRef]
24. Audran, C.; Liotenberg, S.; Gonneau, M.; North, H.; Frey, A.; Tap-Waksman, K.; Vartanian, N.; Marion-Poll, A. Localisation and expression of zeaxanthin epoxidase mRNA in *Arabdopsis* in response to drought stress and during seed development. *Funct. Plant Biol.* **2001**, *28*, 1161–1173. [CrossRef]
25. Schwarz, N.; Armbruster, U.; Iven, T.; Brückle, L.; Melzer, M.; Feussner, I.; Jahns, P. Tissue specific accumulation and regulation of zeaxanthin epoxidase in *Arabidopsis* reflect the multiple functions of the enzyme in plastids. *Plant Cell Physiol.* **2015**, *56*, 346–357. [CrossRef]
26. Thompson, A.; Jackson, A.; Parker, R.; Morpeth, D.R.; Burbidge, A.; Taylor, I.B. Abscisic acid biosynthesis in tomato: Regulation of zeaxanthin epoxidase and 9-cis-epoxycarotenoid dioxygenase mRNAs by light/dark cycles, water stress and abscisic acid. *Plant Mol. Biol.* **2000**, *42*, 833–845. [CrossRef]
27. Wang, N.; Fang, W.; Han, H.; Sui, N.; Li, B.; Meng, Q.W. Overexpression of zeaxanthin epoxidase gene enhances the sensitivity of tomato PSII photoinhibition to high light and chilling stress. *Physiol. Plant.* **2008**, *132*, 384–396. [CrossRef]
28. Zhang, Z.; Wang, Y.; Chang, L.; Zhang, T.; An, J.; Liu, Y.; Cao, Y.; Zhao, X.; Sha, X.; Hu, T.; et al. MsZEP, a novel zeaxanthin epoxidase gene from alfalfa (*Medicago sativa*), confers drought and salt tolerance in transgenic tobacco. *Plant Cell Rep.* **2015**, *35*, 439–453. [CrossRef]
29. Jahns, P.; Latowski, D.; Strzalka, K. Mechanism and regulation of the violaxanthin cycle: The role of antenna proteins and membrane lipids. *Biochim. Biophys. Acta Bioenerg.* **2009**, *1787*, 3–14. [CrossRef]
30. He, H.; Yang, J.; He, Y.; Yang, X.; Fu, C.; Zhang, D.; Dong, J.; Zeb, A.; Qu, J.; Shen, S. Proliferating cell nuclear antigen of *Ulva prolifera* is involved in the response to temperature stress. *J. Oceanol. Limnol.* **2024**, *42*, 1227–1241. [CrossRef]
31. He, Y.; Ma, Y.; Du, Y.; Shen, S. Differential gene expression for carotenoid biosynthesis in a green alga *Ulva prolifera* based on transcriptome analysis. *BMC Genom.* **2018**, *19*, 916. [CrossRef] [PubMed]
32. Gasteiger, E.; Hoogland, C.; Gattiker, A.; Bairoch, A.; Wilkins, M.R.; Duvaud, S.; Appel, R.D. *Protein Identification and Analysis Tools on the ExPASy Server*; Humana Press: Totowa, NJ, USA, 2005. [CrossRef]
33. Kim, M.; Kang, J.; Kang, Y.; Kang, B.S.; Jin, E. Loss of function in zeaxanthin epoxidase of *Dunaliella tertiolecta* caused by a single amino acid mutation within the substrate-binding site. *Mar. Drugs* **2018**, *16*, 418. [CrossRef] [PubMed]
34. Vecchi, V.; Barera, S.; Bassi, R.; Dall'Osto, L. Potential and challenges of improving photosynthesis in algae. *Plants* **2020**, *9*, 67. [CrossRef]
35. Takaichi, S. Carotenoids in algae: Distributions, biosyntheses and functions. *Mar. Drugs* **2011**, *9*, 1101–1118. [CrossRef]
36. He, Y.; Ye, Y.; Shen, S. Effects of light and salinity on carotenoid biosynthesis in *Ulva prolifera*. *Acta Oceanol. Sin.* **2020**, *39*, 50–57. [CrossRef]
37. Rodríguez, F.-E.; Laporte, D.; González, A.; Mendez, K.N.; Castro-Nallar, E.; Meneses, C.; Huidobro-Toro, J.P.; Moenne, A. Copper-Induced increased expression of genes involved in photosynthesis, carotenoid synthesis and C assimilation in the marine alga *Ulva compressa*. *BMC Genom.* **2018**, *19*, 829. [CrossRef]
38. Ruban, A.; Berera, R.; Ilioaia, C.; van Stokkum, I.H.M.; Kennis, J.T.M.; Pascal, A.A.; van Amerongen, H.; Robert, B.; Horton, P.; van Grondelle, R. Identification of a mechanism of photoprotective energy dissipation in higher plants. *Nature* **2007**, *450*, 575–578. [CrossRef]
39. Johnson, M.; Pérez-Bueno, M.; Zia, A.; Horton, P.; Ruban, A. The zeaxanthin-independent and zeaxanthin-dependent qE components of nonphotochemical quenching involve common conformational changes within the photosystem II antenna in *Arabidopsis*. *Plant Physiol.* **2009**, *149*, 1061–1075. [CrossRef]
40. Campbell, D.; Xie, X.; Gao, S.; Gu, W.; Pan, G.; Wang, G. Desiccation induces accumulations of antheraxanthin and zeaxanthin in intertidal macro-alga *Ulva pertusa* (Chlorophyta). *PLoS ONE* **2013**, *8*, e72929. [CrossRef]
41. Gao, S.; Zheng, Z.; Wang, J.; Wang, G.; Cock, M. Slow zeaxanthin accumulation and the enhancement of CP26 collectively contribute to an atypical non-photochemical quenching in macroalga *Ulva prolifera* under high light. *J. Phycol.* **2020**, *56*, 393–403. [CrossRef]
42. Dautermann, O.; Lohr, M. A functional zeaxanthin epoxidase from red algae shedding light on the evolution of light-harvesting carotenoids and the xanthophyll cycle in photosynthetic eukaryotes. *Plant J.* **2017**, *92*, 879–891. [CrossRef]
43. Eilers, U.; Dietzel, L.; Breitenbach, J.; Büchel, C.; Sandmann, G. Identification of genes coding for functional zeaxanthin epoxidases in the diatom *Phaeodactylum tricornutum*. *J. Plant Physiol.* **2016**, *192*, 64–70. [CrossRef] [PubMed]
44. Liu, M.; Ding, W.; Pan, Y.; Hu, H.; Liu, J. Zeaxanthin epoxidase is involved in the carotenoid biosynthesis and light-dependent growth of the marine alga *Nannochloropsis oceanica*. *Biotechnol. Biofuels Bioprod.* **2023**, *16*, 74. [CrossRef]
45. Kim, M.; Kang, Y.; Jin, E. Gene expression analysis of zeaxanthin epoxidase from the marine microalga *Dunaliella tertiolecta* in response to light/dark cycle and salinity. *J. Microbiol. Biotechnol.* **2019**, *29*, 1453–1459. [CrossRef] [PubMed]
46. Kim, H.; Hoang, M.; Jeon, Y.; Wu, G.; Lee, C. Differential down-regulation of zeaxanthin epoxidation in two rice (*Oryza sativa* L.) cultivars with different chilling sensitivities. *J. Plant Biol.* **2017**, *60*, 413–422. [CrossRef]
47. Hoang, M.; Kim, H.; Zulfugarov, I.; Lee, C. Down-Regulation of zeaxanthin epoxidation in vascular plant leaves under normal and photooxidative stress conditions. *J. Plant Biol.* **2020**, *63*, 331–336. [CrossRef]

48. Park, H.-Y.; Seok, H.-Y.; Park, B.-K.; Kim, S.H.; Goh, C.H.; Lee, B.H.; Lee, C.H.; Moon, Y.H. Overexpression of *Arabidopsis* ZEP enhances tolerance to osmotic stress. *Biochem. Biophys. Res. Commun.* **2008**, *375*, 80–85. [CrossRef] [PubMed]
49. Kishimoto, M.; Shimajiri, Y.; Oshima, A.; Hase, A.; Mikami, K.; Akama, K. Functional expression of an animal type-Na^+-ATPase gene from a marine red seaweed *Porphyra yezoensis* increases salinity tolerance in rice plants. *Plant Biotechnol.* **2013**, *30*, 417–422. [CrossRef]

biology

MDPI

Review

Review of Allelopathy in Green Tides: The Case of *Ulva prolifera* in the South Yellow Sea

Yinqing Zeng [1,†], Xinlan Yang [1,†], Zhangyi Xia [1], Runze Chen [1], Faqing He [1], Jianheng Zhang [1,2,3,*] and Peimin He [1,2,3,*]

1 College of Oceanography and Ecological Science, Shanghai Ocean University, Shanghai 201306, China; zyq70442083@gmail.com (Y.Z.); y496898726@163.com (X.Y.); xzy19028@163.com (Z.X.); 18854318218@163.com (R.C.); fqhe0906@163.com (F.H.)
2 Key Laboratory of Exploration and Utilization of Aquatic Genetic Resources, Ministry of Education, Shanghai Ocean University, Shanghai 201306, China
3 Co-Innovation Center of Jiangsu Marine Bio-Industry Technology, Jiangsu Ocean University, Lianyungang 222005, China
* Correspondence: jh-zhang@shou.edu.cn (J.Z.); pmhe@shou.edu.cn (P.H.)
† These authors contributed equally to this work.

Simple Summary: The spread of large green algae in oceans causes green tides, particularly in the South Yellow Sea of China, where *Ulva prolifera* has led to the world's largest green tide events. This review looks at how allelopathy, a process where plants release chemicals to affect the growth and development of other plants, influences green tide dynamics. We focus on four main types of these allelochemicals—fatty acids, aldehydes, phenols, and terpenes—explaining how they influence the algae's growth and behavior. We also discuss new methods for quickly detecting these allelochemicals and how these methods can help monitor green tides. By combining allelochemical detection with advanced technologies like satellite observations and environmental DNA analysis, we can better understand and manage green tides.

Abstract: The proliferation of large green macroalgae in marine environments has led to the occurrence of green tides, particularly in the South Yellow Sea region of China, where *Ulva prolifera* has been identified as the primary species responsible for the world's largest green tide events. Allelopathy among plants is a critical factor influencing the dynamics of green tides. This review synthesizes previous research on allelopathic interactions within green tides, categorizing four extensively studied allelochemicals: fatty acids, aldehydes, phenols, and terpenes. The mechanisms by which these compounds regulate the physiological processes of green tide algae are examined in depth. Additionally, recent advancements in the rapid detection of allelochemicals are summarized, and their potential applications in monitoring green tide events are discussed. The integration of advanced monitoring technologies, such as satellite observation and environmental DNA (eDNA) analysis, with allelopathic substance detection is also explored. This combined approach addresses gaps in understanding the dynamic processes of green tide formation and provides a more comprehensive insight into the mechanisms driving these phenomena. The findings and new perspectives presented in this review aim to offer valuable insights and inspiration for researchers and policymakers.

Keywords: green tide; allelopathy; allelochemicals; *Ulva prolifera*; physiological mechanism; marine monitoring

1. Introduction

A "green tide" is the result of green macroalgae, such as *Ulva prolifera*, growing and spreading quickly in the maritime environment under specific circumstances. Throughout the last fifty years, this phenomenon has been observed in numerous coastal regions of the world [1–3]. Although green tides have the potential to encourage the long-term storage of carbon by releasing huge volumes of dissolved organic carbon that is difficult to degrade into offshore waters, it upsets the balance of the marine ecosystem and burdens the economy [4]. Green tides may also directly worsen the quality of the environment by changing the composition of the phytoplankton community and maybe encouraging the proliferation of some microalgae, including *Aureococcus anophagefferens*, which can then increase brown tides [5]. Increasingly common, these biological occurrences have a significant ecological and economic influence on coastal regions [6–8].

Since 2007, the Yellow Sea region of China has experienced regular instances of green tides dominated by *U. prolifera* [3]. The origin, breakout, and migration routes of the South Yellow Sea green tide have been thoroughly studied by scientists throughout the last ten years [9–11]. Utilizing a combination of remote sensing monitoring and on-site investigations, Liu et al. (2009) proposed that the *Neopyropia* aquaculture areas in northern Jiangsu might be the primary source of the SYS green tide [9], with this viewpoint being widely accepted in subsequent studies [12,13]. The vast rafts dispersed in the northern Jiangsu *Neopyropia* aquaculture regions offer plenty of attachment substrates for the development of huge amounts of green macroalgae [12]. Every year between April and May, green macroalgae affixed to the rafts separate during the retrieval process, and as water temperatures rise, *U. prolifera* takes the front stage. Driven by wind and ocean currents, floating *U. prolifera* then travels northward, creating widespread green tide phenomena (Figure 1) [13]. Due to allelopathic interactions, *U. prolifera* affects the species composition in the marine environment and encourages its growth during green tide occurrences. Still, these allelopathic interactions are fairly little researched, and the present research on the processes behind the formation of green tides is rather incomplete [14–16]. Better knowledge of the allelopathic mechanisms during green tide occurrences and the provision of management strategies based on this knowledge will be essential to effectively manage and reduce the frequency of green tide episodes.

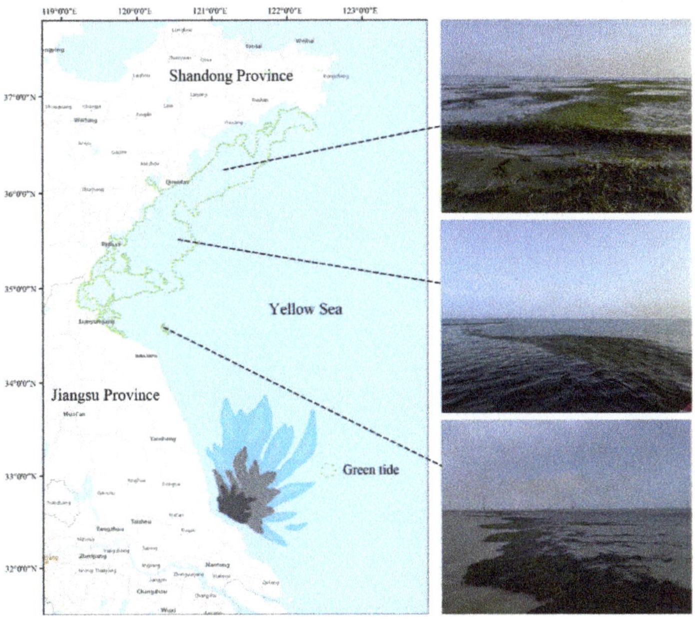

Figure 1. Green tide of *U. prolifera* in the South Yellow Sea in 2022. From May to July each year, large-scale green tides move northwards from Jiangsu to the Shandong Peninsula. Satellite images cited from previous work [17].

Known as allelochemicals, bioactive substances generated by secondary metabolic activities are essential to the communication and signal exchange between plants and the environments around them [18]. These substances can affect the development and reproduction of nearby species, therefore regulating biological interactions within the ecology. More precisely, allelochemicals give plants an edge over other plants in competition and shield them against pathogens and herbivores [19]. Allelochemicals comprise a wide range of intricate chemical structures and biological activities; some examples are alcohols, phenols, terpenoids, alkaloids, and chemicals containing halogen [19,20]. Of them, phenolic chemicals have drawn the attention of a great deal of research since they are essential for controlling plant growth and seed germination [21]. Significant herbicidal effects, inhibition of other plant development, and disruption of other plant physiological activities are all possible with terpenoids and alkaloids [22,23]. Plant root secretion, aqueous solution release, volatilization, and plant residue breakdown are just a few of the ways allelochemicals are produced. These pathways allow allelochemicals to efficiently act on neighboring biomes, thereby influencing their composition and function [24,25]. The release of this family of compounds gives invasive plants a competitive edge in their new surroundings. By limiting native plant development, disrupting microbial symbiosis, and changing the distribution of soil nutrients, these chemicals help invasive species spread and settle [26]. Then, allelochemicals have a wide range of environmental consequences. They demonstrate their essential and significant function in natural ecosystems by influencing not only the relationship between plants but also other ecological communities including soil microbes, mycorrhizal fungi, and invertebrates [27].

Allelopathy among aquatic plants is crucial to the marine ecological environment. The green tide process is greatly aided by these compounds, which may be stimulating or inhibiting, therefore controlling competition and biodiversity in the environment. Citing Teneva et al. (2023) and Chia et al. (2021), cyanobacteria, for instance, release two toxins, microcystin-LR and cylindrospermopsin, which prevent green macroalgae from growing [28,29]. Not only can microalgae interact chemically, but macrophytes and microalgae do too. Seagrasses inhibit the growth of toxic algae by allelopathy, hence preventing their spread [30]. Spent in the intertidal zone, invertebrates like sponges release particular compounds (such as β-sitosterol) to compete for nutrients and living space, which can also impact the physiological processes of other algae [31]. Additionally important is allelopathy among benthic microbes, which through the transmission of chemical signals, affects the composition and ecological efficacy of microbial communities [32]. Significant allelopathic inhibitory effects on *Spartina alterniflora* have been shown for extracts from *U. prolifera*, including valine, hexanedioic acid, and bis(2-ethylhexyl) ester [33]. In ecosystems, allelopathy has a strong and complicated function. It is significant to biodiversity preservation and interspecies competition. It shows how deeply intertwined and dependent ecosystem members are [34,35].

Enhanced by allelopathy, the persistence of green tides in the ocean presents complex ecological and economic problems [36]. Even if allelopathy in marine ecosystems should receive similar attention, particularly during green tide outbreaks, previous research has mostly concentrated on allelopathy in terrestrial settings [37]. This article tries to provide a comprehensive review of works on this topic to highlight the importance of allelopathy in managing species competition and maintaining biodiversity and ecological balance in maritime habitats. In Table 1, we present the key allelochemicals involved in the process of green tide formation. By closely analyzing the effects of allelopathy on the green tide, we hope to add to the future monitoring, prevention, and management plans for the marine environment, which are necessary to ensure its sustainability and health.

Table 1. Key allelochemicals in green tide allelopathy.

Allelochemical Classification	Allelochemicals	Chemical Formula	Chemical Structure	Reference
Fatty acid	Palmitic acid	$C_{16}H_{32}O_2$		[38,39]
	Oleic acid	$C_{18}H_{34}O_2$	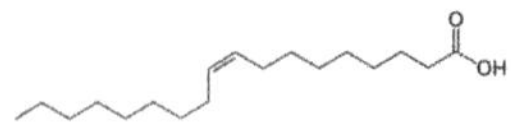	[39]
	Nonanoic acid	$C_9H_{18}O_2$	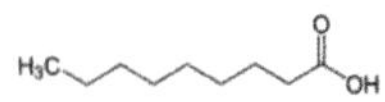	[40]
Aldehyde	Aldehyde 2-trans 4-trans-decadienal	/ /		[41] [41]
	Citral	$C_{10}H_{16}O$	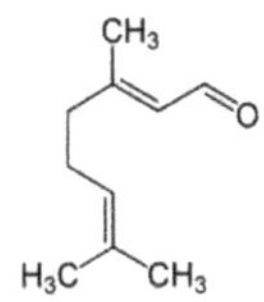	[40]
Phenol	Eugenol	$C_{10}H_{12}O_2$	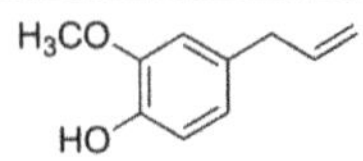	[42]
	Phenol	C_6H_5OH	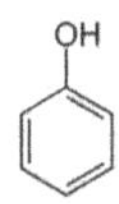	[43]
	Pyrocatechol	$C_6H_4(OH)_2$	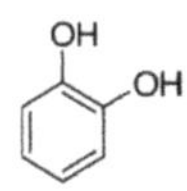	[44]
Terpenoid	Betulinic acid	$C_{30}H_{48}O_3$	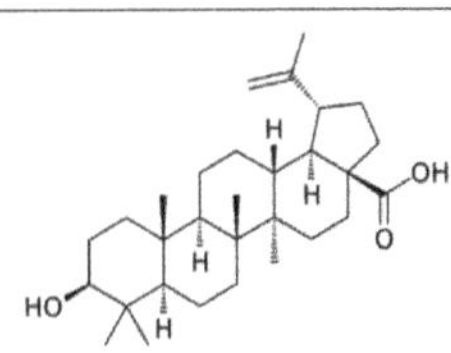	[45]
	Oleanolic acid	$C_{30}H_{48}O_3$	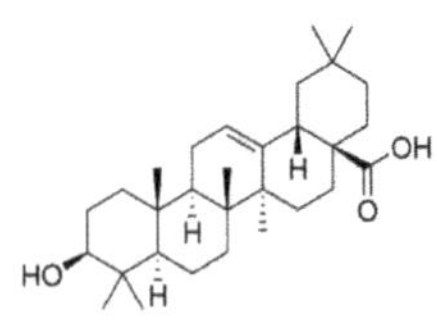	[45]
	Ursolic acid	$C_{30}H_{48}O_3$	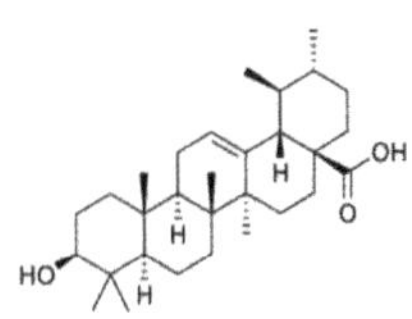	[45]
	Myrcene	$C_{10}H_{16}$	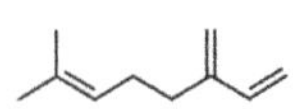	[40]

Table 1. *Cont.*

Allelochemical Classification	Allelochemicals	Chemical Formula	Chemical Structure	Reference
Alcohol	Sterol	/		[46]
	1-Octanol	$C_8H_{18}O$		[40]
Alkaloid	Pyrrolopiperazine-2,5-dione	$C_7H_{10}N_2O_2$	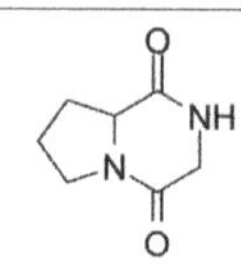	[47]
Amide	N-phenethylacetamide	$C_{10}H_{13}NO$	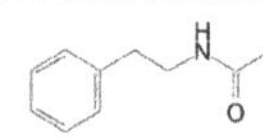	[48,49]
Diketopiperazine derivatives	Cyclo(L-Pro-L-Val)	$C_{10}H_{16}N_2O_2$	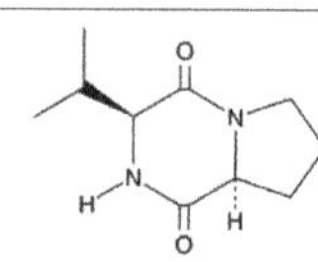	[48,49]
	Cyclo(L-Pro-L-Pro)	$C_{10}H_{14}N_2O_2$	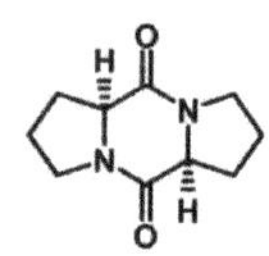	[48,49]
Amino acid	Valine	$C_5H_{11}NO_2$	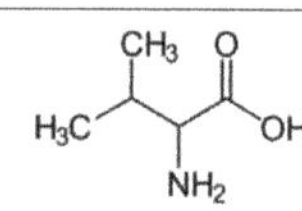	[33]
Carboxylic acid	Hexanedioic acid	$C_6H_{10}O_4$	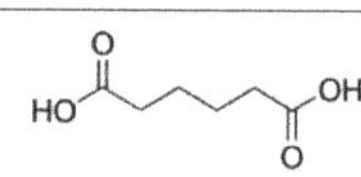	[33]
Esters	Bis(2-ethylhexyl) ester	$C_{24}H_{38}O_4$	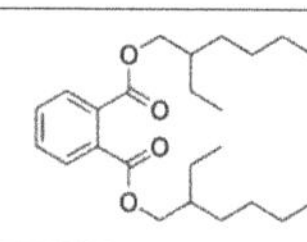	[33]

2. Allelochemicals and Allelopathy Mechanisms in the Green Tide Process

2.1. Fatty Acids

Because fatty acids greatly affect the development and health of both plants and microbes, it is essential to understand their function in ecosystems. To emphasize their significance even further, these substances also function as allelochemicals. Fatty acids play a crucial role in various biological processes, including the regulation of cell structure, the management of oxidative stress responses, and the modulation of gene expression. First of all, changes to the architecture of cells may result from fatty acids. Allelochemicals have been shown in earlier research to influence cell division, elongation, and membrane function [50,51]. For instance, a transmission electron microscopy investigation of the *Heterosigma akashiwo* study revealed that the addition of linoleic acid significantly altered the morphology of algal cells [52]. Furthermore, inside cells, oxidative stress, which is associated with reactive oxygen species (ROS) production, leads to increased lipid peroxidation and can be influenced by fatty acids. As stated by Pinto et al. (2013), this reaction is a typical physiological process [53]. In *H. akashiwo*, linoleic acid boosts the intracellular antioxidant ability to reduce damage [54,55]. Moreover, the way fatty acids regulate gene expression has a significant impact on the body's defenses against metabolic changes and the onset of illnesses. The impacts listed are displayed by epigenetic pathways, as Gniazdowska et al. (2004) have shown [56].

Fatty acids are important participants in the chemical communication network of the intricate undersea ecosystems. As Teneva et al. (2023) have pointed out, important allelopathy between algae and phytoplankton [28]. To an expert on the subject, the significance of fatty acids like hexadecanoic acid and octadecenoic acid, which is from *Ulva linza*, can influence the growth of phytoplankton populations along the shore [39]. And, *U. prolifera*'s spore germination can be inhibited by nonanoic acid [40]. Changes in the local lighting conditions affected the increased production of these fatty acids [57]. Under high light, *U. prolifera* will change their carbon storage and improve their physiological photosynthetic activities. This strategy is mostly shown in the change from C3 to C4 photosynthesis [46]. The biological modification greatly enhances fatty acid synthesis and enables it to continue to be active in harsh conditions [46]. Furthermore, included in this survival strategy is precisely controlled energy expenditure and carbohydrate metabolism [58]. On the other hand, the activity of a triacylglycerol synthetase known as DGAT is increased in the cells of *U. prolifera* when faced with the difficulties of high salinity and temperature. This enzyme helps to accumulate fatty acids and is essential for *U. prolifera*'s ongoing survival [42]. Zhuo et al. (2022) descended further into the genome of *U. prolifera* and discovered that genes involved in fatty acid synthesis in *U. prolifera* became more active in the event of significant salinity fluctuations [59]. Stress tolerance genes are also intimately associated with fatty acid metabolic pathways. These show that fatty acids significantly influence stress and the control of biological responses and that these pathways are essential to the environmental adaptability and growth of algae [59]. Particularly acidifying circumstances significantly affect the levels of polyunsaturated fatty acids and palmitic acid [38].

2.2. Aldehydes

Aldehydes can have a significant impact on the growth and development of nearby plants [60]. These compounds can chemically interact with vital plant macromolecules like proteins and DNA, affecting the plant's general physiology and survival capacity [50]. Aldehyde-related mostly harmful effects include genetic changes and DNA damage [45,61]. The chemical structures of biomolecules control the many aldehyde functions. Aldehydes having α and β-unsaturated carbonyl groups prefer to stick to more flexible biomolecules than others [62]. Targeted and upsetting specific physiological functions in plants, aldehydes can impact cellular activity, metabolism, and respiration. Shirgapure et al. (2020) conducted research demonstrating how environmental variables, including variations in temperature and soil quality, influence the allelopathic effects of aldehydes and other allelochemicals. These abiotic stress factors exert physiological pressure on plants through allelopathic interactions [63]. Plant relationships and adaptation could be changed by outside factors on the synthesis and secretion of allelochemicals [64]. As is well known, plants release volatile organic compounds like aldehydes that strengthen their defenses against diseases and insects. Additionally contributing to the structure and ecological activities of plant communities are these substances' allelopathy [65,66]. Aldehydes are thus crucial to plant allelopathy as well as to the dynamics and overall well-being of plant communities. In specific environmental settings, they engage with biomolecules. This allelopathy emphasizes the significance of aldehydes in maintaining ecological balance and plant interactions [67].

Without aldehydes, green macroalgae cannot survive, and in marine environments, aldehydes also allow green macroalgae to communicate with other living things. Among the phytoplankton, green macroalgae and diatoms release these substances into the water that are necessary for the development of green tides [68]. The hazardous aldehydes 2-trans and 4-trans-decadienal aldehyde are among those generated by some diatoms in both freshwater and marine environments, these aldehydes also reduce the chances of diatom survival, initiating a self-regulating ecological process [41]. Through its regulation of growth rates and induction of cell death, this phenomenon indirectly influences the dynamics of the green tide [69,70]. Variations across species and environmental factors determine how allelopathy affects biological populations. Green macroalgae development has been inhibited by the aldehydes emitted by the red tide dinoflagellate *Karenia brevis*, as laboratory research has shown. It is noteworthy, though, that because of the complex biological diversity found in natural settings, this effect would be less noticeable [71]. Citral was also found to have inhibitory effects on the spore germination of *U. prolifera* [40]. Because secondary compounds produced by the algae might impede the formation of other biological communities, green tide algae accumulate in specific conditions [72].

2.3. Phenols

Mostly produced by the acetic acid and shikimic pathways, phenols have given plants a fundamental environmental stress line [73]. Particularly in the competition amongst species, these compounds have been crucial to the development and evolution of plants. For instance, invasive plants can occasionally force the extinction of current species and, by creating phenolic acid-like components, impact the survival of nearby species [26]. More precisely, phenolic acids have been shown to induce cell death in the root cells of *Pinellia ternata* by causing the excessive accumulation of ROS [74]. Pyrocatechol and other chemicals have been shown to damage plant photosynthetic systems. They interfere with electron transfer, which in turn reduces the photosynthetic capacity in the plants [43,44]. More research shows that by collaborating with other elements, these phenolics can successfully prevent weed growth even in minute amounts within plants. Phenic acids are essential to plant competition in crops like rice since they function as natural weed control agents [24,75]. Strongly herbicidal phenolic chemicals have been discovered to be present in several plants, such as *Artemisia argyi*. The range of phenolic compounds in *Parthenium hysterophorus* has prompted studies to propose that it may be a useful tool for controlling weeds as well as crops [76]. These discoveries have prompted a great deal of research and advancement in bioherbicide technologies [77].

Researchers studying marine ecosystems mainly believe that phenolic allelochemicals are to blame for the proliferating green tide algae. *U. prolifera* is usually floating in the low UV-B radiation levels in the Yellow Sea region. Protecting algae from UV-B exposure is one of the key functions of phenolic compounds, which enables them to better adapt and survive in changing settings [78]. These algae's much greater concentration of phenolic compounds will make survival more likely. Phenols are intermediate products of the ascorbate-glutathione (ASA-GSH) cycle, which is how *U. prolifera* activates its antioxidant system to cope with stress, under low-dose and short-term radiation exposure [48,78]. Meantime, *U. prolifera*'s antioxidant system also helps to control photorespiration metabolism and keeps photosynthesis efficient, which allows it to survive and proliferate in unfavorable environments [79]. Part of the reason for the quick adaption and renewal of *U. prolifera* is its genes, which are essential for the manufacture of phenolic chemicals that allow the green macroalgae to spread quickly in their native habitat [80]. On the other hand, eutrophication and acidification of the ocean have proven green macroalgae to be remarkably adaptive [81]. In these situations, algal physiological regulating ability is increased in part by phenolic substances [38]. Furthermore, some research has demonstrated that green macroalgae can successfully regulate the amount of cyanobacteria by producing phenolic chemicals [82]. Phenolic chemicals may influence species community and mutual competition in the whole aquatic ecosystem in addition to the green macroalgae itself [83]. Li et al. (2021) discovered that eugenol significantly inhibited *U. prolifera* spore germination [40]. The ability of phenolic compounds to control the structure of phytoplankton communities and stifle competitors in these conditions emphasizes the critical function of phenolic compounds in preserving ecological balance and having broad ecological consequences [84].

2.4. Terpenoids

Through the inhibition of the growth of nearby plants, terpenoids have an impact on biodiversity, community organization, and plant competition [85]. For instance, terpenoids can block the development of root systems and seed germination. It is well recognized that several chemicals, such as betulinic acid, oleanolic acid, and ursolic acid, restrict the growth of several plants, such as *Lactuca sativa*, and *Bidens pilosa*. The main mechanism behind this action is damage to photosystem II in the photosynthetic system of the plant [45]. Significant constituents of volatile organic compounds found in plants and terpenoids also influence the competitiveness, resistance to diseases and pests, and developmental patterns of plants [86]. These allelochemicals are essential to plant communication and defense systems since they can obstruct photosynthetic efficiency, root development, and seed germination in several ways [87,88].

Terpenoids are crucial for the growth, reproduction, and environmental adaptation of *U. prolifera*. As important intermediates in the synthesis of carotenoids, these substances are necessary for green tide algae to survive. Two environmental factors affecting this synthesis system and its downstream pathways, which in turn impact carotenoid synthesis, are light and salt. Liu et al. (2023) have clarified the genetic mechanism of the 2-C-methyl-D-erythritol 4-phosphate (MEP) pathway in green macroalgae, which influences carotenoid synthesis and greatly affects the metabolic control of green tide algae in response to environmental changes [58]. *U. prolifera* is fast-growing partly due to its effective carbon fixation ability, which is intimately related to the terpenoid metabolic pathways [80]. Terpenoids are also required for the strong growth of algae, as research has shown, since they are strongly linked to genes linked to stress in green tide algae [89]. The green tide algae synthesis of terpenes is found to be temperature-dependent [90]. A transcription study showed that genes involved in terpenoid synthesis increased dramatically during spore formation [91]. Specifically, green tide algae create terpenoids that are essential for their physiological processes and may also change the amount of nutrients and dissolved oxygen in the surrounding marine environment [68]. The *U. prolifera* spore germination suppression experiment by Li et al. (2021) also revealed notable inhibitory effects of terpenoids, such as myrcene [40]. Terpenoids have been extensively studied ecologically in algae and microorganisms, even if their precise role in allelopathy among marine species is yet unknown [92].

2.5. Other Allelochemicals

Halogen compounds, polysaccharides, alkaloids, and other less studied compounds are similarly significant to the ecological balance in the fields of marine ecology and green tides. Although their exact properties are yet unknown, these chemicals may have a big influence on how ecosystems function and how green tides form [68]. Within the green tide alga *U. prolifera*, Figure 2 demonstrates the known mechanisms underlying the allelopathic effects, which exerted by the major four classes of allelochemicals.

Similar to the above-stated substances, polysaccharides interfere with the physiological processes of neighboring plants, therefore preventing them from developing and germinating. These chemicals can be released into the environment by root secretion, volatilization, or plant breakdown, which influences neighboring plants in the soil rhizosphere and changes the interspecific competition pattern [86]. For example, in marine environments, strong allelopathic effects were shown in polysaccharide and protein complexes produced by the marine phytoplankton *H. akashiwo* [93]. While the controlling mechanism is yet unknown, this complex demonstrates how polysaccharides regulate interspecific competition in the aquatic environment by allelopathy. It can attach itself to the cell surfaces of surrounding algae, therefore stifling the growth of rival algae such as *Skeletonema costatum* [93]. Polysaccharides that exhibit allelopathic effects now clearly cling particularly to the target species' cell surface, drastically restricting their proliferation and altering the number of certain species [94]. Moreover, the marine applications of these polysaccharides have shown considerable potential, such as in aquaculture to strengthen animal immune systems [95]. In-depth research and development of *U. prolifera* polysaccharides can not only help aquaculture thrive and present fresh chances for economic expansion but also provide a remedy for the green tide [96].

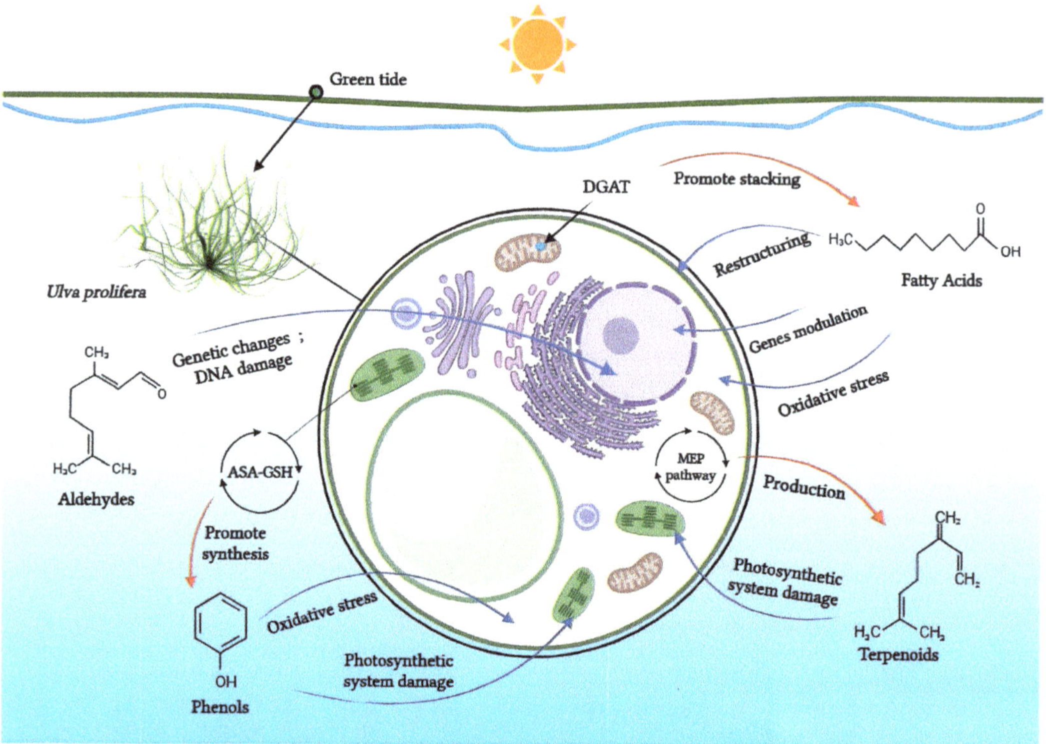

Figure 2. Fatty acids, aldehydes, phenols, and terpenoids in *U. prolifera*: known mechanisms of their allelopathic actions. Allelochemicals' effect on *U. prolifera*'s physiology is shown by blue arrows; algae's physiological processes' effect on allelochemical content is shown by red arrows. (By BioRender).

Meanwhile, *U. prolifera* can produce dimethylsulfoniopropionate, which is a kind of halogenated substance. The production of these compounds can be achieved by modifying the nutrient solubility in gas and water, together with any potential negative effects on the surrounding biological environment [4]. Sterol concentration similarly rises in response to light stress in *U. prolifera* cells, and this buildup of sterols may likewise cause allelopathic consequences [46]. Octanol was also shown to have some inhibitory impact on spore germination in the experiment of allelochemicals inhibiting *U. prolifera* spore; however, the mechanism of action of these allelochemicals has not been further investigated [40]. In agriculture, plant poisons have been shown by the capacity of some specific alkaloids to suppress seed germination and plant growth. For example, indole alkaloid is one of the substances found in barley that stops weeds and other plants from developing [97]. Similarly, alkaloids such as N-phenylacetamide, cyclo (L-Pro-L-Val), and other diketopiperazine derivatives are present in *U. prolifera* and are believed to be one of the ways to regulate green tides because of their inhibition capability against these green macroalgae [49]. Another alkaloid, Pyrrolopiperazine-2,5-dione, derived from *U. prolifera* has also shown resistance to hazardous red tide microalgae in marine environments [47]. And, allelochemicals from *U. prolifera* including valine, hexanedioic acid, and bis(2-ethylhexyl) ester have strong allelopathic inhibitory effects on *S. alterniflora* growth [33].

3. Rapid Detection and Application Prospect of Allelochemicals

Allelochemicals are essential to the process by which green tide algae become the dominating species, even if additional studies are needed to determine their exact mechanism and impacts on the green tide process. A thorough understanding of the mode of action of these chemicals in the natural environment and how they affect the ecological balance of green tide algae and other marine organisms will make new management techniques to effectively control and react to green tide issues and guarantee the ongoing health of marine ecosystems possible. Several sensitive and efficient techniques for chemical detection in water have been established by researchers; these techniques ought to be used for rapid allelochemical in situ detection. Solid phase extraction and ultra-high performance liquid chromatic-tandem mass spectrometry (UPLC-MS/MS) are combined, for example, to enable multiple chemical component monitoring in water with lower detection limits and improved accuracy [98]. Another physical method, including UV absorption-based sensors, can also quickly and accurately identify contaminants in water, such as some phytoallelochemicals [99]. In chemical terms, a less costly and more sensitive assay has been developed by researchers by monitoring glutathione S-transferase activity in aqueous culture systems [100]. Further, the luciferase assay is widely used for field testing since it is very sensitive and easy to use [101]. Using the electrochemical reaction of microorganisms employing microbial fuel cell technology, this biological approach detects harmful compounds in water, including phytoallelochemicals [102]. Currently, existing drone-equipped real-time monitoring systems can effectively collect water samples and quickly perform qualitative and quantitative chemical content analysis [103]. Finally, the use of advanced technologies and algorithms together with a variety of quick probes and automatic alarm systems allows the online biological detection system to provide an instantaneous alarm when anomalies are found, so greatly enhancing the efficacy and precision of water quality monitoring [104–106]. These automated, real-time monitoring systems are suitable for routine testing of water quality and also promise to effectively assess and control environmental hazards via the real-time monitoring of changes in plant allelochemicals in reaction to environmental phenomena such as green tides. In the present monitoring and tracking methods for green tide events, two allelochemicals fatty acids and sterol have been utilized as biomarkers to follow *U. prolifera* settlement sites in the later stages of the events [107–109].

For green tide monitoring, satellite data applications are the most important methods. These technologies make greater use of satellite resources and considerably reduce operational complexity by utilizing the concepts of open data cubes and real-time data analysis [110]. The VGGUnet model shows how well researchers may utilize deep learning to extract features from data and assess biomass [111]. Furthermore, infrared and synthetic aperture radar imaging techniques enhanced the ability to monitor green tides, such as those employed by the Resources 1-02E and Gaofen-3 satellites, though detecting lower-density algae with these techniques is still difficult [112]. Another disruptive monitoring strategy among the ecological monitoring technologies already being developed is environmental DNA (eDNA) technology. This method utilizes the identification of live tissue DNA release in the surroundings, which may come from a variety of organisms, such as microorganisms and organisms with many cells [113]. And, eDNA technology allowed the researchers to identify certain species in the samples, such as macroalgae [114]. EDNA technology with minimal environmental influence provides a cheap, efficient biodiversity analysis [115,116]. In the investigation of early green tides in the Southern Yellow Sea, Zeng et al. (2023b) successfully tracked changes in *U. prolifera* density and distribution during green tides using eDNA technology [117]. An investigation of these DNA samples made it possible to build an association network encompassing *U. prolifera* and other eukaryotic bacteria [117]. Even though such association studies provide new insights into the interactions between organisms, they now mostly rely on ecological niche theories to explain these interactions. Allelopathic studies in green tides will greatly improve our understanding of these interactions.

Whereas satellite technology allows researchers to find and track green tides over a wide area of the sea, eDNA monitoring technology provides an effective way to precisely identify the type and number of organisms in a given area. Particularly crucial during green tides, in situ allelochemical detection can reveal the interactions between algae and other marine life. The integration of satellite technologies and environmental DNA (eDNA) surveillance with rapid allelopathic detection methods represents a significant advancement in ecological monitoring and holds great promise for scientific research and environmental management (Figure 3). Apart from helping scientists to comprehend the growth rates and adaptive strategies of green macroalgae, these data will force a more in-depth analysis of the ecological dynamics and structure and take a front stage compared with traditional nutrient and hydrometeorological factors. All things considered, the coordinated application of these state-of-the-art technologies offers substantial data assistance for studying the ecological processes of green tides.

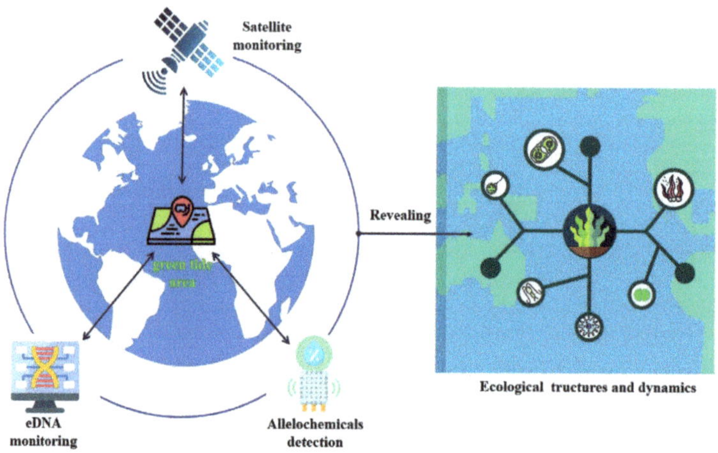

Figure 3. With the confluence of developing technology, ecological adaptation strategies for green tide algae will be better understood. Satellite technology tracks the emergence of green tides regionally; eDNA technology quickly monitors species distribution and community structure; and allelochemical detection greatly fills the gap in the unknown interactions between species. This changes the research emphasis from the traditional nutrient and hydrometeorological factors to the in-depth study of ecological dynamics and structure during green tides.

4. Conclusions

Important consequences of allelopathy are seen in the growth of plants and the condition of aquatic environments. Finding out how various substances affect green macroalgae and their surroundings is very important when researching the green tide phenomena. Presently, the most often used allelochemicals in this field are fatty acids, aldehydes, phenols, and terpenes. These compounds have a range of allelopathic effects, from the breakdown of cell membranes to the inhibition of photosynthesis and enzyme activity. However, marine ecology has not given as much attention to allelochemicals such as polysaccharides and alkaloids, which are ubiquitous in terrestrial habitats. Improvement will come from knowing these allelochemicals and their mechanisms better, and our understanding of green tide occurrence and ocean management options will be enhanced, as will the development of new management methods. In addition, with the gradual understanding of allelopathy in green tides, it has become a reality to develop technologies that can detect these allelopathy substances with high accuracy. Combining these high-precision detection techniques with modern environmental monitoring methods, such as eDNA analysis and satellite remote sensing, can not only improve our insight into the dynamics of green tides but also bring new strategies and perspectives for marine ecosystem research. Such integrated monitoring methods can reveal the complex interaction between green tides and their environment, helping us to better understand the biodiversity, ecological function, and ecological balance of marine ecosystems.

Author Contributions: Conceptualization, Y.Z.; methodology, Y.Z. and X.Y.; software, Y.Z. and X.Y.; validation, Y.Z., X.Y. and Z.X.; formal analysis, Y.Z. and X.Y.; investigation, Y.Z. and X.Y.; resources, Y.Z. and X.Y.; data curation, Y.Z., X.Y. and Z.X.; writing—original draft preparation, Y.Z., X.Y., Z.X., R.C., F.H., J.Z. and P.H.; writing—review and editing, Y.Z., X.Y., Z.X., R.C., F.H., J.Z. and P.H.; visualization, Y.Z., X.Y. and Z.X.; supervision, J.Z. and P.H.; project administration, J.Z. and P.H.; funding acquisition, J.Z. and P.H. All authors have read and agreed to the published version of the manuscript.

Funding: This work was supported by the National Key Research & Development Program of China (grant Nos. 2022YFC3106004 and 2022YFC3106001), Natural Science Foundation of Shanghai (grant No. 21ZR1427400), and National Natural Science Foundation of China (grant No. 41576163).

Institutional Review Board Statement: This study did not require ethical approval.

Informed Consent Statement: Not applicable.

Data Availability Statement: Not applicable.

Conflicts of Interest: The authors declare no conflicts of interest.

References

1. Smetacek, V.; Zingone, A. Green and golden seaweed tides on the rise. *Nature* **2013**, *504*, 84–88. [CrossRef] [PubMed]
2. Zhang, J.; Zhao, P.; Huo, Y.; Yu, K.; He, P. The fast expansion of *Pyropia* aquaculture in "Sansha" regions should be mainly responsible for the *Ulva* blooms in Yellow Sea. *Estuar. Coast. Shelf Sci.* **2017**, *189*, 58–65. [CrossRef]
3. He, P.; Duan, Y.; Qiao, L. Strategy of macroalgae eco-remediation with cases in nearshore China. *J. Appl. Oceanogr.* **2021**, *40*, 557–563.
4. Li, H.; Feng, X.; Xiong, T.; He, C.; Wu, W.; Shi, Q.; Jiao, N.; Zhang, Y. Green tides significantly alter the molecular composition and properties of coastal DOC and perform dissolved carbon sequestration. *Environ. Sci. Technol.* **2022**, *57*, 770–779. [CrossRef] [PubMed]
5. Zhao, J.-Y.; Geng, H.-X.; Zhang, Q.-C.; Li, Y.-F.; Kong, F.-Z.; Yan, T.; Zhou, M.-J.; Yang, D.; Yuan, Y.; Yu, R.-C.J.E.S.; et al. Green tides in the Yellow Sea promoted the proliferation of pelagophyte *Aureococcus anophagefferens*. *Environ. Sci. Technol.* **2022**, *56*, 3056–3064. [CrossRef] [PubMed]
6. Fletcher, R. The occurrence of "green tides"—A review. In *Marine Benthic Vegetation: Recent Changes the Effects of Eutrophication*; Springer: Berlin/Heidelberg, Germany, 1996; pp. 7–43.
7. Morand, P.; Merceron, M. Macroalgal population and sustainability. *J. Coast. Res.* **2005**, *21*, 1009–1020. [CrossRef]
8. Hiraoka, M.; Dan, A.; Shimada, S.; Hagihira, M.; Migita, M.; Ohno, M. Different life histories of *Enteromorpha prolifera* (*Ulvales, Chlorophyta*) from four rivers on Shikoku Island, Japan. *Phycologia* **2003**, *42*, 275–284. [CrossRef]
9. Liu, D.; Keesing, J.K.; Xing, Q.; Shi, P. World's largest macroalgal bloom caused by expansion of seaweed aquaculture in China. *Mar. Pollut. Bull.* **2009**, *58*, 888–895. [CrossRef] [PubMed]
10. Zhang, J.; Huo, Y.; Yu, K.; Chen, Q.; He, Q.; Han, W.; Chen, L.; Cao, J.; Shi, D.; He, P. Growth characteristics and reproductive capability of green tide algae in Rudong coast, China. *J. Appl. Phycol.* **2013**, *25*, 795–803. [CrossRef]
11. Gao, G.; Clare, A.S.; Rose, C.; Caldwell, G.S. Eutrophication and warming-driven green tides (*Ulva rigida*) are predicted to increase under future climate change scenarios. *Mar. Pollut. Bull.* **2017**, *114*, 439–447. [CrossRef]
12. Li, Y.; Song, W.; Xiao, J.; Wang, Z.; Fu, M.; Zhu, M.; Li, R.; Zhang, X.; Wang, X. Tempo-spatial distribution and species diversity of green algae micro-propagules in the Yellow Sea during the large-scale green tide development. *Harmful Algae* **2014**, *39*, 40–47. [CrossRef]
13. Zongling, W.; Mingzhu, F.; Jie, X.; Xuelei, Z.; Wei, S. Progress on the study of the Yellow Sea green tides caused by *Ulva prolifera*. *Acta Oceanol. Sin.* **2018**, *40*, 1–13.
14. Wang, Z.-Y.; Tian, Z.-J.; Li, F.-M.; An, Z.; Hu, H.-Y. Allelopathic effects of large seaweeds on red tide dinoflagellate *Gymnodinium breve*. *Allelopath. J.* **2008**, *22*, 181–188.
15. Xu, D.; Gao, Z.; Zhang, X.; Fan, X.; Wang, Y.; Li, D.; Wang, W.; Zhuang, Z.; Ye, N. Allelopathic interactions between the opportunistic species *Ulva prolifera* and the native macroalga *Gracilaria lichvoides*. *PLoS ONE* **2012**, *7*, e33648. [CrossRef] [PubMed]
16. Fu, M.; Cao, S.; Li, J.; Zhao, S.; Liu, J.; Zhuang, M.; Qin, Y.; Gao, S.; Sun, Y.; Kim, J.K. Controlling the main source of green tides in the Yellow Sea through the method of biological competition. *Mar. Pollut. Bull.* **2022**, *177*, 113561. [CrossRef] [PubMed]
17. Xia, Z.; Liu, J.; Zhao, S.; Sun, Y.; Cui, Q.; Wu, L.; Gao, S.; Zhang, J.; He, P. Review of the development of the green tide and the process of control in the southern Yellow Sea in 2022. *Estuar. Coast. Shelf Sci.* **2024**, *302*, 108772. [CrossRef]

18. Diller, J.G.P.; Hüftlein, F.; Lücker, D.; Feldhaar, H.; Laforsch, C. Allelochemical run-off from the invasive terrestrial plant Impatiens glandulifera decreases defensibility in Daphnia. *Sci. Rep.* **2023**, *13*, 1207. [CrossRef] [PubMed]
19. Liu, J.; Xie, M.; Li, X.; Jin, H.; Yang, X.; Yan, Z.; Su, A.; Qin, B. Main allelochemicals from the rhizosphere soil of *Saussurea lappa* (Decne.) Sch. Bip. and their effects on plants' antioxidase systems. *Molecules* **2018**, *23*, 2506. [CrossRef] [PubMed]
20. Quan, W.; Wang, A.; Li, C.; Xie, L. Allelopathic potential and allelochemical composition in different soil layers of *Rhododendron delavayi* forest, southwest China. *Front. Ecol. Evol.* **2022**, *10*, 963116. [CrossRef]
21. Kong, C.-H.; Xuan, T.D.; Khanh, T.D.; Tran, H.-D.; Trung, N.T. Allelochemicals and signaling chemicals in plants. *Molecules* **2019**, *24*, 2737. [CrossRef]
22. Kato-Noguchi, H. Allelopathy and allelochemicals of *Imperata cylindrica* as an invasive plant species. *Plants* **2022**, *11*, 2551. [CrossRef] [PubMed]
23. Hickman, D.T.; Rasmussen, A.; Ritz, K.; Birkett, M.A.; Neve, P. Allelochemicals as multi-kingdom plant defence compounds: Towards an integrated approach. *Pest Manag. Sci.* **2021**, *77*, 1121–1131. [CrossRef] [PubMed]
24. Rahaman, F.; Shukor Juraimi, A.; Rafii, M.Y.; Uddin, K.; Hassan, L.; Chowdhury, A.K.; Karim, S.M.R.; Yusuf Rini, B.; Yusuff, O.; Bashar, H.K. Allelopathic potential in rice-a biochemical tool for plant defence against weeds. *Front. Plant Sci.* **2022**, *13*, 1072723. [CrossRef] [PubMed]
25. Macías, F.A.; Mejías, F.J.; Molinillo, J.M. Recent advances in allelopathy for weed control: From knowledge to applications. *Pest Manag. Sci.* **2019**, *75*, 2413–2436. [CrossRef] [PubMed]
26. Kalisz, S.; Kivlin, S.N.; Bialic-Murphy, L. Allelopathy is pervasive in invasive plants. *Biol. Invasions* **2021**, *23*, 367–371. [CrossRef]
27. Zeng, R.S.; Wen, Z.; Niu, G.; Schuler, M.A.; Berenbaum, M.R. Allelochemical induction of cytochrome P450 monooxygenases and amelioration of xenobiotic toxicity in *Helicoverpa zea*. *J. Chem. Ecol.* **2007**, *33*, 449–461. [CrossRef] [PubMed]
28. Teneva, I.; Velikova, V.; Belkinova, D.; Moten, D.; Dzhambazov, B. Allelopathic Potential of the Cyanotoxins Microcystin-LR and Cylindrospermopsin on Green Algae. *Plants* **2023**, *12*, 1403. [CrossRef] [PubMed]
29. Chia, M.A.; Bittencourt-Oliveira, M.d.C. Allelopathic interactions between phytoplankton species alter toxin production, oxidative response, and nitrogen fixation. *Hydrobiologia* **2021**, *848*, 4623–4635. [CrossRef]
30. Zhang, Z.; Liu, Y.; Yuan, L.; Weber, E.; van Kleunen, M. Effect of allelopathy on plant performance: A meta-analysis. *Ecol. Lett.* **2021**, *24*, 348–362. [CrossRef]
31. Singh, A.; Thakur, N.L. Allelopathic interaction among rocky intertidal invertebrates: Sponge *Cinachyrella* cf. *cavernosa and Zooxanthellate zoanthids Zoanthus sansibaricus*. *Hydrobiologia* **2021**, *848*, 4647–4659.
32. Allen, J.L.; Ten-Hage, L.; Leflaive, J. Allelopathic interactions involving benthic phototrophic microorganisms. *Environ. Microbiol. Rep.* **2016**, *8*, 752–762. [CrossRef] [PubMed]
33. Zhao, Y.; Li, Y.; Jin, Y.; Cui, T.; Wang, S.; Kong, F. The inhibitory effects of *Ulva prolifera* extracts on early growth of *Spartina alterniflora* and the underlying mechanisms. *J. Environ. Manag.* **2022**, *319*, 115639. [CrossRef] [PubMed]
34. Gostyńska, J.; Pankiewicz, R.; Romanowska-Duda, Z.; Messyasz, B. Overview of allelopathic potential of *Lemna minor* L. obtained from a shallow eutrophic lake. *Molecules* **2022**, *27*, 3428. [CrossRef] [PubMed]
35. Felpeto, A.B.; Roy, S.; Vasconcelos, V.M. Allelopathy prevents competitive exclusion and promotes phytoplankton biodiversity. *Oikos* **2018**, *127*, 85–98. [CrossRef]
36. Gross, E.M. Allelopathy of aquatic autotrophs. *Crit. Rev. Plant Sci.* **2003**, *22*, 313–339. [CrossRef]
37. Semmouri, I.; Janssen, C.R.; Asselman, J. Allelopathy in macroalgae: Ecological principles, research opportunities and pitfalls reviewed. *J. Appl. Phycol.* **2024**, *36*, 441–458. [CrossRef]
38. Cai, J.; Ni, J.; Chen, Z.; Wu, S.; Wu, R.; He, C.; Wang, J.; Liu, Y.; Zhou, W.; Xu, J. Effects of ocean acidification and eutrophication on the growth and photosynthetic performances of a green tide alga *Ulva prolifera*. *Front. Mar. Sci.* **2023**, *10*, 1145048. [CrossRef]
39. Lv, M.; Yuan, M.; Wang, Y.; Tang, X.; Zhao, Y. Allelopathic effects of *Ulva linza* on marine phytoplankton and identification of the allelochemicals. *Environ. Sci. Pollut. Res.* **2021**, *28*, 45714–45723. [CrossRef] [PubMed]
40. Li, N.; Zhang, J.; Hu, C. Study of allelopathic effect on spore attachment of *Ulva prolifera*. *Environ. Sci. Technol.* **2021**, *44*, 26–34.
41. Żak, A.; Kosakowska, A. Allelopathic influence of cyanobacteria *Microcystis aeruginosa* on green algae *Chlorella vulgaris*. In *Insights on Environmental Changes: Where the World is Heading*; Springer: Cham, Switzerland, 2014; pp. 141–150.
42. Zhang, X.; Chi, X.; Wang, Y.; Zhang, J.; Zhang, Y.; Xu, D.; Fan, X.; Liang, C.; Ye, N. Characterization of a broad substrates specificity acyl-CoA: Diacylglycerol acyltransferase 1 from the green tide alga *Ulva prolifera*. *Acta Oceanol. Sin.* **2020**, *39*, 42–49. [CrossRef]
43. Matorin, D.; Plekhanov, S.; Bratkovskaya, L.; Yakovleva, O.; Alekseev, A. The effect of phenols on the parameters of chlorophyll fluorescence and reactions of P 700 in green algae *Scenedesmus quadricauda*. *Biophysics* **2014**, *59*, 374–379. [CrossRef]
44. Zaytseva, T.; Medvedeva, N.; Mamontova, V. Peculiarities of the effect of octyl-and nonylphenols on the growth and development of microalgae. *Inland Water Biol.* **2015**, *8*, 406–413. [CrossRef]
45. Xie, Y.; Tian, L.; Han, X.; Yang, Y. Research advances in allelopathy of volatile organic compounds (VOCs) of plants. *Horticulturae* **2021**, *7*, 278. [CrossRef]
46. Gu, K.; Liu, Y.; Jiang, T.; Cai, C.; Zhao, H.; Liu, X.; He, P. Molecular response of *Ulva prolifera* to short-term high light stress revealed by a multi-omics approach. *Biology* **2022**, *11*, 1563. [CrossRef]
47. Sun, Y.-y.; Wang, H.; Guo, G.-l.; Pu, Y.-f.; Yan, B.-l.; Wang, C.-h. Isolation, purification, and identification of antialgal substances in green alga *Ulva prolifera* for antialgal activity against the common harmful red tide microalgae. *Environ. Sci. Pollut. Res.* **2016**, *23*, 1449–1459. [CrossRef] [PubMed]

48. Li, N.; Tong, M.; Glibert, P.M. Effect of allelochemicals on photosynthetic and antioxidant defense system of *Ulva prolifera*. *Aquat. Toxicol.* **2020**, *224*, 105513. [CrossRef] [PubMed]
49. Li, N.; Zhang, J.; Zhao, X.; Wang, P.; Tong, M.; Glibert, P.M. Allelopathic Inhibition by the Bacteria *Bacillus cereus* BE23 on Growth and Photosynthesis of the Macroalga *Ulva prolifera*. *J. Mar. Sci. Eng.* **2020**, *8*, 718. [CrossRef]
50. Mushtaq, W.; Siddiqui, M.B.; Hakeem, K.R.; Mushtaq, W.; Siddiqui, M.B.; Hakeem, K.R. Mechanism of Action of Allelochemicals. In *Allelopathy: Potential for Green Agriculture*; Springer: Cham, Switzerland, 2020; pp. 61–66.
51. Salek, M.; Hosseini Hooshiar, S.; Salek, M.; Poorebrahimi, M.; Jafarnejad, S. Omega-3 fatty acids: Current insights into mechanisms of action in systemic lupus erythematosus. *Lupus* **2023**, *32*, 7–22. [CrossRef] [PubMed]
52. Wang, R.; Liu, Q. Responses of bloom-forming *Heterosigma akashiwo* to allelochemical linoleic acid: Growth inhibition, oxidative stress and apoptosis. *Front. Mar. Sci.* **2022**, *8*, 793567. [CrossRef]
53. Pinto, M.E.A.; Silva, N.L.; de Siqueira, E.P.; dos Santos, F.J.L.; dos Santos Lima, L.A.R. Efeito alelopático dos ésteres metílicos dos ácidos graxos obtidos de óleos de milho e girassol. *BBR—Biochem. Biotechnol. Rep.* **2013**, *2*, 44–48. [CrossRef]
54. González-Becerra, K.; Ramos-López, O.; Barrón-Cabrera, E.; Riezu-Boj, J.I.; Milagro, F.; Martínez-López, E.; Martínez, J. Fatty acids, epigenetic mechanisms and chronic diseases: A systematic review. *Lipids Health Dis.* **2019**, *18*, 178. [CrossRef]
55. Gniazdowska, A.; Bogatek, R.J. Allelopathic interactions between plants. Multi site action of allelochemicals. *Acta Physiol. Plant.* **2005**, *27*, 395–407. [CrossRef]
56. Gniazdowska, A.; Oracz, K.; Bogatek, R. Allelopatia-nowe interpretacje oddziaływań pomiędzy roślinami. *Kosmos* **2004**, *53*, 207–217.
57. Ali, H.A.; Al-Darraji, M.N.; Ali, S.F. The Allelopathic in the Diatoms. *Rafidain J. Sci.* **2021**, *30*, 30–38. [CrossRef]
58. Liu, Q.; Cui, R.; Lin, J.; Kang, Z.; Zhou, X. Competition relations between selected microalgae and bloom-forming *Ulva prolifera*. *J. Mar. Biol. Assoc.* **2023**, *103*, e43. [CrossRef]
59. Zhuo, J.; Wang, H.; Du, Y.; Shi, M.; Huan, L.; Wang, G. Transcriptomic Analysis of *Ulva prolifera* in Response to Salt Stress. *Water* **2022**, *15*, 63. [CrossRef]
60. Xie, M.-Z.; Shoulkamy, M.I.; Salem, A.M.; Oba, S.; Goda, M.; Nakano, T.; Ide, H. Aldehydes with high and low toxicities inactivate cells by damaging distinct cellular targets. *Mutat. Res. Fundam. Mol. Mech. Mutagen.* **2016**, *786*, 41–51. [CrossRef] [PubMed]
61. Das, C.; Dey, A.; Bandyopadhyay, A. Allelochemicals: An emerging tool for weed management. In *Evidence Based Validation of Traditional Medicines: A Comprehensive Approach*; Springer: Singapore, 2021; pp. 249–259.
62. Kaiira, M.; Chemining'wa, G.; Ayuke, F.; Baguma, Y.; Atwijukire, E. Allelopathic potential of compounds in selected crops. *J. Agric. Sci.* **2021**, *13*, 192. [CrossRef]
63. Shirgapure, K.; Ghosh, P. Allelopathy a tool for sustainable weed management. *Arch. Curr. Res. Int.* **2020**, *20*, 17–25. [CrossRef]
64. Chou, C.-H. Allelopathic compounds as naturally occurring herbicides. In Proceedings of the 15th Asian-Pacific Weeds Control Conference, Tsukuba, Japan, 24–28 July 1995; pp. 107–115.
65. LoPachin, R.M.; Gavin, T. Molecular mechanisms of aldehyde toxicity: A chemical perspective. *Chem. Res. Toxicol.* **2014**, *27*, 1081–1091. [CrossRef] [PubMed]
66. Ambika, S. Multifaceted attributes of allelochemicals and mechanism of allelopathy. In *Allelopathy: Current Trends and Future Applications*; Springer: Berlin/Heidelberg, Germany, 2012; pp. 389–405.
67. Scavo, A.; Restuccia, A.; Mauromicale, G. Allelopathy: Principles and basic aspects for agroecosystem control. In *Sustainable Agriculture Reviews 28: Ecology for Agriculture*; Springer: Cham, Switzerland, 2018; pp. 47–101.
68. Van Alstyne, K.L.; Nelson, T.A.; Ridgway, R.L. Environmental chemistry and chemical ecology of "green tide" seaweed blooms. *Integr. Comp. Biol.* **2015**, *55*, 518–532. [CrossRef] [PubMed]
69. Brannon, M.A.; Bartsch, A.F. Influence of growth substances on growth and cell division in green algae. *Am. J. Bot.* **1939**, *26*, 271–279. [CrossRef]
70. Jalliffier-Merlon, E.; Marty, J.-C.; Denant, V.; Saliot, A. Phytoplanktonic sources of volatile aldehydes in the river Rhône estuary. *Estuar. Coast. Shelf Sci.* **1991**, *32*, 463–482. [CrossRef]
71. Prince, E.K.; Myers, T.L.; Kubanek, J. Effects of harmful algal blooms on competitors: Allelopathic mechanisms of the red tide dinoflagellate *Karenia brevis*. *Limnol. Oceanogr.* **2008**, *53*, 531–541. [CrossRef]
72. Hajnal-Jafari, T.I.; Đurić, S.S.; Stamenov, D.R. Influence of green algae Chlorella vulgaris on initial growth of different agricultural crops. *Zb. Matice Srp. Prir. Nauk.* **2016**, *130*, 29–33. [CrossRef]
73. Misra, D.; Dutta, W.; Jha, G.; Ray, P. Interactions and regulatory functions of phenolics in soil-plant-climate nexus. *Agronomy* **2023**, *13*, 280. [CrossRef]
74. He, Z.; Wang, Y.; Yan, Y.; Qin, S.; He, H.; Mao, R.; Liang, Z. Dynamic analysis of physiological indices and transcriptome profiling revealing the mechanisms of the allelopathic effects of phenolic acids on *Pinellia ternata*. *Front. Plant Sci.* **2022**, *13*, 1039507. [CrossRef]
75. Pardo-Muras, M.; Puig, C.G.; Pedrol, N. Complex synergistic interactions among volatile and phenolic compounds underlie the effectiveness of allelopathic residues added to the soil for weed control. *Plants* **2022**, *11*, 1114. [CrossRef]
76. Bashar, H.K.; Juraimi, A.S.; Ahmad-Hamdani, M.S.; Uddin, M.K.; Asib, N.; Anwar, M.P.; Rahaman, F.; Haque, M.A.; Hossain, A. Evaluation of allelopathic effects of *Parthenium hysterophorus* L. methanolic extracts on some selected plants and weeds. *PLoS ONE* **2023**, *18*, e0280159. [CrossRef] [PubMed]

77. Chen, L.; Li, J.; Zhu, Y.; Zhao, T.; Guo, L.; Kang, L.; Yu, J.; Du, H.; Liu, D. Weed suppression and molecular mechanisms of isochlorogenic acid A isolated from *Artemisia argyi* extract via an activity-guided method. *J. Agric. Food Chem.* **2022**, *70*, 1494–1506. [CrossRef]
78. Zhong, Y.; Xu, J.; Zhao, X.; Qu, T.; Guan, C.; Hou, C.; Tang, X.; Wang, Y. Balancing Damage via Non-Photochemical Quenching, Phenolic Compounds and Photorespiration in *Ulva prolifera* Induced by Low-Dose and Short-Term UV-B Radiation. *Int. J. Mol. Sci.* **2022**, *23*, 2693. [CrossRef] [PubMed]
79. Yu, Y.; Lin, J.; Jiang, J.; Hu, S.; Kang, C.K.; Xu, N.; Li, Y. Environmental history affects the growth and photosynthesis of a green-tide macroalgae *Ulva prolifera*. *Aquac. Res.* **2022**, *53*, 2509–2517. [CrossRef]
80. He, Y.; Shen, S.; Yu, D.; Wang, Y.; Yin, J.; Wang, Z.; Ye, Y. The *Ulva prolifera* genome reveals the mechanism of green tides. *J. Oceanol. Limnol.* **2021**, *39*, 1458–1470. [CrossRef]
81. Del Mondo, A.; Smerilli, A.; Ambrosino, L.; Albini, A.; Noonan, D.M.; Sansone, C.; Brunet, C. Insights into phenolic compounds from microalgae: Structural variety and complex beneficial activities from health to nutraceutics. *Crit. Rev. Biotechnol.* **2021**, *41*, 155–171. [CrossRef]
82. Pei, Y.; Liu, L.; Hilt, S.; Xu, R.; Wang, B.; Li, C.; Chang, X. Root exudated algicide of *Eichhornia crassipes* enhances allelopathic effects of cyanobacteria *Microcystis aeruginosa* on green algae. *Hydrobiologia* **2018**, *823*, 67–77. [CrossRef]
83. Huang, S.; Zhu, J.; Zhang, L.; Peng, X.; Zhang, X.; Ge, F.; Liu, B.; Wu, Z. Combined effects of allelopathic polyphenols on *Microcystis aeruginosa* and response of different chlorophyll fluorescence parameters. *Front. Microbiol.* **2020**, *11*, 614570. [CrossRef] [PubMed]
84. Zhang, C.; Wang, X.; Ma, Z.; Luan, Z.; Wang, Y.; Wang, Z.; Wang, L. Removal of phenolic substances from wastewater by algae. A review. *Environ. Chem. Lett.* **2020**, *18*, 377–392. [CrossRef]
85. Xu, Y.; Chen, X.; Ding, L.; Kong, C.-H. Allelopathy and allelochemicals in grasslands and forests. *Forests* **2023**, *14*, 562. [CrossRef]
86. Ooka, J.K.; Owens, D.K. Allelopathy in tropical and subtropical species. *Phytochem. Rev.* **2018**, *17*, 1225–1237. [CrossRef]
87. Muzell Trezzi, M.; Vidal, R.A.; Balbinot Junior, A.A.; von Hertwig Bittencourt, H.; da Silva Souza Filho, A.P. Allelopathy: Driving mechanisms governing its activity in agriculture. *J. Plant Interact.* **2016**, *11*, 53–60. [CrossRef]
88. Cheng, F.; Cheng, Z. Research progress on the use of plant allelopathy in agriculture and the physiological and ecological mechanisms of allelopathy. *Front. Plant Sci.* **2015**, *6*, 160714. [CrossRef] [PubMed]
89. Wang, Y.; Liu, F.; Wang, M.; Moejes, F.W.; Bi, Y. Characterization and transcriptional analysis of one carbonic anhydrase gene in the green-tide-forming alga *Ulva prolifera* (*Ulvophyceae, Chlorophyta*). *Phycol. Res.* **2020**, *68*, 90–97. [CrossRef]
90. He, Y.; Ye, Y.; Shen, S. Effects of light and salinity on carotenoid biosynthesis in *Ulva prolifera*. *Acta Oceanol. Sin.* **2020**, *39*, 50–57. [CrossRef]
91. He, Y.; Ma, Y.; Du, Y.; Shen, S. Differential gene expression for carotenoid biosynthesis in a green alga *Ulva prolifera* based on transcriptome analysis. *BMC Genom.* **2018**, *19*, 916. [CrossRef] [PubMed]
92. Nelson, T.; Lee, D.; Smith, B.; Prins, R. Are 'Green TIDES'Harmful Algal Blooms? Allelopathic Properties of Extracts from *Ulva Fenestrata* and *Ulvaria Obscura*. *J. Phycol.* **2002**, *38*, 28–29. [CrossRef]
93. Qiu, X.; Yamasaki, Y.; Shimasaki, Y.; Gunjikake, H.; Honda, M.; Kawaguchi, M.; Matsubara, T.; Nagasoe, S.; Etoh, T.; Matsui, S. Allelopathy of the raphidophyte *Heterosigma akashiwo* against the dinoflagellate *Akashiwo sanguinea* is mediated via allelochemicals and cell contact. *Mar. Ecol. Prog. Ser.* **2012**, *446*, 107–118. [CrossRef]
94. Deng, R.; Wang, F.; Wang, L.; Xiong, L.; Shen, X.; Song, H. Advances in plant polysaccharides as antiaging agents: Effects and signaling mechanisms. *J. Agric. Food Chem.* **2023**, *71*, 7175–7191. [CrossRef] [PubMed]
95. Li, M.; Zhou, Y.; Hu, C. One Step Catalytic Conversion of Polysaccharides in *Ulva prolifera* to Lactic Acid and Value-Added Chemicals. *Catalysts* **2023**, *13*, 262. [CrossRef]
96. Ge, H.; Ni, Q.; Chen, Z.; Li, J.; Zhao, F. Effects of short period feeding polysaccharides from marine macroalga, *Ulva prolifera* on growth and resistance of *Litopenaeus vannamei* against *Vibrio parahaemolyticus* infection. *J. Appl. Phycol.* **2019**, *31*, 2085–2092. [CrossRef]
97. Lebecque, S.; Crowet, J.-M.; du Jardin, P.; Delory, B.; Lins, L.; Deleu, M.; Fauconnier, M.-L. Interactions of allelochemicals with plant plasma membrane: A case study with alkaloids from barley. In Proceedings of the 41st New Phytologist Symposium, Nancy, France, 11–13 April 2018.
98. Greer, B.; McNamee, S.E.; Boots, B.; Cimarelli, L.; Guillebault, D.; Helmi, K.; Marcheggiani, S.; Panaiotov, S.; Breitenbach, U.; Akçaalan, R. A validated UPLC–MS/MS method for the surveillance of ten aquatic biotoxins in European brackish and freshwater systems. *Harmful Algae* **2016**, *55*, 31–40. [CrossRef]
99. Ma, J.; Meng, F.; Zhou, Y.; Wang, Y.; Shi, P. Distributed water pollution source localization with mobile UV-visible spectrometer probes in wireless sensor networks. *Sensors* **2018**, *18*, 606. [CrossRef]
100. Hellou, J.; Ross, N.W.; Moon, T.W. Glutathione, glutathione S-transferase, and glutathione conjugates, complementary markers of oxidative stress in aquatic biota. *Environ. Sci. Pollut. Res.* **2012**, *19*, 2007–2023. [CrossRef] [PubMed]
101. Denisov, I.; Lukyanenko, K.; Yakimov, A.; Kukhtevich, I.; Esimbekova, E.; Belobrov, P. Disposable luciferase-based microfluidic chip for rapid assay of water pollution. *Luminescence* **2018**, *33*, 1054–1061. [CrossRef]
102. Chouler, J.; Di Lorenzo, M. Water quality monitoring in developing countries; can microbial fuel cells be the answer? *Biosensors* **2015**, *5*, 450–470. [CrossRef]

103. De Fazio, R.; Dinoi, L.M.; De Vittorio, M.; Visconti, P. A sensor-based drone for pollutants detection in eco-friendly cities: Hardware design and data analysis application. *Electronics* **2021**, *11*, 52. [CrossRef]
104. Jurdak, R.; Elfes, A.; Kusy, B.; Tews, A.; Hu, W.; Hernandez, E.; Kottege, N.; Sikka, P. Autonomous surveillance for biosecurity. *Trends Biotechnol.* **2015**, *33*, 201–207. [CrossRef] [PubMed]
105. Dörnhöfer, K.; Oppelt, N. Remote sensing for lake research and monitoring–Recent advances. *Ecol. Indic.* **2016**, *64*, 105–122. [CrossRef]
106. Sagan, V.; Peterson, K.T.; Maimaitijiang, M.; Sidike, P.; Sloan, J.; Greeling, B.A.; Maalouf, S.; Adams, C. Monitoring inland water quality using remote sensing: Potential and limitations of spectral indices, bio-optical simulations, machine learning, and cloud computing. *Earth-Sci. Rev.* **2020**, *205*, 103187. [CrossRef]
107. Geng, H.; Yu, R.; Yan, T.; Zhang, Q.; Kong, F. Using sterol biomakers to trace deposition areas of floating green algae after green tides. *Oceanol. Limnol. Sin.* **2018**, *49*, 1094–1102.
108. Geng, H.; Yu, R.; Zhang, Q.; Yan, T.; Kong, F.; Zhou, M. Tracing the settlement region of massive floating green algae in the Yellow Sea. *J. Oceanol. Limnol.* **2019**, *37*, 1555–1565. [CrossRef]
109. Liu, J.; Yang, X.; Li, J.; Wen, Q.; Liu, W.; Song, W.; He, P.; Zhang, J. Research progress on settled mature *Ulva prolifera* during the outbreak of green tide in the Yellow Sea. *Environ. Pollut. Control* **2020**, *42*, 614–618.
110. Malthus, T.J.; Lehmann, E.; Ho, X.; Botha, E.; Anstee, J. Implementation of a satellite based inland water algal bloom alerting system using analysis ready data. *Remote Sens.* **2019**, *11*, 2954. [CrossRef]
111. Shang, W.; Gao, Z.; Gao, M.; Jiang, X. Monitoring green tide in the Yellow Sea using high-resolution imagery and deep learning. *Remote Sens.* **2023**, *15*, 1101. [CrossRef]
112. Song, Q.; Ma, C.; Liu, J.; Wei, H. Quantifying ocean surface green tides using high-spatial resolution thermal images. *Opt. Express* **2022**, *30*, 36592–36602. [CrossRef] [PubMed]
113. Ortega, A.; Geraldi, N.R.; Díaz-Rúa, R.; Ørberg, S.B.; Wesselmann, M.; Krause-Jensen, D.; Duarte, C.M. A DNA mini-barcode for marine macrophytes. *Mol. Ecol. Resour.* **2020**, *20*, 920–935. [CrossRef] [PubMed]
114. Zeng, Y.; Wang, X.; Liu, J.; Cao, J.; Sun, Y.; Zhao, S.; Chen, Z.; Kim, J.K.; Zhang, J.; He, P. Harnessing the power of eDNA technology for macroalgal ecological studies: Recent advances, challenges, and future perspectives. *Algal Res.* **2023**, *77*, 103340. [CrossRef]
115. Rees, H.C.; Maddison, B.C.; Middleditch, D.J.; Patmore, J.R.; Gough, K.C. The detection of aquatic animal species using environmental DNA–a review of eDNA as a survey tool in ecology. *J. Appl. Ecol.* **2014**, *51*, 1450–1459. [CrossRef]
116. Ruppert, K.M.; Kline, R.J.; Rahman, M.S. Past, present, and future perspectives of environmental DNA (eDNA) metabarcoding: A systematic review in methods, monitoring, and applications of global eDNA. *Glob. Ecol. Conserv. Genet.* **2019**, *17*, e00547. [CrossRef]
117. Zeng, Y.; Chen, Z.; Cao, J.; Li, S.; Xia, Z.; Sun, Y.; Zhang, J.; He, P. Revolutionizing early-stage green tide monitoring: eDNA metabarcoding insights into *Ulva prolifera* and microecology in the South Yellow Sea. *Sci. Total Environ.* **2024**, *912*, 169022. [CrossRef]

biology

MDPI

Article

Enhanced Soil Fertility and Carbon Sequestration in Urban Green Spaces through the Application of Fe-Modified Biochar Combined with Plant Growth-Promoting Bacteria

Guoyao Niu [1], Chiquan He [1,*], Shaohua Mao [1], Zongze Chen [1], Yangyang Ma [1] and Yi Zhu [2,*]

1 School of Environmental and Chemical Engineering, Shanghai University, Shanghai 200444, China; guoyaoniu@163.com (G.N.); shmao1234@163.com (S.M.); czzsuccess@163.com (Z.C.); 16609453783@163.com (Y.M.)

2 Key Laboratory of National Forestry and Grassland Administration on Ecological Landscaping of Challenging Urban Sites, Shanghai Engineering Research Center of Landscaping on Challenging Urban Sites, Shanghai Academy of Landscape Architecture Science and Planning, Shanghai 200232, China

* Correspondence: cqhe@shu.edu.cn (C.H.); zhuyi0816@163.com (Y.Z.)

Simple Summary: This study investigated the effects of plant growth promoting bacteria (*Bacillus clausii*) and Fe-modified biochar on soil fertility increases and mechanisms of carbon sequestration. Additionally, the impact on C-cycling-related enzyme activity and the bacterial community was also explored. The study results demonstrate that in comparison to the individual application of FeB and BC, the FeBBC treatment significantly relieves soil alkalization and enhances soil alkali-hydro nitrogen content and aggregate stability (particle size > 0.25 mm), thereby contributing to improved soil fertility and ecological function. Additionally, all biochar treatments exhibit higher soil organic carbon, thereby increasing organic carbon sequestration, particularly in the FeBBC treatment. Compared to a single ecological restoration method, FeBBC treatment can improve soil fertility and carbon sequestration, providing important reference values for urban green space soil ecological restoration projects.

Abstract: The soil of urban green spaces is severely degraded due to human activities during urbanization, and it is crucial to investigate effective measures that can restore the ecological functions of the soil. This study investigated the effects of plant growth promoting bacteria (*Bacillus clausii*) and Fe-modified biochar on soil fertility increases and mechanisms of carbon sequestration. Additionally, the effects on C-cycling-related enzyme activity and the bacterial community were also explored. Six treatments included no biochar or *Bacillus clausii* suspension added (CK), only *Bacillus clausii* suspension (BC), only biochar (B), only Fe-modified biochar (FeB), biochar combined with *Bacillus clausii* (BBC), and Fe-modified biochar combined with *Bacillus clausii* (FeBBC). Compared with other treatments, the FeBBC treatment significantly decreased soil pH, alleviated soil alkalization, and increased the alkali-hydro nitrogen content in the soil. Compared to the individual application of FeB and BC, the FeBBC treatment significantly improved aggregates' stability and positively improved soil fertility and ecological function. Additionally, compared to the individual application of FeB and BC, the soil organic carbon (SOC), particulate organic carbon (POC), and soil inorganic carbon (SIC) contents for the FeBBC-treated soil increased by 28.46~113.52%, 66.99~434.72%, and 7.34~10.04%, respectively. In the FeBBC treatment, FeB can improve soil physicochemical properties and provide bacterial attachment sites, increase the abundance and diversity of bacterial communities, and promote the uniform distribution of carbon-related bacteria in the soil. Compared to a single ecological restoration method, FeBBC treatment can improve soil fertility and carbon sequestration, providing important reference values for urban green space soil ecological restoration.

Keywords: carbon sequestration; enzyme activity; Fe-modified biochar; soil fertility; soil restoration

1. Introduction

Urban green spaces refer to land areas in urban environments where natural and artificial vegetation are the predominant forms of existence [1]. The soil in urban green spaces is a vital constituent of terrestrial ecosystems, playing a crucial role in mitigating climate change and sequestering carbon in the soil [2]. The soil in urban green spaces can increase its nutrient content and enhance its carbon storage capacity through atmospheric carbon dioxide precipitation or decomposition of plant waste [3]. Nevertheless, during the rapid urbanization process, the soil of these urban green spaces is commonly affected by multiple direct or indirect human disturbances or obstacles, resulting in problems such as compaction, high pH, and low soil nutrient and organic matter content, profoundly impacting its ability to fulfill ecological functions [4,5]. Hence, investigating measures to alleviate the degradation of soil fertility and enhance soil carbon sequestration in urban green spaces has emerged as a crucial research focus.

Biochar is a solid carbonaceous material produced by converting carbon rich biomass through thermochemical methods under anaerobic or hypoxic conditions. Previous studies have demonstrated that adding biochar to soil effectively improves soil fertility and increases carbon sequestration [6,7]. Recent research has focused on the functional modification of biochar to enhance its potential applications in carbon sequestration, soil fertility improvement, and environmental pollution remediation [8,9]. Wen et al. [10] indicated that, compared with biochar, iron-modified biochar significantly reduces the pH value and has more potential in alleviating soil alkalization. Wu et al. [11] found that Fe (III)-biochar has a larger specific surface area, pores, and more functional groups, which helps enhance its ability to restore soil ecological functions. Furthermore, Zhang et al. [12] found that the application of Fe_3O_4-modified biochar in soil resulted in higher organic carbon content and carbon-nitrogen ratio compared to the addition of raw biochar and Fe_3O_4 alone. Based on these findings, taking advantage of Fe-modified biochar for soil recovery in urban green spaces has great potential [9,13].

Plant Growth-Promoting Bacteria (PGPB) can promote plant growth through direct and indirect effects, thereby increasing soil carbon sequestration [14]. PGPB can directly promote plant growth through the synthesis of growth hormones (IAA) or through the effects of phosphorus solubilization, nitrogen fixation, and potassium solubilization [15]. In addition, PGPB can indirectly promote plant growth by improving soil structure and inhibiting the growth of pathogens [16]. *Bacillus* sp. is the bacterium that is most frequently utilized for promoting plant growth. Some studies have demonstrated that the inoculation of *Bacillus* sp. can effectively increase soil nutrient and carbon sequestration [17]. *Bacillus clausii* is a microorganism that has been recognized for its ability to promote plant growth. Li et al. [18] confirmed the production of extracellular polysaccharides by *Bacillus clausii* isolated from saline-alkali soil in Xinjiang. The *Bacillus clausii* B8 strain isolated from soil by Oulebsir-Mohandkaci et al. [19] can produce siderophore, HCN, and IAA, leading to a significant improvement in rapeseed seed germination rate. Although *Bacillus clausii* has been recognized for its plant growth-promoting attributes, its potential role as a beneficial bacterium in improving soil fertility and carbon sequestration within the context of urban green spaces is yet to be comprehensively elucidated.

Ecological remediation of soils by plants has been widely used and is often attributed to changes in plant and soil microbial communities [20]. However, the comparison of the effects between biotic and abiotic soil ecological remediation materials has not been fully studied. Previous research has demonstrated that biochar can serve as a shelter for bacteria by providing a favorable pore structure and specific surface area, while also supplying essential nutrients for bacterial growth [21,22]. Furthermore, biochar has been observed to enhance the physicochemical properties of soil, stimulate microbial activity, and thus positively affect the soil carbon pool [23]. For example, Jabborova et al. [24] indicated that co-inoculation of biochar and rhizobacteria significantly increased soil SOC and nutrient content. Ren et al. [25] indicated that the combined application of PGPR and biochar significantly increased the soil bacterial diversity index. However, most research mainly focused on the combined biotechnology of biochar and PGPB. In contrast, the effect of Fe-modified biochar combined with *Bacillus clausii* on improving soil fertility in urban green spaces and carbon sequestration is still unclear.

To fill this knowledge gap, this study aimed to investigate the effects of applied *Bacillus clausii*, biochar, Fe-modified biochar, biochar combined with *Bacillus clausii*, and Fe-modified biochar combined with *Bacillus clausii* on soil fertility in urban green spaces and carbon sequestration and explored the changes in soil C-cycling-related enzyme activity and the bacterial community. We hypothesized that (1) compared to their individual application, the combination of Fe-modified biochar and *Bacillus clausii* will result in improving soil fertility; (2) the combination of Fe-modified biochar and *Bacillus clausii* could significantly enhance the ability of soil carbon sequestration; and (3) the combination of Fe-modified biochar and *Bacillus clausii* can improve enzyme activity and bacterial community composition related to carbon cycling in the soil.

2. Materials and Methods

2.1. Preparation of Soil Samples, Biochar, Bacillus clausii, and Ryegrass Seeds

The soil samples were collected from the green spaces of Shanghai University (31°19′34″ N; 121°24′22″ E). Using a multipoint mixing method, four sampling points were used to collect soil samples from the surface layers (upper 20 cm). The drying method was used to measure the soil moisture content of the field soil samples immediately upon their return to the laboratory. Before being used for the incubation experiment, ensure that the soil is thoroughly mixed and dried using air conditions and passed through a 2 mm sieve. The physicochemical properties of the soil are shown in Table S1.

The biochar (B) was provided by Shanghai Carbon Suo Era Environmental Technology Co., Ltd., Shanghai, China. The biochar was derived from apple wood and obtained using a continuous operation reactor at a temperature of 400 °C for 5 h. The preparation of Fe-modified biochar (FeB) is based on the modification method mentioned by Liu et al. [26]. All the biochar samples were passed through a 2 mm sieve before being used. *Bacillus clausii* CICC 21104 was provided by the China Center of Industrial Culture Collection (http://www.china-cicc.org/cicc/detail2/?sid=3305, accessed on 14 January 2023), Beijing, China. The ryegrass seeds (*Lolium perenne* L.) were provided by Beijing Hejia Eco-technology Co., Ltd., Beijing, China.

2.2. Bacterial Inoculum Preparation

Nutrient broth (CM0002, China Center of Industrial Culture Collection, Beijing, China) culture medium was the medium used for culture of *Bacillus clausii*. *Bacillus clausii* was inoculated into sterile medium to cultivate and expand at 37 °C for 48 h. We regulated the suspension of *Bacillus clausii* to 1×10^8 CFU mL^{-1} and then used it as a standard inoculum.

2.3. Experimental Design

On 1 July 2023, an indoor potted plant experiment was conducted at the artificial climate laboratory of the School of Environmental and Chemical Engineering, Shanghai University. The six treatments are shown in Table 1. Each treatment had three replicates of pots. Firstly, we filled 1200 g soil into pots with an inside diameter of 12 cm and a height of 22 cm. For the CK treatment, no biochar or *Bacillus clausii* suspension was added. For the *Bacillus clausii* (BC) treatment, we diluted 10 mL of *Bacillus clausii* suspension with sterile water to 100 mL, and then fully mixed with the soil in the pot. For the biochar (B) or Fe-modified biochar (FeB) treatment, we fully mixed the biochar or Fe-modified biochar with the soil in the pot at a rate of 2% (*w*/*w*), respectively. For the biochar combined with *Bacillus clausii* (BBC) or Fe-modified biochar combined with *Bacillus clausii* (FeBBC) treatment, we fully mixed the same above-mentioned doses of biochar or Fe-modified biochar and *Bacillus clausii* suspension with the soil in the pot. To ensure consistency, add the same doses of sterile water for treatments without *Bacillus clausii* inoculation.

Table 1. Experimental treatments settings.

Samples	Biochar (%)	*Bacillus clausii* Suspension (1×10^8 CFU mL^{-1}) (mL)	*Lolium perenne* L.
CK	×	×	√
BC	×	10	√
B	2	×	√
FeB	2	×	√
BBC	2	10	√
FeBBC	2	10	√
Soil (g)		1200	
Lolium perenne L. (Number)		50	
Incubation cycle (d)		56	

Note: CK: control; BC: *Bacillus clausii*; B: biochar; FeB: Fe-modified biochar; BBC: biochar combined with *Bacillus clausii*; FeBBC: Fe-modified biochar combined with *Bacillus clausii*.

The ryegrass seeds (*Lolium perenne* L.) underwent sterilization using a 10% H_2O_2 solution for 20 min. After sterilization, the seeds were thoroughly rinsed with deionized water until clean and soaked in deionized water for 2 h prior to sowing. All pots were watered with deionized water every two days, and the soil moisture content was maintained at about 60% field soil moisture content. After ten days of growth, 50 seedlings of the same growth were retained in each pot. The incubation experiment lasted 56 days at a controlled temperature of 25 ± 2 °C.

2.4. Samples Analysis

2.4.1. Characterization of *Bacillus clausii* and Biochar Samples

The pH, element contents (C, N, and H), surface area, pore volume, and pore diameter were selected to characterize the basic properties of biochar samples. A scanning electron microscope with energy dispersive spectrometer (SEM-EDS), Fourier transform infrared (FTIR) spectra, X-ray diffraction (XRD), and X-ray Photoelectron Spectroscopy (XPS) were used to analyze the structural features of biochar samples.

The growth promoting characteristics of *Bacillus clausii* were evaluated, including IAA production, ammonia production, phosphate solubilization, nitrogen fixation, siderophore production, and the production of carbonic anhydrase. Inoculate *Bacillus clausii* into the liquid B4 medium for biomineralization experiments to evaluate the effect of *Bacillus clausii* on carbonate precipitation. The FTIR spectra, XRD, and SEM-EDS were used to analyze the surface morphological characteristics and the composition of the mineralization products.

2.4.2. Analysis of Plant and Soil Samples

Soil and plant samples were taken after the incubation experiment to evaluate the effect of soil treatments on plant biomass and soil. A portion of the freshly collected soil was stored in a refrigerator at −20 °C for the analysis of C-cycling enzyme activity. To analyze the diversity and abundance of bacterial communities, a portion of fresh soil was gathered in sterilized bags and kept it in a refrigerator at −80 °C. The remaining fresh soil was dried under air conditions and then analyzed for soil properties after being sieved.

The basic soil properties pH, alkali-hydro nitrogen, available phosphorus, available potassium, available Fe, free Fe oxide (Fed), and soil aggregates were measured to evaluate the effects of soil treatment on soil fertility and structure. The contents of soil organic carbon (SOC), particle organic carbon (POC), $KMnO_4$-oxidized organic carbon (EOC), dissolved organic carbon (DOC), and soil inorganic carbon (SIC) were measured to evaluate the effects of soil treatment on soil carbon sequestration. This study selected invertase and β-glucosidase to assess the impact of soil treatment on C-cycling enzyme activity. The Supplementary Information provides detailed information on the measured methods.

2.4.3. Analytic Method for Bacterial Community Analysis

Soil samples were processed using the MagAttract® PowerSoil® Pro DNA Kit (QIAGEN, Hilden, Germany) following the manufacturer's instructions. To analyze bac-

terial diversity, the V3-V4 hypervariable region of the 16S rRNA genes were amplified with universal primers 338F (5′-ACTCCTACGGGAGGCAGCAG-3′) and 806R (5′-GGACTACHVGGGT
WTCTAAT-3′). The raw sequencing data have been deposited into the NCBI Sequence Read Archive (SRA) database (Accession Number: PRJNA1069871).

2.5. Statistical Analysis

To assess significant differences in plant biomass, soil physicochemical properties, soil nutrient content, soil carbon component content, and enzyme activity were examined using a one-way analysis of variance (ANOVA) with Duncan's multiple range test ($p < 0.05$). The results were presented as the mean ± standard error of the mean of three replicates. The Pearson correlation analysis method was employed to explore the relationship between soil and environmental factors. A co-occurrence network was constructed using Gephi to visualize the associations between the diversity of different bacterial orders and ecological aspects. A correlation between two nodes was considered to be statistically robust if the Spearman's correlation coefficient was over 0.8 or less than −0.8 and the p-value was less than 0.05. The software of Excel 2016 and Origin 2022 was used to produce the tables and figures.

3. Results

3.1. Changes in Characteristics of Fe-Modified Biochar

The physicochemical properties of biochar are listed in Table S2. The results showed that the pH and elemental content (C, N, and H) of FeB were significantly decreased. In addition, the specific surface area and Fe content of FeB increased to 9.35 m^2 g^{-1} and 22.63 g kg^{-1}, respectively. SEM analysis clearly revealed the porous structure of biochar, and after Fe-modification, nanoparticles were observed to be loaded into the pores of FeB (Figure S2). FTIR spectroscopy revealed the presence of new absorption peaks at 650 cm^{-1} and 460 cm^{-1} in FeB (Figure S1a). XRD analysis demonstrated the presence of characteristic peaks of Fe_2O_3, Fe_3O_4, and FeO on the surface of FeB (Figure S1b). After modification, the C-O peak of FeB disappeared, while the peaks corresponding to C-C, C=O, and O-C=O increased (Figure S1d,e).

3.2. Plant Growth-Promoting Characteristics and Biomineralization of *Bacillus clausii*

The results showed that *Bacillus clausii* has various plant growth promoting characteristics, such as indole-3-acetic acid (IAA) production, phosphorus solubilization, nitrogen fixation, siderophores production, and carbonic anhydrase production (Table S3). In the biomineralization experiment (Figure S3), the SEM-EDS analysis of the mineralized product (MP) revealed that the main elements of the MPs were C, O, and Ca (Figure S4c,d). The absorption peaks at 875 cm^{-1}, 1401 cm^{-1}, and 1436 cm^{-1} were the absorption peaks of hexagonal calcium carbonate and calcite, respectively (Figure S4a) [27]. XRD spectrum showed (Figure S4b) that the diffraction peak at $2\theta = 29.4°$ corresponds to the (104) crystal plane of calcite [28].

3.3. Changes in Soil Physicochemical Properties and Fertility of FeB Combined with BC

The effects of various treatments on the physicochemical properties of the soils differ significantly (Table 2). The soil pH significantly decreased in the BC, FeB, and FeBBC treatments. However, the soil pH significantly increased in the B and BBC treatments. Compared with other treatments, in the FeBBC treatments, both the aboveground weight of *Lolium perenne* L. and the alkali-hydro nitrogen content in the soil remained consistently high (Figure S5). However, FeBBC treatment had a negative effect on both available P and K compared to the BBC treatment. Furthermore, the FeB treatment exhibited a significant decrease in soil available Fe content when compared to the CK. The available Fe and Fed content in the FeBBC-treated soil is significantly increased compared to the BBC treatment, indicating that the combination of FeB and BC positively affects the available Fe and

Fed content. After 56 days of incubation, compared with other treatments, there was a significant increase in the specific gravity of soil aggregates (particle size > 0.25 mm) that was observed for the FeBBC treatment (Figure 1f).

Table 2. Soil physicochemical properties after incubation for 56 days.

Samples	pH	AP (mg kg^{-1})	AN (mg kg^{-1})	AK (mg kg^{-1})	AFe (mg kg^{-1})	Fed (mg kg^{-1})
CK	8.15 ± 0.02 ^{b}	7.36 ± 0.82 ^{c}	17.97 ± 1.07 ^{d}	81.31 ± 2.75 ^{c}	13.54 ± 0.50 ^{b}	5.63 ± 0.13 ^{ab}
BC	8.12 ± 0.02 ^{c}	7.59 ± 0.23 ^{bc}	20.77 ± 0.81 ^{bc}	93.02 ± 2.53 ^{b}	19.15 ± 1.35 ^{a}	4.40 ± 0.68 ^{c}
B	8.33 ± 0.01 ^{a}	8.73 ± 0.60 ^{a}	18.67 ± 2.65 ^{cd}	124.34 ± 9.21 ^{a}	12.64 ± 0.13b ^{cd}	4.64 ± 0.55 ^{c}
FeB	7.86 ± 0.02 ^{d}	6.14 ± 0.48 ^{d}	22.63 ± 1.46 ^{b}	77.77 ± 3.80 ^{c}	11.94 ± 0.70 ^{cd}	5.65 ± 0.46 ^{ab}
BBC	8.32 ± 0.02 ^{a}	8.35 ± 0.26 ^{ab}	21.93 ± 0.40 ^{b}	117.46 ± 6.10 ^{a}	11.31 ± 0.75 ^{d}	4.98 ± 0.21 ^{bc}
FeBBC	7.82 ± 0.01 ^{e}	6.75 ± 0.13 ^{cd}	25.43 ± 1.07 ^{a}	80.51 ± 7.34 ^{c}	12.76 ± 0.42 ^{bc}	5.86 ± 0.14 ^{a}

Note: AP, available P; AN, alkali-hydro nitrogen; AK, available K; AFe, available Fe; Fed, free Fe oxides; CK: control; BC: *Bacillus clausii*; B: biochar; FeB: Fe-modified biochar; BBC: biochar combined with *Bacillus clausii*; FeBBC: Fe-modified biochar combined with *Bacillus clausii*.

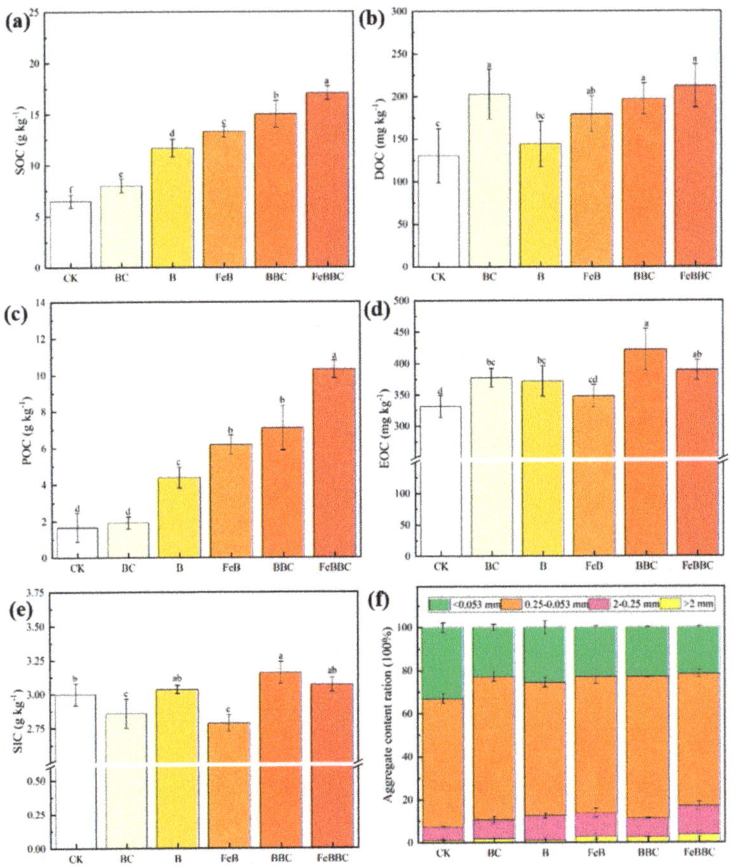

Figure 1. Mean contents of soil organic carbon (SOC) (**a**), dissolved organic carbon (DOC) (**b**), particulate organic carbon (POC) (**c**), $KMnO_4$-oxidized organic carbon (EOC) (**d**), soil inorganic carbon (SIC) (**e**), and aggregate content ration (**f**) in the different treatments. Different letters indicate significant differences between the treatments and control ($p < 0.05$). CK: control; BC: *Bacillus clausii*; B: biochar; FeB: Fe-modified biochar; BBC: biochar combined with *Bacillus clausii*; FeBBC: Fe-modified biochar combined with *Bacillus clausii*.

3.4. Changes in Soil Carbon Components for FeB Combined with BC

The combination of FeB and BC resulted in elevated levels of SOC content compared to other treatments (Figure 1a). Compared with BC or unmodified biochar (B and BBC) treatments, the higher content of POC was observed for Fe-modified biochar treatments (FeB and FeBBC) (Figure 1c). Furthermore, including BC, FeB, BBC, and FeBBC treatments resulted in significantly higher levels of DOC (Figure 1b). Regarding EOC, the Fe-modified biochar treatments (FeB and FeBBC) exhibited a slight decrease in EOC content compared to the unmodified biochar treatments (B and BBC). However, this difference was not statistically significant (Figure 1d). Among all the treatments, BC and FeB treatments showed lower SIC content, whereas BBC treatment exhibited higher SIC content. Furthermore, when comparing FeB with FeBBC treatment, a significant increase in SIC content was observed in the latter (Figure 1e).

3.5. Changes in Soil C-Cycling-Related Enzyme Activity and Soil Bacterial Diversity for FeB Combined with BC

Among all the treatment groups, the FeB and FeBBC treatments exhibited lower levels of β-glucosidase (Figure 2a). Similar trends were observed in terms of invertase activity among all treatments. The B and FeBBC treatments displayed higher levels of invertase activity, while no significant differences were observed between the other treatment groups (Figure 2b).

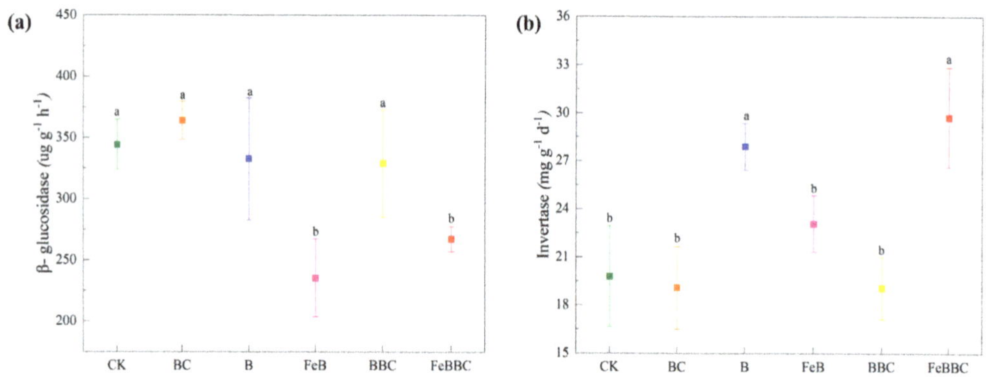

Figure 2. Changes in soil C-cycling-related enzyme activity in different treatments: β-glucosidase (**a**) and invertase (**b**). Different lowercase letters indicate significant differences between the treatment groups ($p < 0.05$). CK: control; BC: *Bacillus clausii*; B: biochar; FeB: Fe-modified biochar; BBC: biochar combined with *Bacillus clausii*; FeBBC: Fe-modified biochar combined with *Bacillus clausii*.

The changes in the alpha diversity of soil bacterial communities are shown in Table 3. The results of this study suggested that the gene sequence utilized accurately reflects the composition of bacterial communities in the soil samples. This conclusion is supported by the high Goods coverage, ranging from 99.59% to 100.00%. The FeBBC treatment resulted in a diverse bacterial community, as demonstrated in Table 3. However, the BBC treatment provided the least diversity in the community. Likewise, the number of ASVs correlates with the trend of similarity in the alpha diversity index. The Venn diagram analysis revealed variations in the ASVs among different treatments (Figure S6). The number of treatment-specific ASVs ranged from 1684 to 2292, and the highest value was observed for the FeBBC treatment, which suggests that the FeBBC treatment improved the environment for microbial survival. Furthermore, a total of 533 ASVs were found to be shared among all treatments, as depicted in Figure S6. These core bacterial groups primarily included *Proteobacteria*, *Acidobacteriota*, *Actinobacteruota*, *Chloroflexi*, and *Firmicutes*.

Table 3. Changes in the richness and diversity of soil microbial communities.

Sample	ASVs	Ace	Chao	Goods Coverage (%)	Shannon	Simpson
CK	3916	3994.39	3937.67	99.59	7.47	0.0012
BC	3690	3734.58	3702.95	99.73	7.59	0.0009
B	3536	3556.61	3538.84	99.87	7.52	0.0009
FeB	3453	3453.00	3453.00	100.00	7.47	0.0011
BBC	3424	3461.45	3431.95	99.78	7.41	0.0012
FeBBC	4077	4113.06	4081.85	99.80	7.61	0.0009

Note: ASVs: Amplicon sequence variants; Ace: Ace index; Chao: Species richness estimator; Goods coverage: Microbial analysis depth; Shannon: The Shannon index; Simpson: The Simpson index. CK: control; BC: *Bacillus clausii*; B: biochar; FeB: Fe-modified biochar; BBC: biochar combined with *Bacillus clausii*; FeBBC: Fe-modified biochar combined with *Bacillus clausii*.

3.6. Changes in Bacterial Community Composition for FeB Combined with BC

The relative abundance of dominant phyla in bacterial communities and the heatmaps of 30 dominant bacterial orders are shown in Figure 3, respectively. The focus of this study will be on taxonomic groups that are closely linked to carbon decomposition/fixation in soil. The result showed that the dominant C-cycling-related bacterial taxa included the Proteobacteria, Acidobacteriota, Actinobacteriota, Chloroflexi, Firmicutes, Myxococcota, and Gemmatimonadota. Compared with unmodified biochar, the relative abundance of Proteobacteria, Firmicutes, and Gemmatimonadota increased with Fe-modified biochar treatment. Actinobacteriota and Chloroflexi were less abundant in the FeBBC-treated soil than in the BBC-treated soil. Moreover, the abundance of Acidobacteriota for the unmodified biochar treatment was higher than that for the Fe-modified biochar treatment. The distribution of C-cycling-related bacteria treated with FeBBC was uniform compared to adding FeB or BC alone. Notably, the FeBBC treatment exhibited a higher relative abundance of *Rhizobiales*, *Burkholderiales*, *Sphingomonadales*, and *Gemmatimonadales* than the BBC treatment. Additionally, the relative abundance of *Myxococcales* in the FeBBC-treated soil decreased in comparison to the FeB treatment (Figure 3b).

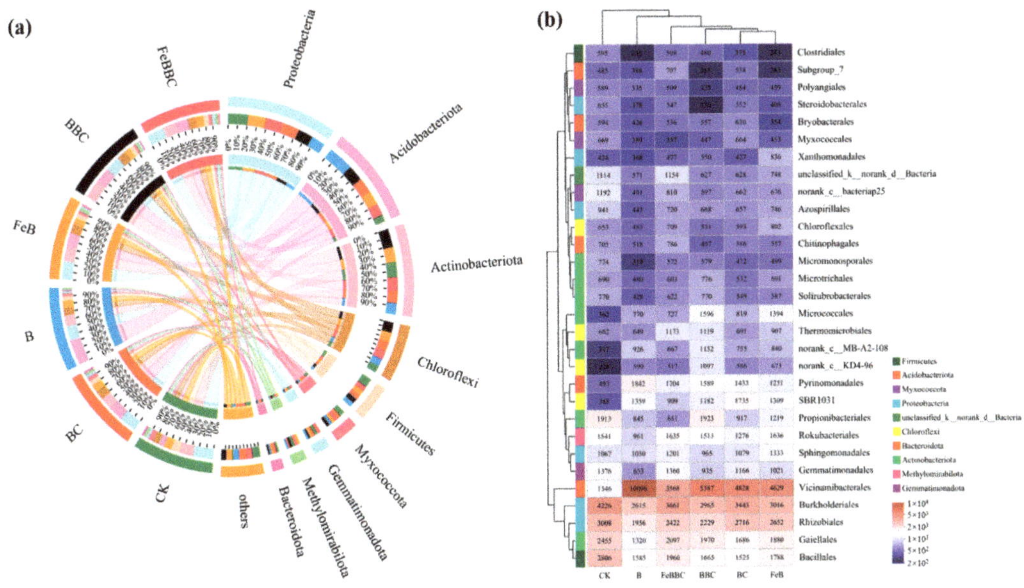

Figure 3. The relative abundance (**a**) of dominant species of bacterial communities. Heatmap diagram (**b**) of the dominant 30 bacterial orders in different treatments. CK: control; BC: *Bacillus clausii*; B: biochar; FeB: Fe-modified biochar; BBC: biochar combined with *Bacillus clausii*; FeBBC: Fe-modified biochar combined with *Bacillus clausii*.

3.7. Exploring the Correlations between Environmental Factors and Bacterial Communities

Pearson correlation analysis (Figure 4a) revealed significant positive correlations between soil pH and available P, available K, and β-glucosidase activity. Moreover, SOC demonstrated significant positive correlations with the aboveground weight of *Lolium perenne* L., particulate organic carbon, and invertase activity. As illustrated in Figure 4b, the co-occurring network analysis revealed significant negative correlations between soil pH and *Chloroflexales* and *Sphingomonadales*. A significant negative correlation between SOC and *Myxococcales* was also revealed.

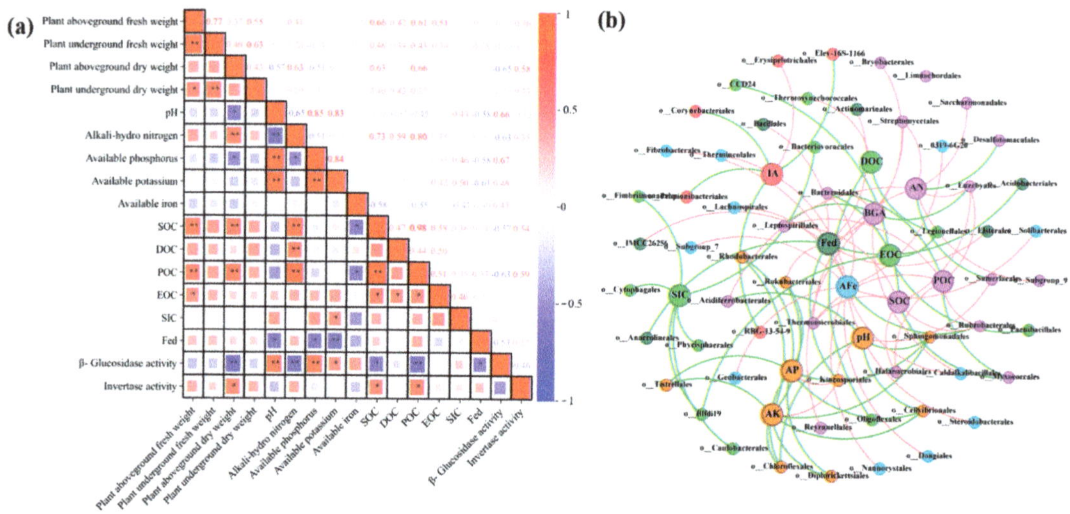

Figure 4. Pearson correlations (**a**) between soil environmental factors among all treatments. * and ** indicate significant correlations at $p < 0.05$ and 0.01, respectively. Co-occurring network (**b**) of the dominant 200 bacterial orders and environmental factors based on correlation analysis. The nodes in the network are colored based on modularity class. The connections stand for a strong (Spearman's $\rho > 0.8$) and significant ($p < 0.05$) correlations. The red and green lines represent positive and negative correlations, respectively. Abbreviations: AP, available P; AN, alkali-hydro nitrogen; AK, available K; AFe, available Fe; IA, invertase activity; BGA, β-glucosidase activity.

4. Discussion

4.1. Characteristics of Biochar and Fe-Modified Biochar

The main reason for the significant decrease in the pH value of FeB was the hydrolysis of Fe^{3+} on the surface of FeB [10]. This indicated that adding acidic FeB to soil will positively impact the pH, alleviating soil alkalization. In addition, an increase in the specific surface area of FeB and the porous structure observed via SEM indicates that FeB had more adsorption sites, which was beneficial for the survival of bacteria. FTIR spectroscopy revealed the presence of new absorption peaks at 650 cm^{-1} and 460 cm^{-1}, indicating the successful loading of Fe_2O_3 and Fe_3O_4, which is also confirmed in the XRD analysis [29]. XPS spectroscopy shows that the changes in different functional groups on the surface of FeB indicate that the increased aromaticity of FeB and alterations in its surface characteristics can impact its interaction with various soil environmental factors, consequently influencing the composition of bacterial communities within the soil.

4.2. Characteristics of *Bacillus clausii*

The results in Table S3 indicate that *Bacillus clausii* exhibited positive traits for plant growth promotion. The mineralization experiment's products were mainly carbonates, as determined with SEM-EDS, FTIR, and XRD characterization analysis. The findings of this study provide evidence that *Bacillus clausii* is capable of producing carbonic anhydrase, an enzyme that facilitates the conversion of CO_2 into carbonate. This enzymatic activity is advantageous for the sequestration of inorganic carbon in soil, contributing to its overall carbon storage capacity.

4.3. *Effect of FeB Combined with BC Improves Soil Physicochemical Properties and Fertility*

In this study, the high pH of biochar itself and the hydrolysis of soluble alkaline substances led to an increase in soil pH in the B and BBC treatments, exacerbating soil alkalization. In accordance with our first hypothesis, both FeB and BC treatments had a positive effect on reducing the soil pH value. Moreover, compared with an individual application, the combined application of FeB and BC resulted in a more significant decrease in soil pH value. The hydrolysis of Fe^{3+} on the surface of Fe-modified biochar released a large amount of H^+, and consequently reduced soil pH [10]. In addition, this phenomenon may also be attributed to FeB providing a favorable habitat for *Bacillus clausii*, which in turn secretes organic acids. These organic acids not only contribute to the reduction in soil pH but also facilitate the dissolution of inorganic phosphates that are present [30].

In this study, the aboveground fresh and dry weight of *Lolium perenne* L. were significantly increased in the FeBBC treatment (Figure S5). This is because the addition of Fe-modified biochar provided a larger contact area for plant roots to absorb nutrients from the soil, and improved plant growth under the action of plant growth promoting bacteria. Compared with BBC or FeB treatments, the highest alkali-hydro nitrogen content in soil was observed with the FeBBC treatment. This may be attributed to the ability of FeB to regulate the C/N ratio in the soil by altering the soil pH, thereby affecting the availability of nitrogen in the soil. Additionally, as shown in Table S3, *Bacillus clausii* in the FeBBC treatment exhibited characteristics such as IAA production and nitrogen fixation, promoting plant growth and enhancing the effectiveness of nitrogen in the soil [31]. Comparatively, the utilization of FeBBC led to a significant decrease in the availability of soil P and K in comparison to the BBC treatment. These findings align with previous studies that observed reduced soil phosphorus availability upon adding Fe-modified biochar [11]. The decrease in the soil availability of P may have been related to the combination of phosphorus anions and iron hydroxides, which reduces the bioavailability of phosphorus in the soil. In addition, the low pH value of the soil and the adsorption of Fe-modified biochar may also be the reasons for the decrease in soil nutrients (P and K). The application of iron-modified biochar may inhibit the activity of iron reducing bacteria [32]. In addition, the adsorption of biochar and the binding of available Fe and phosphorus reduced the content of available Fe in the FeB treatment soil [33]. Compared with the BBC treatment, adding FeB and BC also led to a significant increase in available Fe content in the soil, which was attributed to the role of the siderophore produced by *Bacillus clausii* (Table S3).

After 56 days of incubation, compared to the BBC treatment, the FeBBC treatment exhibited a significant increase in the proportion of aggregates (particle size > 0.25 mm). The result showed a negative correlation between pH value and Fed (Figure 4a), indicating that soil pH influenced the formation of Fed. The introduction of exogenous Fe oxides through Fe-modified biochar played a crucial role in forming macroaggregates in the soil. It possesses the ability to bind with soil organic carbon, resulting in the formation of organic iron composite colloids [34,35]. In addition, our research results indicated that the higher aboveground biomass of plants and the richness and diversity of the soil bacterial communities in the FeBBC treatment increased the production of bacterial and plant viscous substances, consequently promoting the formation of aggregates. Therefore, the results of this study indicated that the combination of FeB and BC had great potential in alleviating soil alkalization, increasing soil alkali-hydro nitrogen content, and improving soil aggregate structure.

4.4. Effect of FeB Combined with BC Improves Soil Carbon Sequestration

It is well-established that adding biochar to the soil has been proven to significantly increase SOC content by providing exogenous organic carbon [36]. The results of this study were consistent with the previous studies, and our findings demonstrated higher SOC content in all biochar treatments. In accordance with our second hypothesis, compared with individual application, the combination of FeB and BC significantly enhanced the content of SOC in soil, and SOC reached its highest values for the FeBBC treatment. The highest content of SOC in the FeBBC treatment may be attributed to the high specific surface area of FeB and the creation of a suitable environment for microbial survival by providing nutrients, increasing the colonization of bacterial communities, and promoting plant growth, thus increasing SOC content. In our study, the result showed a notable positive correlation between SOC and the aboveground weight of *Lolium perenne* L. (Figure 4a), confirming that *Lolium perenne* L. growth contributes to the accumulation of SOC.

Compared with CK treatment, BC treatment showed no significant difference in POC. However, the higher POC content observed with FeBBC treatment can be attributed to the high surface activity of the Fe oxide on the surface of Fe-modified biochar, which can increase the carbon content. It also strengthens the bonding between soil clay particles and organic molecules and forms water-stable soil aggregates, preventing microorganisms' rapid degradation of POC [26]. It is worth noting that higher DOC content was observed in Fe-modified biochar treatments (Figure 1b). This observation may be attributed to the biological pathway of Fe^{3+} reduction, which had been previously shown to increase DOC content [37]. After 56 days of BC treatment, the concentrations of DOC and EOC continued to increase. This may be attributed to the fact that BC could promote the secretion of a large amount of low-molecular weight organic compounds by ryegrass roots and enhanced bacterial activity, and thus accelerated the decomposition of soil organic matter. In terms of EOC, after modification, the active functional groups on the surface of FeB were removed (Figure S1d,e), and the pore volume and pore diameter were decreased (Table S2). This may be the reason for the decrease in EOC content in Fe-modified biochar treatments compared to unmodified biochar treatments.

The higher SIC content with the BBC treatment may be explained by its higher soil pH (Figure 1e). The significant decrease in SIC content with the BC and FeB treatments may be attributed to the growth of *Lolium perenne* L. Evidence has shown that plant roots absorb inorganic carbon from the soil during their growth process and convert it into organic carbon to meet plant growth needs [38]. The combination of FeB and BC resulted in a higher SIC content than individual applications. This may have occurred because the added *Bacillus clausii* inoculant produced carbonic anhydrase, which was beneficial for converting CO_2, generated by atmospheric or biological respiration into carbonates. In addition, a short-term study found that carbonic anhydrase had the highest activity at pH 7.5, and the presence of Fe^{3+} enhanced the activity of carbonic anhydrase [39]. Therefore, the results of this study indicated that compared with individual applications, the combination of FeB and BC can significantly increase the content of SOC, POC, and SIC, promoting soil carbon sequestration.

4.5. Effect of FeB Combined with BC Improves C-Cycling-Related Enzyme Activity and Bacterial Community

The activity of enzymes related to carbon cycling in soil serves as a crucial indicator reflecting changes in soil organic matter (SOM). The decrease in β-glucosidase activity observed with the FeB and FeBBC treatments may be attributed to the co-localization of carbon and microorganisms on the surface of FeB in the soil, improving carbon utilization efficiency and reducing the production demand for β-glucosidase [40]. Compared with CK, there was no significant difference in the invertase activity between the BC treatment and FeB treatment, while the invertase activity with the FeBBC treatment significantly increased. The higher bacterial community richness may be responsible for the increase in invertase activity during FeBBC treatment. Furthermore, it has been established that a soil environment with a neutral to slightly alkaline pH promotes increased bacterial and enzymatic activity [41].

Bacterial communities play a crucial role in urban soil ecosystems, and changes in their structure can significantly impact the cycling of elements urban green spaces. In this study, it was observed that the richness and diversity of bacterial communities were significantly lower in all treatments except for the FeBBC treatment. This finding aligns with a study conducted by Zheng et al. [42], which also reported a reduction in bacterial richness with the addition of biochar. In addition, the competition between the inoculated *Bacillus clausii* and the local microbial community is weak and the colonization rate is limited [43]. The dramatically higher richness and diversity of bacterial communities in the FeBBC treatment, which may be attributed to the fact that positively charged Fe oxides on the surface of Fe-modified biochar can combine with negatively charged bacteria through electrostatic attraction, making them less prone to leaching in the soil and increasing the bacterial colonization rate.

Relative to the unmodified biochar treatment, the dominant C-cycling-related bacterial taxa increased under Fe-modified biochar treatment included Proteobacteria, Firmicutes, and Gemmatimonadota. The relative abundance of *Proteobacteria* is the highest in soils that have high levels of labile organic carbon. The order *Rhizobiales* (Alphaprotobacteria) is genetically capable of degrading various plant organic compounds [44]. Compared with unmodified biochar, the higher relative abundance of *Rhizobiales* with FeBBC treatment indicated that the addition of FeBBC had a good promoting effect on plant growth. The order *Sphingomonadales* and *Burkholderiales* belong to the phylum Proteobacteria, which can utilize various organic compounds or more refractory compounds like aromatics in the soil as a carbon source [45,46]. Meanwhile, the result indicated that the abundance of *Sphingomonadales* negatively correlated with the soil pH (Figure 4b). These observations suggested that *Sphingomonadales* demonstrated higher activity in neutral and slightly alkaline environments. *Firmicutes* bacteria are particularly responsive to an increase in recalcitrant carbon sources in the soil [26]. Therefore, the input of stable organic carbon may promote the rapid growth of *Firmicutes* in FeBBC-treated soil. The colonization of *Bacillus clausii* was indicated by the relative abundance of *Bacillales* belonging to *Firmicutes* increasing during FeBBC treatment. *Gemmatimonadales* contain diverse genes associated with organic carbon metabolism; they can utilize carbon and play a significant role in soil carbon fixation [47]. Therefore, the increase in relative abundance of carbon cycling-related bacteria indicated that Fe-modified biochar treatment was more effective at improving soil environment (pH and fertility) than unmodified biochar treatment.

The Acidobacteriota is a crucial bacterial community in soil that plays a significant role in the degradation of complex organic compounds [48]. The Actinobacteriota includes various microorganisms such as *Micromonosporales* and *Gaiellales*. The order *Micromonosporales* could encode multiple enzymes that aid in the degrading of organic C compounds [44]. The order of *Gaiellales* had been proven to be involved in the degradation of polycyclic aromatic hydrocarbons [49]. They play an important role in the degradation of various labile and stubborn carbon compounds. The lower relative abundance of Acidobacteriota and Actinobacteriota with FeBBC treatment indicated potential carbon sequestration and reduction in carbon dioxide emissions. According to several studies, Chloroflexi were typically found in anaerobic or nutrient-deficient environments [50]. The degrading overall of complex organic compounds is associated with Chloroflexi's relative abundance in agricultural soils [47]. In addition, the result indicated that the abundance of *Chloroflexales* had negatively correlated with the soil pH (Figure 4b), The lower relative abundance of Chloroflexi with FeBBC treatment indicated that the addition of FeB combined with BC could improve the soil environment and could enhance soil fertility and aeration. The results indicated a significant negative correlation between *Myxococcales* and SOC. It has been confirmed that the extracellular hydrolytic enzymes secreted by *Myxococcales* play a crucial role in promoting the decomposition of complex carbon sources [47]. The lower relative abundance of *Myxococcales* with FeBBC treatment indicated that it was beneficial for reducing soil carbon mineralization. Therefore, the FeBBC treatment has a higher rate of soil carbon sequestration because Acidobacteriota, Actinobacteriota, Chloroflexi, and the order *Myxococcales* have lower relative abundance than the BBC treatment.

5. Conclusions

In conclusion, this study innovatively combined Fe-modified biochar (FeB) with *Bacillus clausii* (BC) to restore the ecological function of urban green soil. The results showed that the combination of FeB and BC effectively alleviates soil alkalization, improves soil fertility, and increases soil carbon sequestration. Additionally, the combination of FeB and BC significantly increased the activity of invertase and the richness and diversity of bacterial communities in the soil, promoting the uniform distribution of carbon cycle-related bacteria in the soil. The results clearly indicate that the combined application of FeB and BC has greater potential in improving soil fertility and increasing soil carbon sequestration in urban green spaces compared to their individual applications. These results provide a reference solution for the future restoration of soil ecological functions and provide a solid theoretical basis for restoring the ecological functions of urban green space soil and achieving sustainable management of urban or agricultural soil.

Supplementary Materials: The following supporting information can be downloaded at: https://www.mdpi.com/article/10.3390/biology13080611/s1, Figure S1: FTIR (a) and XRD (b) of biochar and Fe-modified biochar samples. Fe2p (c) XPS spectrum of Fe-modified biochar. (d,e) are the C1s spectra of biochar and Fe-modified biochar, respectively; Figure S2: SEM-EDS images of biochar and Fe-modified biochar: SEM results of biochar (a) and Fe-modified biochar (b), and the corresponding EDS results of biochar (c) and Fe-modified biochar (d); Figure S3: Biomineralization experiments of carbonic anhydrase from *Bacillus clausii*. The experimental treatment with *Bacillus clausii* is represented by (a,c), while the control treatment without *Bacillus clausii* is represented by (b,d); Figure S4: FTIR (a), XRD (b), and SEM-EDS images of mineralization products (MPs). SEM (c) results for MPs, and the corresponding EDS (d) results for MPs; Figure S5: Effects of plant fresh (a) and dry weights (b) in different treatments. CK: Control; BC: *Bacillus clausii*; B: Biochar; FeB: Fe-modified biochar; BBC: Biochar combined with *Bacillus clausii*; FeBBC: Fe-modified biochar combined with *Bacillus clausii*; Figure S6: Venn diagram of exclusive and shared bacterial ASVs with different treatments. CK: Control; BC: *Bacillus clausii*; B: Biochar; FeB: Fe-modified biochar; BBC: Biochar combined with *Bacillus clausii*; FeBBC: Fe-modified biochar combined with *Bacillus clausii*; Table S1: The physicochemical properties of soil; Table S2: The physicochemical properties of biochar (B) and Fe-modified biochar (FeB); Table S3: Plant growth-promoting bacterial traits of *Bacillus clausii*; Text S1: Methods of plant characterization; Text S2: Methods of soil characterization; Text S3: Methods of biochar characterization; Text S4: Methods of *Bacillus clausii* characterization.

Author Contributions: G.N.; Conceptualization, Methodology, Validation, Investigation, Writing—Original Draft. C.H.; Conceptualization, Resources, Data Curation, Writing—Review and Editing, Funding acquisition. S.M.; Investigation, Writing—Review and Editing. Z.C.; Investigation, Writing—Review and Editing. Y.M.; Investigation. Y.Z.; Resources, Investigation, Project administration. All authors have read and agreed to the published version of the manuscript.

Funding: This study was financially supported by the Shanghai Municipal Agricultural and Rural Committee Extension Project (2022) 2-6 and the Shanghai Greening and Urban Appearance Management Bureau's research project G190201.

Institutional Review Board Statement: Not applicable.

Informed Consent Statement: Not applicable.

Data Availability Statement: The datasets generated during and/or analyzed during the current study are available from the corresponding author on reasonable request.

Conflicts of Interest: All authors declare that they have no known competing financial interests or personal relationships that could have appeared to influence the work reported in this paper.

Abbreviations

AFe, available Fe; AK, available K; AN, alkali-hydro nitrogen; AP, available P; ASVs, amplicon sequence variants; B, biochar; BBC: biochar combined with *Bacillus clausii*; BC, *Bacillus clausii*; BGA, β-glucosidase activity; Chao: species richness estimator; CK, control; DOC, dissolved organic carbon; EOC, $KMnO_4$-oxidized organic carbon; FeB, Fe-modified biochar; FeBBC: Fe-modified biochar combined with *Bacillus clausii*; Fed, free Fe oxides; FTIR, Fourier transform infrared; IA, invertase activity; IAA, indole-3-acetic acid; MP, mineralized product; POC, particulate organic carbon; PGPB, Plant growth-promoting bacteria; SEM-EDS, scanning electron microscope with energy dispersive spectrometer; SOC, soil organic carbon; SIC, soil inorganic carbon; XPS, X-ray Photoelectron Spectroscopy; XRD, X-ray diffraction.

References

1. Feng, X.J.; Sun, X.Y.; Li, S.Y.; Zhang, J.D.; Hu, N. Relationship study among soils physicochemical properties and bacterial communities in urban green space and promotion of its composition and network analysis. *Agron. J.* **2021**, *113*, 515–526. [CrossRef]
2. O'Riordan, R.; Davies, J.; Stevens, C.; Quinton, J.N.; Boyko, C. The ecosystem services of urban soils: A review. *Geoderma* **2021**, *395*, 115076. [CrossRef]
3. Vasenev, V.; Kuzyakov, Y. Urban soils as hot spots of anthropogenic carbon accumulation: Review of stocks, mechanisms and driving factors. *Land Degrad. Dev.* **2018**, *29*, 1607–1622. [CrossRef]
4. Li, G.; Sun, G.X.; Ren, Y.; Luo, X.S.; Zhu, Y.G. Urban soil and human health: A review. *Eur. J. Soil Sci.* **2018**, *69*, 196–215. [CrossRef]
5. Ungaro, F.; Maienza, A.; Ugolini, F.; Lanini, G.M.; Baronti, S.; Calzolari, C. Assessment of joint soil ecosystem services supply in urban green spaces: A case study in Northern Italy. *Urban For. Urban Green.* **2022**, *67*, 127455. [CrossRef]
6. Kumar, A.; Singh, E.; Mishra, R.; Lo, S.-L.; Kumar, S. A green approach towards sorption of CO_2 on waste derived biochar. *Environ. Res.* **2022**, *214*, 113954. [CrossRef] [PubMed]
7. Zhao, P.; Palviainen, M.; Koster, K.; Berninger, F.; Bruckman, V.J.; Pumpanen, J. Effects of Biochar on Fluxes and Turnover of Carbon in Boreal Forest Soils. *Soil Sci. Soc. Am. J.* **2019**, *83*, 126–136. [CrossRef]
8. Nath, H.; Sarkar, B.; Mitra, S.; Bhaladhare, S. Biochar from Biomass: A Review on Biochar Preparation Its Modification and Impact on Soil Including Soil Microbiology. *Geomicrobiol. J.* **2022**, *39*, 373–388. [CrossRef]
9. Wang, J.; Wang, S. Preparation, modification and environmental application of biochar: A review. *J. Clean. Prod.* **2019**, *227*, 1002–1022. [CrossRef]
10. Wen, E.; Yang, X.; Chen, H.; Shaheen, S.M.; Sarkar, B.; Xu, S.; Song, H.; Liang, Y.; Rinklebe, J.; Hou, D.; et al. Iron-modified biochar and water management regime-induced changes in plant growth, enzyme activities, and phytoavailability of arsenic, cadmium and lead in a paddy soil. *J. Hazard. Mater.* **2021**, *407*, 124344. [CrossRef]
11. Wu, L.; Zhang, S.; Wang, J.; Ding, X. Phosphorus retention using iron (II/III) modified biochar in saline-alkaline soils: Adsorption, column and field tests. *Environ. Pollut.* **2020**, *261*, 114223. [CrossRef] [PubMed]
12. Zhang, Y.; Zhao, C.; Chen, G.; Zhou, J.; Chen, Z.; Li, Z.; Zhu, J.; Feng, T.; Chen, Y. Response of soil microbial communities to additions of straw biochar, iron oxide, and iron oxide–modified straw biochar in an arsenic-contaminated soil. *Environ. Sci. Pollut. Res.* **2020**, *27*, 23761–23768. [CrossRef]
13. Bao, Z.; Shi, C.; Tu, W.; Li, L.; Li, Q. Recent developments in modification of biochar and its application in soil pollution control and ecoregulation. *Environ. Pollut.* **2022**, *313*, 120184. [CrossRef]
14. Ajijah, N.; Fiodor, A.; Pandey, A.K.; Rana, A.; Pranaw, K. Plant Growth-Promoting Bacteria (PGPB) with Biofilm-Forming Ability: A Multifaceted Agent for Sustainable Agriculture. *Diversity* **2023**, *15*, 112. [CrossRef]
15. Khatoon, Z.; Huang, S.; Rafique, M.; Fakhar, A.; Kamran, M.A.; Santoyo, G. Unlocking the potential of plant growth-promoting rhizobacteria on soil health and the sustainability of agricultural systems. *J. Environ. Manag.* **2020**, *273*, 111118. [CrossRef]
16. Li, H.; Qiu, Y.; Yao, T.; Ma, Y.; Zhang, H.; Yang, X. Effects of PGPR microbial inoculants on the growth and soil properties of Avena sativa, Medicago sativa, and Cucumis sativus seedlings. *Soil Tillage Res.* **2020**, *199*, 104577. [CrossRef]
17. Colo, J.; Hajnal-Jafari, T.I.; Duric, S.; Stamenov, D.; Hamidovic, S. Plant Growth Promotion Rhizobacteria in Onion Production. *Pol. J. Microbiol.* **2014**, *63*, 83–88. [CrossRef]
18. Li, X.; Chen, X.; Wang, H.; Cai, Q. Identification of An Alkalophilic Strain and Activities of Its Extracelluar Polysaccharide. *Pharm. Biotechnol.* **2008**, *15*, 370–374.
19. Oulebsir-Mohandkaci, H.; Benzina-Tihar, F.; Hadjouti, R. Exploring biofertilizer potential of plant growth-promoting rhizobacteria *Bacillus clausii* strain B8 (MT305787) on *Brassica napus* and *Medicago sativa*. *Not. Bot. Horti Agrobot. Cluj-Napoca* **2021**, *49*, 12484. [CrossRef]
20. Bhanse, P.; Kumar, M.; Singh, L.; Awasthi, M.K.; Qureshi, A. Role of plant growth-promoting rhizobacteria in boosting the phytoremediation of stressed soils: Opportunities, challenges, and prospects. *Chemosphere* **2022**, *303*, 134954. [CrossRef]
21. Zhu, X.; Chen, B.; Zhu, L.; Xing, B. Effects and mechanisms of biochar-microbe interactions in soil improvement and pollution remediation: A review. *Environ. Pollut.* **2017**, *227*, 98–115. [CrossRef]
22. Khadem, A.; Raiesi, F.; Besharati, H.; Khalaj, M.A. The effects of biochar on soil nutrients status, microbial activity and carbon sequestration potential in two calcareous soils. *Biochar* **2021**, *3*, 105–116. [CrossRef]

23. Sarfraz, R.; Hussain, A.; Sabir, A.; Ben Fekih, I.; Ditta, A.; Xing, S. Role of biochar and plant growth promoting rhizobacteria to enhance soil carbon sequestration—A review. *Environ. Monit. Assess.* **2019**, *191*, 251. [CrossRef] [PubMed]
24. Jabborova, D.; Wirth, S.; Kannepalli, A.; Narimanov, A.; Desouky, S.; Davranov, K.; Sayyed, R.Z.; El Enshasy, H.; Malek, R.A.; Syed, A.; et al. Co-Inoculation of Rhizobacteria and Biochar Application Improves Growth and Nutrientsin Soybean and Enriches Soil Nutrients and Enzymes. *Agronomy* **2020**, *10*, 1142. [CrossRef]
25. Ren, H.; Li, Z.; Chen, H.; Zhou, J.; Lv, C. Effects of Biochar and Plant Growth-Promoting Rhizobacteria on Plant Performance and Soil Environmental Stability. *Sustainability* **2022**, *14*, 10922. [CrossRef]
26. Liu, S.; Kong, F.; Li, Y.; Jiang, Z.; Xi, M.; Wu, J. Mineral-ions modified biochars enhance the stability of soil aggregate and soil carbon sequestration in a coastal wetland soil. *Catena* **2020**, *193*, 104618. [CrossRef]
27. Srivastava, S.; Bharti, R.K.; Verma, P.K.; Thakur, I.S. Cloning and expression of gamma carbonic anhydrase from *Serratia* sp. ISTD04 for sequestration of carbon dioxide and formation of calcite. *Bioresour. Technol.* **2015**, *188*, 209–213. [CrossRef]
28. Zheng, T.; Qian, C. Influencing factors and formation mechanism of $CaCO_3$ precipitation induced by microbial carbonic anhydrase. *Process Biochem.* **2020**, *91*, 271–281. [CrossRef]
29. Lu, J.; Yang, Y.; Liu, P.; Li, Y.; Huang, F.; Zeng, L.; Liang, Y.; Li, S.; Hou, B. Iron-montmorillonite treated corn straw biochar: Interfacial chemical behavior and stability. *Sci. Total Environ.* **2020**, *708*, 134773. [CrossRef]
30. Khalil, H.M.A.; Hassan, R.M. Raising the Productivity and Fiber Quality of Both White and Colored Cotton Using Eco-Friendly Fertilizers and Rice Straw. *Int. J. Plants Res.* **2015**, *5*, 122–135.
31. Ren, H.; Lv, C.; Fernández-García, V.; Huang, B.; Yao, J.; Ding, W. Biochar and PGPR amendments influence soil enzyme activities and nutrient concentrations in a eucalyptus seedling plantation. *Biomass Convers. Biorefin.* **2019**, *11*, 1865–1874. [CrossRef]
32. Hu, J.; Dong, H.; Xu, Q.; Ling, W.; Qu, J.; Qiang, Z. Impacts of water quality on the corrosion of cast iron pipes for water distribution and proposed source water switch strategy. *Water Res.* **2018**, *129*, 428–435. [CrossRef] [PubMed]
33. Manirakiza, N.; Şeker, C.; Negiş, H. Effects of Woody Compost and Biochar Amendments on Biochemical Properties of the Wind Erosion Afflicted a Calcareous and Alkaline Sandy Clay Loam Soil. *Commun. Soil Sci. Plant Anal.* **2021**, *52*, 487–498. [CrossRef]
34. Liu, L.; Zhang, S.; Chen, M.; Fei, C.; Zhang, W.; Li, Y.; Ding, X. Fe-modified biochar combined with mineral fertilization promotes soil organic phosphorus mineralization by shifting the diversity of phoD-harboring bacteria within soil aggregates in saline-alkaline paddy soil. *J. Soils Sediments* **2022**, *23*, 619–633. [CrossRef]
35. Cai, L.; Yang, Y.; Chong, Y.; Xiong, J.; Wu, J.; Ai, X.; Guo, Q.; Yuan, Y.; Li, Z. Higher Soil Aggregate Stability in Subtropical Coniferous Plantations Than Natural Forests Due to Microbial and Aggregate Factors. *Forests* **2022**, *13*, 2110. [CrossRef]
36. Smith, J.L.; Collins, H.P.; Bailey, V.L. The effect of young biochar on soil respiration. *Soil Biol. Biochem.* **2010**, *42*, 2345–2347. [CrossRef]
37. Guo, Z.; Kang, Y.; Wu, H.; Li, M.; Hu, Z.; Zhang, J. Enhanced removal of phenanthrene and nutrients in wetland sediment with metallic biochar: Performance and mechanisms. *Chemosphere* **2023**, *327*, 138523. [CrossRef]
38. He, C.; Wang, X.; Wang, D.; Zhao, Z.; Wang, F.; Cheng, L.; Feng, H.; Zhang, P. Impact of *Spartina alterniflora* invasion on soil bacterial community and associated greenhouse gas emission in the Jiuduansha wetland of China. *Appl. Soil Ecol.* **2021**, *168*, 104168. [CrossRef]
39. Sharma, T.; Sharma, A.; Xia, C.L.; Lam, S.S.; Khan, A.A.; Tripathi, S.; Kumar, R.; Gupta, V.K.; Nadda, A.K. Enzyme mediated transformation of CO_2 into calcium carbonate using purified microbial carbonic anhydrase. *Environ. Res.* **2022**, *212*, 113538. [CrossRef]
40. Tian, J.; Wang, J.; Dippold, M.; Gao, Y.; Blagodatskaya, E.; Kuzyakov, Y. Biochar affects soil organic matter cycling and microbial functions but does not alter microbial community structure in a paddy soil. *Sci. Total Environ.* **2016**, *556*, 89–97. [CrossRef]
41. Khan, M.N.; Li, D.; Shah, A.; Huang, J.; Zhang, L.; Núñez-Delgado, A.; Han, T.; Du, J.; Ali, S.; Sial, T.A.; et al. The impact of pristine and modified rice straw biochar on the emission of greenhouse gases from a red acidic soil. *Environ. Res.* **2022**, *208*, 112676. [CrossRef]
42. Zheng, H.; Liu, D.; Liao, X.; Miao, Y.; Li, Y.; Li, J.; Yuan, J.; Chen, Z.; Ding, W. Field-aged biochar enhances soil organic carbon by increasing recalcitrant organic carbon fractions and making microbial communities more conducive to carbon sequestration. *Agric. Ecosyst. Environ.* **2022**, *340*, 108177. [CrossRef]
43. Ji, C.; Liu, Z.; Hao, L.; Song, X.; Wang, C.; Liu, Y.; Li, H.; Li, C.; Gao, Q.; Liu, X. Effects of Enterobacter cloacae HG-1 on the Nitrogen-Fixing Community Structure of Wheat Rhizosphere Soil and on Salt Tolerance. *Front. Plant Sci.* **2020**, *11*, 111094. [CrossRef] [PubMed]
44. Ibrahim, M.M.; Guo, L.; Wu, F.; Liu, D.; Zhang, H.; Zou, S.; Xing, S.; Mao, Y. Field-applied biochar-based MgO and sepiolite composites possess CO_2 capture potential and alter organic C mineralization and C-cycling bacterial structure in fertilized soils. *Sci. Total Environ.* **2022**, *813*, 152495. [CrossRef]
45. Shivaji, S.; Ray, M.K.; Shyamala Rao, N.; Saisree, L.; Jagannadham, M.V.; Seshu Kumar, G.; Reddy, G.S.N.; Bhargava, P.M. *Sphingobacterium antarcticus* sp. nov., a Psychrotrophic Bacterium from the Soils of Schirmacher Oasis, Antarctica. *Int. J. Syst. Evol. Microbiol.* **1992**, *42*, 102–106. [CrossRef]
46. Nicolitch, O.; Feucherolles, M.; Churin, J.L.; Fauchery, L.; Turpault, M.P.; Uroz, S. A microcosm approach highlights the response of soil mineral weathering bacterial communities to an increase of K and Mg availability. *Sci. Rep.* **2019**, *9*, 14403. [CrossRef]
47. Ibrahim, M.M.; Zhang, H.; Guo, L.; Chen, Y.; Heiling, M.; Zhou, B.; Mao, Y. Biochar interaction with chemical fertilizer regulates soil organic carbon mineralization and the abundance of key C-cycling-related bacteria in rhizosphere soil. *Eur. J. Soil Biol.* **2021**, *106*, 103350. [CrossRef]

48. Wang, S.; Li, T.; Zheng, Z.; Chen, H.Y.H. Soil aggregate-associated bacterial metabolic activity and community structure in different aged tea plantations. *Sci. Total Environ.* **2019**, *654*, 1023–1032. [CrossRef]
49. Lin, W.; Fan, F.; Xu, G.; Gong, K.; Cheng, X.; Yuan, X.; Zhang, C.; Gao, Y.; Wang, S.; Ng, H.Y.; et al. Microbial community assembly responses to polycyclic aromatic hydrocarbon contamination across water and sediment habitats in the Pearl River Estuary. *J. Hazard. Mater.* **2023**, *457*, 131762. [CrossRef]
50. Yamada, T.; Sekiguchi, Y. Cultivation of Uncultured *Chloroflexi* Subphyla: Significance and Ecophysiology of Formerly Uncultured *Chloroflexi* 'Subphylum I' with Natural and Biotechnological Relevance. *Microbes Environ.* **2009**, *24*, 205–216. [CrossRef]

biology

MDPI

Review

Typical Marine Ecological Disasters in China Attributed to Marine Organisms and Their Significant Insights

Lulu Yao [1], Peimin He [1,*], Zhangyi Xia [2], Jiye Li [3] and Jinlin Liu [4,5,*]

1 College of Oceanography and Ecological Science, Shanghai Ocean University, Shanghai 201306, China; llyao166@163.com
2 College of Ocean and Earth Science, Xiamen University, Xiamen 361102, China; xzy19028@163.com
3 Key Laboratory of Ecological Prewarning of Bohai Sea of Ministry of Natural Resources, North China Sea Environmental Monitoring Center of State Oceanic Administration, Qingdao 266033, China; lijiye@ncs.mnr.gov.cn
4 State Key Laboratory of Marine Geology, Tongji University, Shanghai 200092, China
5 Project Management Office of China, National Scientific Seafloor Observatory, Tongji University, Shanghai 200092, China
* Correspondence: jlliu@tongji.edu.cn (J.L.); pmhe@shou.edu.cn (P.H.)

Simple Summary: In China, in addition to green tides and red tides, outbreaks of Cnidaria (jellyfish organisms), Annelida (*Urechis unicinctus* Drasche, 1880), Mollusca (*Philine kinglipini* S. Tchang, 1934), Arthropoda (*Acetes chinensis* Hansen, 1919), and Echinodermata (Asteroidea organisms (commonly known as starfish), Ophiuroidea organisms, and *Acaudina molpadioides* Semper, 1867) not only damage marine resources, tourism, coastal industries, and navigation but also profoundly impact ecosystems near nuclear plants, beaches, and infrastructures, endangering human lives. This article suggests measures to mitigate future disasters, emphasizing enhanced monitoring and integration into China's marine ecological disaster monitoring system. This review is conducive to enhancing researchers' comprehension and reflection regarding biological disasters in marine ecosystems.

Abstract: Owing to global climate change or the ever-more frequent human activities in the offshore areas, it is highly probable that an imbalance in the offshore ecosystem has been induced. However, the importance of maintaining and protecting marine ecosystems' balance cannot be overstated. In recent years, various marine disasters have occurred frequently, such as harmful algal blooms (green tides and red tides), storm surge disasters, wave disasters, sea ice disasters, and tsunami disasters. Additionally, overpopulation of certain marine organisms (particularly marine faunas) has led to marine disasters, threatening both marine ecosystems and human safety. The marine ecological disaster monitoring system in China primarily focuses on monitoring and controlling the outbreak of green tides (mainly caused by outbreaks of some *Ulva* species) and red tides (mainly caused by outbreaks of some diatom and dinoflagellate species). Currently, there are outbreaks of Cnidaria (Hydrozoa and Scyphozoa organisms; outbreak species are frequently referred to as jellyfish), Annelida (*Urechis unicinctus* Drasche, 1880), Mollusca (*Philine kinglipini* S. Tchang, 1934), Arthropoda (*Acetes chinensis* Hansen, 1919), and Echinodermata (Asteroidea organisms, Ophiuroidea organisms, and *Acaudina molpadioides* Semper, 1867) in China. They not only cause significant damage to marine fisheries, tourism, coastal industries, and ship navigation but also have profound impacts on marine ecosystems, especially near nuclear power plants, sea bathing beaches, and infrastructures, posing threats to human lives. Therefore, this review provides a detailed introduction to the marine organisms (especially marine fauna species) causing marine biological disasters in China, the current outbreak situations, and the biological backgrounds of these outbreaks. This review also provides an analysis of the causes of these outbreaks. Furthermore, it presents future prospects for marine biological disasters, proposing corresponding measures and advocating for enhanced resource utilization and fundamental research. It is recommended that future efforts focus on improving the monitoring of marine biological disasters and integrating them into the marine ecological disaster monitoring system. The aim of this review is to offer reference information and constructive suggestions for enhancing future monitoring, early warning systems, and prevention efforts related to marine ecological disasters in support of the healthy development and stable operation of marine ecosystems.

Keywords: marine ecological disasters; Cnidaria; Annelida; Mollusca; Arthropoda; Echinodermata; harmful algal blooms; marine management

1. Introduction

Currently, there is a global increase in frequent marine disasters, resulting in significant damage to nearshore and terrestrial ecosystems [1–3]. Simultaneously, these disasters disrupt fishing activities and tourism development, with some also causing destruction of residents' livelihoods and productive resources, thereby impacting regional economic development [4–7]. For instance, storm surges and tsunamis [8,9] not only pose threats to human lives and property but also result in the destruction of marine and terrestrial habitats. This leads to the influx of land-based pollutants into the ocean or seawater encroaches on land areas. Consequently, freshwater resources essential for human survival become contaminated, exacerbating ecological crises in coastal regions and leading to the collapse and reorganization of social-ecological systems [10].

China is among the countries most affected by marine disasters [11,12], and these disasters are primarily attributed to physical and biological factors. The key disasters included in China's marine disaster monitoring system encompass storm surge disasters, wave disasters, sea ice disasters, tsunami disasters, red tide events, and green tide events [13]. For instance, in 2023, the storm surge disaster (seven disaster incidents) resulted in the largest direct economic loss for China, totaling 2.4805027 billion yuan and accounting for 99% of the total direct economic loss from all disasters that year. Five wave disaster incidents led to a direct economic loss of 26.25 million yuan and caused eight fatalities or disappearances [13]. In the long term, both physical and biological factors can lead to significant indirect economic losses, particularly ecological and economic impacts.

The continuous outbreaks of marine biological disasters in China have led to significant direct and indirect economic losses that cannot be overlooked. In recent years, there has been a frequent occurrence of green tides (Figure 1a) and red tides (Figure 1b), which have become the primary biological disasters monitored in China's marine disaster monitoring system [14–18]. Red tides are characterized by sudden increases in population or aggregation of marine plankton (Figure 1b), resulting in water color changes (Figure 1b) or harm to other marine organisms [19]. In 2023, a total of 46 red tide blooms were discovered in China's marine areas, with a total outbreak area of 1466 square kilometers; among them, 29 were toxic or harmful, covering an area of 1118 square kilometers; also, the East China Sea experienced the most occurrences and the largest total outbreak area [13]. Green tides exhibit large-scale outbreaks (Figure 1a), widespread distribution, and long duration. The Yellow Sea green tide outbreak is particularly typical, with macroalgae such as *Ulva prolifera* O.F. Müller, 1778 being the main species involved [18]. In 2023, the maximum covered area by the Yellow Sea green tide was recorded at 998 square kilometers, while its maximum distribution area reached 61,159 square kilometers [13], causing significant impacts and unquantifiable economic losses as well as algal prevention costs. Currently, the primary impact from marine disasters caused by biological factors (Figure 1) is on the loss to the marine ecological economy; nevertheless, it remains challenging to effectively assess this specific economic loss scale accurately. Objectively speaking, this issue requires sustained high-level attention along with improvements to disaster assessment standards.

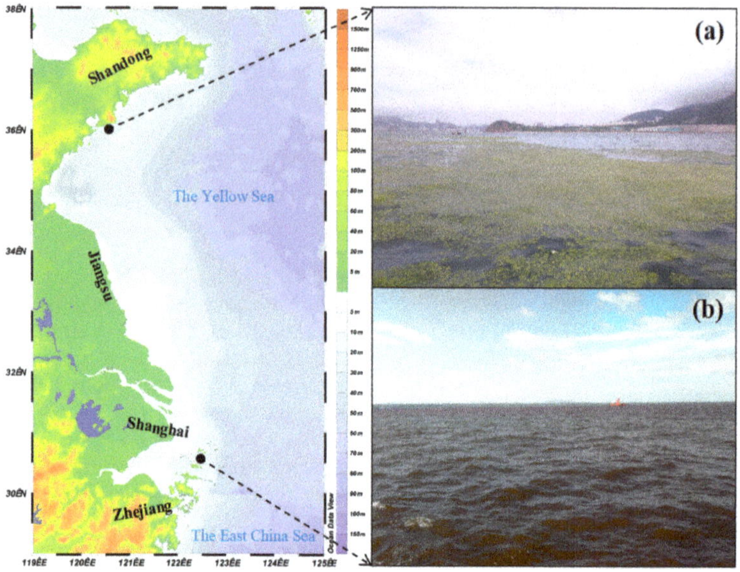

Figure 1. The large-scale outbreak of green tides and red tides occurred in China. (**a**) A green tide event occurred in the Yellow Sea in July 2019; (**b**) A red tide event occurred in the East China Sea in June 2019.

The current China's marine disaster monitoring system primarily focuses on microplankton and macroalgae [13]. However, there is minimal attention given to common marine animals that can cause ecological disasters, and they have not been included in China's regular marine disaster monitoring system to date. The sustainable development of marine environmental protection and ecological systems is increasingly crucial for the development of coastal economies. The frequent outbreaks of jellyfish and starfish in the China Sea [20,21] have had significant negative impacts on fishery farming, ecological restoration, tourism, as well as the service functions of the marine ecological system. This has led to a series of economic and social problems, making it a long-term marine ecological disaster that urgently needs to be studied and solved in China. Additionally, small-scale outbreak phenomena are also being observed with *Acetes chinensis* Hansen, 1919; Ophiuroidea organisms; *Acaudina molpadioides* Semper, 1867; *Urechis unicinctus* Drasche, 1880; and *Philine kinglipini* S. Tchang, 1934 [22–24], gradually affecting human coastal infrastructure and the service functions of marine ecological systems. These phenomena have started attracting attention from researchers and the Chinese government.

Therefore, this review focuses on the fundamental biological background of the outbreaks of common marine animals—Cnidaria (jellyfish organisms), Annelida (*U. unicinctus*), Mollusca (*P. kinglipini*), Arthropoda (*A. chinensis*), and Echinodermata (Asteroidea organisms, Ophiuroidea organisms, and *A. molpadioides*) in China. Simultaneously, a detailed introduction of the current outbreak status of the aforementioned species in the China Sea is provided, with particular emphasis on summarizing and discussing the underlying causes of these outbreaks. The aim is to offer reference information and constructive suggestions for future disaster monitoring, early warning systems, and prevention efforts in order to support the healthy and stable development of the marine ecosystem.

2. Cnidaria—The Biological Background of Jellyfish and the Marine Ecological Disasters It Causes in China

Jellyfish, an important group of gelatinous plankton in the phylum Cnidaria, encompasses five categories: Hydromedusae, Siphonophore, Scyphomedusae, Cubomedusae, and

Ctenophore [25]. Jellyfish originated in the Cambrian period, and jellyfish have survived five mass extinctions and continue to thrive in the global ocean [26]. Large jellyfish species are prevalent in the waters of China, Japan, and Korea [27]. Globally, more than 1400 species of jellyfish have been identified [28], with approximately 420 species reported in China's nearshore waters [26,29]. Jellyfish possess a translucent and pliable body structure, characterized by a high water content and the absence of skeletal support [30]. They exhibit an alternation of generations in reproduction, and their demise can exert a substantial influence on biogeochemical processes and the cycling of life-source elements. Deceased jellyfish may serve as sustenance for micro- and small scavengers while also providing dissolved organic matter for microorganisms [31]. This influx of organic matter has the potential to impact plankton distribution and biomass, thereby playing a crucial role in maintaining marine ecosystem equilibrium and facilitating nutrient cycling.

Jellyfish are rich in minerals and proteins, with low fat, cholesterol, and calorie content, making them a natural delicacy in certain Asian countries [32]. The collagen present in jellyfish exhibits potential in preventing infectious pathogens and diseases and promoting health by eliciting immune responses [33,34]. Research has shown that collagen from *Nemopilema nomurai* Kishinouye, 1922, enhances immunity [35], while quniumucin extracted from various jellyfish also shows promising potential as a novel marine resource [33]. In the United States, salted and dried *Stomolophus meleagris* Agassiz, 1860 are utilized for the production of novel marine gelatin powders, serving as gelling agents, thickeners, and binders in food applications. They are increasingly recognized as a valuable source of collagen within the emerging fishery and food industry [32]. Owing to their extensive health benefits, jellyfish are progressively establishing themselves as an essential marine biological resource for nutritional supplements, cosmetics, and functional foods, demonstrating significant potential in both the food and health industries.

Surprisingly, jellyfish blooms have become a global phenomenon. Since 1995, numerous countries and regions, including the Baltic Sea, Japan, South Korea, India, Australia, the United States, Israel, and Scotland, have witnessed significant impacts from jellyfish outbreaks [36–38]. Jellyfish have proliferated and formed large aggregations, leading to blockages in the cooling water intakes of nuclear power plants. This has resulted in forced shutdowns of facilities such as the Madras Atomic Power Station in the Bay of Bengal and the Kashiwazaki Kariwa Nuclear Power Station in Japan. Similar incidents have also occurred at coastal power plants in South Korea, the United States, Israel, and Scotland [37]. Furthermore, jellyfish outbreaks have caused substantial economic losses within the fishing industry. For instance, Japan suffered losses of USD 20 million in 2003 and USD 250 million in 2005 due to *N. nomurai* outbreaks. Additionally, the Peruvian purse seiners of Ilo lost over USD 200,000 during 2008–2009 [38,39]. Moreover, jellyfish blooms have resulted in significant ecological harm, such as the incursion of Ctenophores along the Mediterranean coast of Spain and Israel in 2009, leading to extensive damage to the coastal ecosystem and fisheries [40]. In 2011, there was an outbreak of *Cephea cephea* Forskål, 1775, in the Red Sea, which served as an opportunistic food source for coral reef fish and disrupted the community structure of marine ecosystems [41]. In addition, approximately 52.8 million jellyfish stranded on Cabo Beach in 2012, with *Crambione mastigophora* Maas, 1903—a dominant species—exhibiting exceptionally high predation rates on bivalve larvae, potentially posing a serious threat to local oyster farming [42]. Moreover, a survey conducted among fishermen in the northern Adriatic Sea revealed that 70% of them had encountered jellyfish entanglement in their nets, resulting in reduced catches and economic losses amounting to up to 8.2 million euros per year [39].

In the Bohai Sea, Yellow Sea, and East China Sea (Figure 2), jellyfish outbreaks occur almost annually, leading to significant adverse effects on marine ecosystems and human livelihoods. Major species involved in these outbreaks include *N. nomurai*, *Aurelia aurita* Linnaeus, 1758, and *Cyanea nozakii* Kishinouye, 1891 [36]. Among these species, *N. nomurai* and *C. nozakii* primarily impact aquaculture and fishing by preying on the eggs or larvae of economically valuable fish and shrimp species [39]. This results in a decrease in fishing yields, leading to the obstruction or damage of fishing nets [43], thereby impacting fishing catches. Between 2000 and 2003, the annual catch volume in the northern East China Sea and the Yellow Sea's fishery decreased by 64%, while the annual catch volume of *N. nomurai* increased by 250% [36]. In 2003, an outbreak of *Rhopilema esculentum* Kishinouye, 1891, occurred in the East China Sea, with an average biomass of 1555 kg/ha and a maximum biomass of 15,000 kg/ha; during this period, there was a reduction of about 20% in the catch volume of economically important species *Pseudosciaena polyactis* Bleeker, 1877. In 2004, an outbreak of *C. nozakii* occurred in Liaodong Bay, resulting in an approximately 80% reduction in edible *R. esculentum* and direct economic losses amounting to approximately USD 80 million [27,36].

Figure 2. The outbreak of jellyfish occurred in the China Sea. On 21 June 2021, the jellyfish bloom was observed in the offshore waters of the Yellow Sea (**a**,**b**); on 26 June 2021 (**c**) and 3 June 2024 (**d**), the jellyfish bloom was observed in the East China Sea. The red or white dashed arrows indicate the location of the jellyfish that have been spotted.

Aurelia aurita has impacted the normal operation of some power plants in China. For instance, in August 2008, 20 to 50 tons of *A. aurita* were removed from the air intake screen at the Weihai coastal power plant [36]. In the same year, an outbreak of *A. aurita* blocked the rotary filter net at the seafront pumping station of the Qinhuangdao Power Company, leading to a unit shutdown and threatening power supply for Olympic venues [44]. Similarly, on 7–8 July 2010, over 10 tons of *A. aurita* obstructed the air intake screen at the Qingdao coastal power plant [36]. Additionally, in 2009, a large quantity of *A. aurita* blocked the seawater intake at Huadian Qingdao Power Generation Co., Ltd. (Qingdao,

China), endangering its circulating water system. It took thirty employees over two days of continuous work to restore normal operations [45]. In addition, in 2013, an invasion by numerous jellyfish threatened safe operations at Liaoning Hongyanhe Nuclear Power Plant by blocking its cooling water intake [45]. Since 2014, the Hongyanhe Nuclear Power Plant has faced persistent challenges with jellyfish infestation at its intake, particularly from *N. nomurai* and *A. aurita*, posing a high risk of impacting the plant's cooling source [46]. In 2014, an influx of jellyfish into the inlet area of the circulating water filtration system resulted in the shutdown of Units 1 and 2. Subsequently, in 2015, a significant population of jellyfish obstructed the intake due to breaches in both first and third barrier nets [37].

Furthermore, toxic jellyfish have increasingly appeared at coastal bathing areas, posing a threat to seaside tourism by endangering tourists. Species like *N. nomurai*, *C. nozakii*, *A. aurita*, *Physalia physalis* Linnaeus, 1758, *Pelagia noctiluca* Forsskål, 1775, and *R. esculentum* are notorious for their propensity to sting humans [36]. The stings from toxic jellyfish can result in symptoms such as skin redness, swelling, and burning pain, and in severe cases may lead to skin necrosis, cardiac and neurotoxic reactions, or even fatalities [36,45]. Since 1983, over 2000 incidents of jellyfish stings have been reported along China's coast, resulting in 13 fatalities [44]. For example, in 2013 near Qinhuangdao City, *A. aurita* proliferated with a maximum density of 1500×10^4 ind./km^2, severely disrupting tourist activities and prohibiting swimming in the sea [44]. More than a thousand individuals sought medical attention due to jellyfish stings, which tragically resulted in the death of a child at Daihe Beach [36]. Meanwhile, in 2019, there were also jellyfish outbreaks reported in the vicinity of Haikou City in Hainan Province and near Maoming City in Guangdong Province [47].

Jellyfish outbreaks not only cause significant damage to marine fisheries, tourism, coastal industries, and ship navigation, but they also have a profound impact on marine ecosystems. The proliferation of jellyfish in large numbers leads to competition with other aquatic organisms for food, predation on marine organism larvae, and their stings can cause the mortality of certain marine organisms, resulting in a decline in ecosystem biodiversity [48]. Furthermore, the decomposition of jellyfish during their demise contributes to the accumulation of organic matter in the ocean, which promotes the rapid proliferation of toxic microorganisms. This process creates a low-oxygen zone at the ocean floor and releases acidic substances that harm marine organisms [49]. These impacts severely compromise the health of the marine environment and disrupt ecosystem stability.

The proliferation of jellyfish can be attributed to a combination of factors, including the impact of global climate change, alterations in environmental conditions, and human activities [50,51]. Firstly, global climate change has resulted in elevated sea temperatures, expanding the suitable temperature range for polyp budding and promoting their proliferation, consequently leading to increased jellyfish populations [30,49,51]. Climate change also induces variations in salinity, temperature, and other marine parameters that may influence jellyfish reproduction and distribution patterns. For instance, tropical jellyfish may migrate to subtropical and temperate regions due to these changes [52,53]. Secondly, environmental changes encompass diminished predator populations, overfishing practices, and water eutrophication. The decline in predators directly contributes to an upsurge in jellyfish numbers by reducing predation pressure and competition from other species [36]. Increased nutrient input leads to heightened plankton biomass, which serves as a food source for jellyfish. Simultaneously, eutrophication results in algae sedimentation and decomposition, causing oxygen depletion at the water column bottom—conditions unsuitable for many organisms but favorable for sustained jellyfish outbreaks due to their tolerance [54]. Furthermore, coastal nuclear power plants constructed by humans emit heat, potentially contributing to the increased survival rate of the polyp stage and the growth rate of the medusa stage of jellyfish [55]. Artificial structures in the ocean offer numerous attachment opportunities for the polyp stage of jellyfish, facilitating asexual reproduction through budding in certain jellyfish species [38]. Additionally, solid surfaces or shells (such as balanidae, Mytilidae, Vermetidae, and Hiatellidae) serve as attachment points for the polyp stage. The extensive cultivation of *Ruditapes philippinarum* A. Adams and Reeve,

1850, by human activities has altered subsea substrate and structure, providing ample attachment space and substrate for the polyp stage of jellyfish [56], thereby promoting jellyfish proliferation. Maritime shipping can introduce non-native species via ballast water and facilitate widespread dispersal of jellyfish [40]; a survey conducted in 2018 revealed that *A. aurita* reached peak populations in areas with high levels of human activity such as port construction and aquaculture in the Bohai Sea [52]. Moreover, the enforcement of China's fishing ban system has led to a reduction in fishing activities and minimized human interference and mortality of jellyfish, thereby creating favorable conditions for the rapid proliferation of jellyfish [57]. Also, the surge in jellyfish populations may also represent an adaptive response to expand their distribution range and seek new habitats to cope with adverse environments [26].

3. Annelida—The Biological Background of *Urechis unicinctus* Drasche, 1880 and the Marine Ecological Disasters It Causes in China

The *U. unicinctus* is commonly known as 'Hai Chang' in China and belongs to the phylum Annelida, class Echiurida, order Xenopneusta, family Urchidae, and was originally classified in the phylum Echiurioidea [58,59]. This species is widely distributed along the North Pacific coast, including the coasts of the Yellow Sea and Bohai Sea in China, Japan, the Korean Peninsula, and the Far East Region of Russia [60–62]. *Urechis unicinctus* typically inhabits muddy or sandy intertidal and sublittoral zones (Figure 3a) with 'U'-shaped burrows and exhibits strong burrowing ability [63,64]. Its biological disturbance significantly enhances dissolved inorganic nitrogen (DIN) diffusion from sediments to water bodies by altering sediment structure through burrowing and feeding behavior [65,66], thereby promoting nutrient cycling in marine ecosystems [67]. Additionally, *U. unicinctus* feeds on phytoplankton, microzooplankton, and debris in water bodies to promote organic matter decomposition and transformation while reducing suspended particulate matter content. This contributes to purifying water quality and maintaining a clean and stable marine environment with significant ecological implications [68].

Although a small outbreak of *U. unicinctus* has only been reported in China, fortunately, *U. unicinctus* possesses significant potential for application in the fields of food and pharmaceuticals. Its meat is not only delectable but also abundant in protein and various essential amino acids [69], rendering it a high-value sea food with substantial market demand. Furthermore, its rich content of flavor amino acids can be utilized for the production of seafood seasonings [70–72]. Moreover, within the medical domain, *U. unicinctus* harbors diverse biologically active peptides and substances that exhibit anti-thrombotic, anti-tumor, antioxidant, antibacterial, and immune-regulatory properties [73,74]. The *U. unicinctus*-derived fibrinoclase demonstrates direct degradation of fibrin and fibrinogen while activating plasminogen without inducing acute bleeding side effects in animal models [75]. Additionally, antioxidant peptides VTSALVGPR, IGLGDEGLRR, and TKIRNEISDLNER isolated from the internal organs of *U. unicinctus* display promising antioxidant capabilities suitable for incorporation into health foods and beverages [76]. The active peptide DDL derived from *U. unicinctus* exhibits erectile function improvement properties [77]. Furthermore, glycosaminoglycan found in *U. unicinctus* manifests hypoglycemic effects by enhancing the body's antioxidant capacity and tissue repair ability while increasing glucose tolerance and insulin sensitivity [78]. Lastly, due to its capability to tolerate and metabolize high concentrations of exogenous sulfides toxic to other organisms, *U. unicinctus* plays a crucial role in improving the quality of polluted sediments [63,73,79].

The recent widespread occurrence of *U. unicinctus* on the beaches of Yantai City, Shandong Province, China, has garnered significant attention. The reproduction and growth of *U. unicinctus* are seasonal, typically occurring in large numbers during spring and autumn with strong wind and wave conditions [63]. The large quantities of *U. unicinctus* washed onto the beach cover the sandy area, attracting residents and tourists to collect them; however, accidents can occur when individuals are swept into the sea by beachside winds and waves. Furthermore, due to its high tolerance for various heavy metals and easy accumulation in its body, consuming *U. unicinctus* in large quantities may pose a potential food safety issue [80,81]. It is noteworthy that *Listriolobus brevirostris* Chen and Yeh, 1958, which shares similar habits with *U. unicinctus*, has been found near multiple nuclear power plants, posing a threat to their safe operation as its *grappler method risk index* exceeds 50% [82,83]. Future massive outbreaks of *U. unicinctus* may clog seawater intake for nearby nuclear power plants, presenting a safety hazard.

Researchers generally believe that the significant outbreak of *U. unicinctus* is attributed to strong winds, turbulent seas, and extreme weather conditions. The sudden changes in sea temperature and disturbances in ocean sediments caused by extreme weather alter the habitat of *U. unicinctus*, compelling them to migrate from sedimentary bottoms to shallow waters and beaches. Additionally, the southward movement of a robust cold front brings about substantial cooling and strong northerly winds, potentially leading to the displacement of *U. unicinctus* onto the shore. Moreover, in Shandong Province's Bajiao Bay, where the primary outbreak occurred, its trumpet-shaped geographical formation concentrates wave energy as they approach the coast. This, combined with high tides and strong winds, facilitates the stranding of *U. unicinctus* living on the seafloor. Furthermore, substrate conditions significantly impact their growth and reproduction; larvae exhibit higher survival rates in muddy environments compared with sandy ones [61,84]. Bajiao Bay's bay-mouth has a gentle slope with flat beaches consisting mainly of silt-deposited layers conducive for *U. unicinctus'* development. The convergence water zone formed by the Yi River estuary provides abundant food for these organisms, while their mass stranding may also be linked to nocturnal breeding behavior [62]. Also, in recent years, local government initiatives releasing various marine species, including *U. unicinctus* larvae, have aimed at enhancing biological community structure and restoring marine resources but inadvertently led to rapid proliferation, disrupting local biodiversity and ecological equilibrium.

4. Mollusca—The Biological Background of *Philine kinglipini* S. Tchang, 1934, and the Marine Ecological Disasters It Causes in China

Philine kinglipini, commonly known as 'Bai Ni Ma' in China, belongs to the phylum Mollusca, class Gastropoda, subclass Opisthobranchia, order Cephalaspidea, and family Onchidiidae, with close genetic relationships to the family Aglajidae [23]. This species is mainly found in the Bohai Sea, Yellow Sea, and East China Sea, inhabiting sandy beaches in the lower layer of high tidal regions to muddy bottom substrates several tens of meters deep. As an omnivorous animal, *P. kinglipini* prefers feeding on cultivated mollusks such as clams and oysters. Its appearance resembles that of a snail, with a long oval-shaped shell that is thin and fragile with a white translucent color. Additionally, it has a well-developed radula and three differently sized rhomboid calcareous gizzards [23,85]. In terms of its ecological role, this species falls between nutritional levels II and III within the ecosystem [86–88].

Outbreaks of *P. kinglipini* occur only in China, with limited economic value, and its secreted slime can induce allergic reactions in humans. In recent years, outbreaks of *P. kinglipini* in the Qingdao Bay area of Shandong Province, China, have posed a significant threat to local shellfish aquaculture (Figure 3b) [89]. The species not only consumes large areas of bivalve larvae but also secretes slime, leading to localized oxygen depletion in the seabed and exacerbating shellfish mortality. In 2022, clam larvae in some aquaculture areas of Qingdao Bay decreased by two-thirds due to outbreaks of *P. kinglipini* and other species, resulting in losses of several million RMB for aquaculture farmers. Furthermore, biological surveys have revealed that *P. kinglipini* tends appears in high numbers in areas with abundant shellfish populations, such as its prevalence in Jiaozhou Bay [90,91], a frequency of up to 66.7% in the western Liaodong Bay [92], dominance on Nanjiu Island, which is primarily reserved for shellfish and other organisms [93], and an increasing abundance trend in the southern part of the Yellow Sea with densities reaching up to 55 ind/m^2 [94]. *Philine kinglipini* exhibits a larger biomass in June in the sea off Yantai City, Shandong [95];

a survey conducted in 2022 revealed that *P. kinglipini* is the predominant species in the sea area near the Dao Island Archipelago in the Yellow Sea [96]. It is evident that *P. kinglipini* has exhibited significant breeding and dispersal patterns across multiple marine regions, potentially posing a threat to marine ecosystems.

Figure 3. The outbreak of *Urechis unicinctus* Drasche, 1880; *Philine kinglipini* S. Tchang, 1934; and *Acetes chinensis* Hansen, 1919, occurred in China. *Urechis unicinctus* was discovered in the seafloor habitats of Jiaozhou Bay on 1 August 2019 (**a**); an outbreak of *P. kinglipini* occurred in the seawaters of Jiaozhou Bay in 2022 [89] (**b**); *Acetes chinensis* were captured in the Southern Yellow Sea in 2024 (**c**). Among them, the white dashed arrow indicates the sediment collected from the seafloor, and the red dashed arrow indicates *U. unicinctus*.

The outbreak of *P. kinglipini* may be attributed to marine eutrophication, global climate change, and degradation of coastal environments, among other factors. In recent years, rapid human development and exploitation of marine resources have accelerated coastal degradation and disrupted the ecological balance in nearshore waters. The increased occurrence of marine biological outbreaks in China's waters could be linked to starfish and jellyfish outbreaks that have disrupted the food chain balance [23]. Furthermore, global warming has favored the growth of temperature-tolerant *P. kinglipini*, while sedimentary mud deposits in certain areas also provide favorable conditions for its growth. The rapid expansion of shellfish aquaculture has provided a rich food source for *P. kinglipini*, accelerating its growth and reproductive capacity. Simultaneously, overfishing of economically important fish species has led to a decline in *P. kinglipini*'s predators (such as *Lateolabrax maculatus* McClelland, 1844, *Sebastes schlegelii* Hilgendorf, 1880, and *Hexagrammos otakii* Jordan and Starks, 1895) [97]. These combined effects have resulted in a significant increase in *P. kinglipini* populations, posing a threat to the stability of marine ecosystems and sustainable development of shellfish aquaculture.

5. Arthropoda—The Biological Background of *Acetes chinensis* Hansen, 1919, and the Marine Ecological Disasters It Causes in China

The species *A. chinensis*, belonging to Arthropoda, Malacostraca, Decapoda, and Sergestidae, *Acetes* [98], is a significant economic resource in China (Figure 3c) and is widely distributed in the near-shore waters of the Northwest Pacific region, including China, Japan, and South Korea [99]. Specifically, *A. chinensis* exhibits abundant resources in the coastal areas of the Bohai Sea, Yellow Sea, East China Sea, and northern South China Sea [100]. Its role as a primary consumer feeding on phytoplankton makes it crucial for marine ecosystems. Additionally, its significance as food for various marine organisms contributes to the material cycle and energy flow within marine environments [98,101,102].

Acetes chinensis is prone to outbreaks near China's nuclear power plants, while it also holds considerable economic and nutritional significance. Its unique amino acid structure provides potential for the development of flavor substances, such as shrimp-flavored seafood seasonings [103]. Furthermore, *A. chinensis* is abundant in essential macronutrients and micronutrients, particularly vitamin B_5, vitamin E, and calcium, making it suitable for use as a calcium supplement food and for the development of other health foods [104,105]. In terms of medicinal properties, the active peptide components in *A. chinensis* have demonstrated the ability to stimulate cellular and humoral immune systems for immune regulation enhancement [106]. Additionally, its neuraminidase inhibitory peptide exhibits anti-influenza virus effects and can serve as raw material for new anti-influenza drug development [107], while its ACE inhibitory peptide shows hypotensive effects [108–110].

However, *A. chinensis* poses a significant biological threat to the safety of cold source water for coastal nuclear power plants, as its 'explosive' aggregations have caused multiple blockages in intake screens. For instance, in January 2016, a large number of *A. chinensis* bloomed in the intake screen of the No. 2 reactor at Ling'ao Nuclear Power Plant, leading to filter blockage and cessation of the circulating water pump, resulting in an emergency shutdown [111,112]. In January 2019, there was an outbreak of *A. chinensis* near the intake screen of Daya Bay Nuclear Power Plant, with over 20 tons captured [113]; biological surveys revealed that *A. chinensis* was abundant in Daya Bay and accounted for over 99% of trawl catches near the nuclear power plant area, posing a threat to safe operations at Daya Bay Nuclear Power Plant [114]. In March 2020, the No. 4 reactor unit at Yangjiang Nuclear Power Plant was shut down due to massive blockage in the seawater circulation water filtration system by *A. chinensis*, which caused pump tripping and triggered a reactor emergency shutdown. After salvage and system restoration, the unit was reconnected to the grid the next day but experienced repeated failures due to subsequent 'explosions' of *A. chinensis* populations [111,115].

Multiple incidents of *A. chinensis* blockages in offshore nuclear power plants have been observed, primarily due to the combined effects of various factors. Firstly, in the context of global warming, changes in sea temperature directly impact the seasonal spawning migration and foraging activities of *A. chinensis* [101,114,116]. A significant drop in winter temperatures may force *A. chinensis* to migrate to warmer waters, and the vicinity of nuclear power plants often has higher water temperatures, increasing the likelihood of a large concentration of *A. chinensis* in the area [117]. Secondly, eutrophication and overfishing have disrupted the food chain. Human activities and meteorological factors such as storms, rainfall, and cold surges cause changes in the nutrient environment of the sea, leading to rapid changes in phytoplankton communities and extensive reproduction that provide abundant food resources for *A. chinensis* [117]. Simultaneously, overfishing has reduced predators of *A. chinensis*, further contributing to a potential population explosion. The construction of artificial reefs generates upwelling and counter-rotating currents that bring nutrients from the seabed to reach the sea surface, which may also affect distribution and population dynamics [102]. Additionally, dynamic changes in water flow patterns along with tidal movements sweep *A. chinensis* into nuclear power plant intake areas [114,115]. As more units are gradually put into operation at most nuclear power plant facilities sharing a common intake structure, it increases water flow rates near intakes, raising risks for ocean life being swept into them [118]. Due to *A. Chinensis*'s weak swimming ability and reproductive pattern closely tied to environmental conditions, causing sharp increases during specific seasons ultimately leads to biological blockage issues at nuclear power plant intakes [114].

6. Echinodermata

6.1. The Biological Background of Asteroidea and the Marine Ecological Disasters It Causes in China

The Asteroidea, the member of the Echinodermata phylum, is currently classified into five orders: Spinulosa, Forcipulata, Paxillosida, Valvatide, and Platyasterida [119–121]. There are approximately 1500 species worldwide [120–122], with over 100 species found in China; also, fossil remains of Asteroidea, Benthopectinidae species, have been discovered in Sichuan Province of China [123]. Meanwhile, Xiao et al. [124] identified nine new record species of Asteroidea in Chinese seawaters, including *Henricia arcystata* Fisher, 1917; *Henricia densispina* Sladen, 1878; *Henricia ohshimai* Hayashi, 1935; *Henricia exigua* Hayashi, 1940; and *Henricia pacifica* Hayashi, 1940, etc. [125].

Asteroidea, a carnivorous predator, inhabits sandy bottoms, muddy bottoms, rocky reefs, or coral reefs [126]. It plays a crucial role in regulating the population of its prey and maintaining local ecosystem equilibrium. Moreover, Asteroidea has the ability to absorb carbon from seawater for exoskeleton formation and, upon death, contribute to carbon sequestration in the ocean [121]. Additionally, it exhibits both sexual and asexual reproduction methods while possessing regenerative capabilities in its arms. In response to predation and environmental stressors, Asteroidea undergoes autotomy as an adaptive mechanism for survival and reproduction within dangerous habitats [120].

Asteroidea organisms are rich in lipids, proteins, polysaccharides, saponins, sterols, carotenoids, and cerebrosides, making them a promising source of functional health food processing ingredients [127]. They also contain a diverse range of pharmacologically active compounds, including polyunsaturated fatty acids with anti-hyperlipidemia, anti-hypertension, anti-platelet aggregation, and anti-thrombosis effects, as well as enhanced memory and improved autoimmune system disorders effects [128]. Furthermore, the asterosaponin of *Asterias rollestoni* Bell, 1881, exhibits significant inhibitory activity against various cancer cells [129]. Additionally, *Craspidaster hesperus* Muller and Troschel, 1840, *Stellaster equestris* Bruzelius, 1805, and *Asterina limboonkengi* G.A. Smith, 1927, have been traditionally used in Chinese folk medicine for treating goiter and rheumatism [122]. In conclusion, Asteroidea organisms demonstrate great potential for functional foods and pharmaceutical ingredients.

However, numerous marine ecosystems worldwide are under serious threat from biological outbreaks of Asteroidea. *Acanthaster planci* Linnaeus, 1758, has caused damage to coral reef ecosystems in the Indian Ocean and Pacific [130]. It is estimated that Asteroidea species have periodically erupted in the Great Barrier Reef for at least 7000 years [131]. Between 1985 and 2012, *A. planci* outbreaks resulted in a 42% mortality rate of corals in the Great Barrier Reef, leading to significant reductions in coral cover and biodiversity, as well as declines in fish density that feed on corals and structural and functional instability of the coral reef ecosystem [132–134]. Additionally, periodic outbreaks of *Asterias amurensis* Lutken, 1871, have affected nearshore waters of Australia, France, Japan, and Vietnam, posing threats to ecosystems while also causing damage to fisheries and marine aquaculture with substantial economic losses [132,135,136]. Furthermore, larvae of Asteroidea species can be transported globally through ocean currents and ship transport. For example, *A. amurensis* was transported to Tasmania, Australia, where it became an invasive species, causing significant damage to local fisheries and marine benthic ecosystems [121,135].

In various marine regions of China, there have been multiple large-scale outbreaks of Asteroidea biological species (Figure 4a,b), with the outbreak numbers reaching 150,000–720,000 per hectare [137]. They are commonly referred to as 'underwater locusts' due to their significant population and destructive impact [138]. The highest concentrations of Asteroidea in China are found in the Yellow Sea, central Bohai Sea area, and the South China Sea. Among them, *A. amurensis* and *Asterina pectinifera* Muller and Troschel, 1842 mainly pose a threat to bivalve mollusks in marine aquaculture areas [126,138], while *A. planci*, *Acanthaster* cf. *Solaris*, and *Acanthaster solaris* Schreber, 1793 cause damage to coral reef ecosystems [139,140]. In marine aquaculture areas, the rapid increase in Asteroidea species leads to high density and large-scale aggregation, resulting in significant harm to fishery production. In 2006, the outbreak density of *A. amurensis* reached up to 300 ind/m^2 in multiple aquaculture areas in Laoshan District and Huangdao District, Shandong Province, China [138], leading to a sharp decline in economically important species output; in 2007, the affected area within *R. philippinarum* aquaculture region in Jiaozhou Bay reached up to 60%, with a mortality rate as high as 80%, resulting in a loss of over 100,000 tons of output [138,141,142]; in 2008, an outbreak of *A. pectinifera* resulted in a 50% reduction in *Apostichopus japonicus* Selenka, 1867 production in the cultivation area [126]; in 2012, *A. amurensis* invaded the *Chlamys farreri* K.H. Jones and Preston, 1904 cultivation area in Qingdao City, leading to a decline of over 80% in *C. farreri* production [126]. In 2020, the average density of Asteroidea species outbreaks reached as high as 50 ind/m^2 and peaked at 200 ind/m^2 in the bottom-cultivated areas of *Crassostrea gigas* Thunberg, 1793, and *R. philippinarum* in Jiaozhou Bay, Shandong Province. This affected an area of approximately 66,666.7 hectares and caused direct economic losses exceeding 100 million RMB. Additionally, Asteroidea species outbreaks were observed in the waters of Jiaozhou Bay, Shandong Province, both in February and July 2022 [138].

In the marine areas of China where coral reefs thrive, the outbreak of *A. planci* exhibits a distinct periodicity. Its larvae consume phytoplankton, while adult *A. planci* can consume hundreds of square centimeters of coral reefs in a single night, causing severe damage to the structure and function of coral reefs and leading to a significant decline in biodiversity [143]. During the five-year period from 2006 to 2010, the outbreak of *A. planci* resulted in a decrease in coral coverage in the Xisha Islands from 60% to below 5%, resulting in the destruction of over 95% of live corals [134,136,140]. Between 2017 and 2021, the outbreak of *A. planci* led to a significant decline in coral coverage in the Nansha Islands, dropping from 33.0% to 0.9% [144–146]. In June 2019, the average density of *A.* cf. *solaris* in the northern South China Sea at Dongsha Atoll was recorded as high as 630 ind/ha [139]. By April 2021, the average density of *A.* cf. *solaris* in the central Nansha Islands had risen to as high as 630 ind/ha, with peak densities reaching up to 1920 ind/ha [139,144]. The destruction of coral reefs in the South China Sea will not only lead to a decline in biodiversity and deterioration of fishery resources but also pose a more serious threat by accelerating coastal reef erosion and endangering national security around China's Xisha Islands and Nansha Islands, among others [142].

There are many factors that contribute to the mass outbreak of Asteroidea organisms, mainly influenced by the species' growth characteristics, environmental conditions, predator numbers, and human activities. Asteroidea organisms have extremely strong reproductive capabilities, producing up to 65 million eggs per season [133] and up to 200 million eggs per year [145]. Additionally, global warming can cause an increase in sea temperature, which improves the survival rate of Asteroidea larvae and significantly increases the predation strength of adults [137,140,147]. Nutrient content in the sea may be increased through human activities, starfish excretion [148], and nutrients brought by storms or typhoons, further promoting the growth of phytoplankton and providing sufficient food and nutrition for Asteroidea larvae, thereby stimulating the growth and dispersal of the Asteroidea population [142,148,149]. Furthermore, overfishing of *Charonia tritonis* Linnaeus, 1758, *Chaetodon*, and Pomacentridae, Asteroidea's predators, may increase the survival rate of Asteroidea larvae and adults, potentially leading to a massive outbreak of the species [130,150]. In addition, human activities such as the development of shellfish seeding technology and the expansion of aquaculture areas not only provide sufficient food for Asteroidea organisms but also destroy the original benthic marine ecosystem and increase the likelihood of Asteroidea outbreaks [120]. Meanwhile, in the management of marine resources, the ratio of the boundaries to the area of small marine protected areas is disproportionately large, which allows Asteroidea to enter protected areas rich in coral and leads to mass outbreaks, putting the increasingly fragmented and degraded coral reef ecosystem at greater risk [151].

6.2. The Biological Background of Ophiuroidea and the Marine Ecological Disasters It Causes in China

Ophiuroidea, a member of the phylum Echinodermata of invertebrate animals [152], is often misidentified by the general public as a starfish due to its morphological similarities with Asteroidea organisms. The current classification divides Ophiuroidea into five suborders: Ophiomyxina, Laemophiurina, Gnathophiurina, Chilophiurina, and Euryalidae. Originating in the Cambrian period 500 million years ago, Ophiuroidea experienced a mass extinction in the Permian period before gradually recovering in the Triassic period [153–155]. Most existing Ophiacanthidae organisms originated from deep sea environments, and species differentiation occurred no later than the end of the Triassic period [156]. There are at least 2064 globally recognized species of Ophiuroidea [155], with China alone hosting at least 211 species [154,155,157,158]. Fossils of Ophiuroidea have been discovered in inland or coastal provinces such as Hunan, Fujian, Guangdong, and Shaanxi, including *Ciliophiurina hunanensis* Lin, 1988; *Ophioderma huaanensis* Wu, 1980; *Ophioderma schistovertebrata* sp. nov.; *Ophiaulax bijieensis* sp. nov.; *Syntomospina kaiyangensis* sp. nov.; *Ophiolepis gulinensis* sp. nov.; and *Ophoderma qingchangensis* sp. nov. [159–161].

Recently, Chen et al. [162] have identified two new record species, *Ophiurothamnus discycla* H.L. Clark, 1911, and *Ophiurothamnus clausa* Lyman, 1878, on Huangyan Seamount in the central South China Sea; Li et al. [163] have discovered a new species named *Histampica haimaensis* sp. nov. in the cold seeps of the South China Sea; and Nethupul et al. [164–166] have found seven new species, including *Asteroschema domogranulatum* sp. nov., *Asteroschema shenhaiyongshii* sp. nov., *Ophiacantha aster* sp. nov., *Ophiomoeris petalis* sp. nov., *Ophiopristis shenhaiyongshii* sp. nov., and *Ophiophthalmus serratus* sp. nov. in the South China Sea, the seamount in the Northwest Pacific Ocean and the Mariana Trench. Additionally, recent years have seen discoveries of *Asteroschema* cf. *Bidwillae* McKnight, 2000; *Asteroschema rubrum* Lyman, 1879; *Asteroschema tubiferum* Matsumoto, 1911; *Asteroschema salix* Lyman, 1879; *Asteroschema* cf. *Lissum* H.L. Clark, 1939; and *Ophiotreta eximia* Koehler, 1904 [164–166]. These studies and discoveries further enrich our understanding of Ophiuroidea diversity and contribute to specimen preservation.

Ophiuroidea outbreaks mainly occur in China's mariculture zones and waters near nuclear power plants. However, Ophiuroidea organisms (Figure 4c) play a crucial role in the economy, medicine, and environment, exerting a positive influence on human health and environmental sustainability. They act as primary or secondary consumers in the food chain by feeding on organic debris and small benthic organisms, thereby transferring energy to higher trophic levels. Additionally, they are significant contributors to the marine carbon cycle and nutrient cycling in sediments [158,167]. Due to their sensitivity to environmental changes, Ophiuroidea organisms can serve as indicators of global climate change and human activities' impact. They can also be utilized for monitoring aquaculture activities' effects on the marine environment and the risk of marine eutrophication [158]. Furthermore, Ophiuroidea organisms contain valuable compounds such as saponins, polysaccharides, and calcium with potential medicinal properties including anti-tumor, anti-viral, and anti-bacterial effects, along with anti-inflammatory properties. Moreover, they provide analgesic effects while enhancing calcium supplementation, immune system enhancement, and cardiovascular health. This provides abundant resources for research and development in the medical field [154,158,168,169].

However, the marine ecosystem is currently facing numerous challenges due to the impact of climate change and other factors, which in turn affect the ecological characteristics and population dynamics of Ophiuroidea organisms. Climate change has been shown to influence the reproductive activities of Ophiuroidea organisms [170]. Additionally, overfishing has resulted in a decline in the number of large carnivorous fish, leading to disruptions in the food chain and nutrient transmission at lower food chains. Furthermore, extensive expansion of marine aquaculture has caused significant damage to the marine benthic system, contributing to a rapid increase in Ophiuroidea populations [155]. In areas with abundant food sources, Ophiuroidea organism densities can reach up to 1200 individuals per square meter [158], potentially leading to sudden outbreaks and ecological disasters. Surveys have revealed that Ophiuroidea organisms are abundant in the Yellow Sea Cold Water Mass region [171,172] and have become dominant species in multiple aquaculture areas [173], posing a potential risk for outbreaks.

Additionally, *Amphioplus laevis* Lyman, 1874 presents a potential risk of obstructing the cooling systems of nuclear power plants in Fujian Province, Guangdong Province, and Hainan Province. Specifically, the *grappler method risk index* for blocking at Fuqing Nuclear Power Plant is 32.6%, while it is 45.8% for Daya Bay Nuclear Power Plant and Lingao Nuclear Power Plant, and 36.9% for Hainan Changjiang Nuclear Power Plant [174]. Ophiuroidea species pose a threat to the safe and stable operation of nuclear power plants. Furthermore, the Zhangzi Island Seafood Cultivation Base is the largest marine cultivation area for *Mizuhopecten yessoensis* Jay, 1857 in China; Ophiuroidea organisms (Figure 4d) are natural predators of *M. yessoensis*. The larvae of Ophiuroidea organisms, known as ophiopluteus, are abundant and easily dispersed, and the cultivated *M. yessoensis* in this area provides a rich food resource for their breeding and growth. Meanwhile, the increased numbers of *Ophiopholis mirabilis* Duncan, 1879, *Ophiura sarsii* subsp. *Vadicola* Djakonov, 1954, and *Siegophiura sladeni* Duncan, 1879, have led to clogging of fishing nets and underwater facilities, causing damage to fishing gear, reducing fishing efficiency, and resulting in economic loss [155]. This has further impacted the stability of benthic community structure and consequently affected the number and distribution of fishery resources.

Moreover, the alterations in the Yellow Sea Cold Water Mass impact the Ophiuroidea species, leading to their formation of high-density populations in the southern waters of the Zhangzi Island area during the summer season (Figure 4d). Among them, the predominant species, *O. sarsii*, preys on bivalve organisms such as *M. yessoensis* larvae [175], while *O. mirabilis* primarily feeds on suspended matter in the water, competing for food resources with *M. yessoensis* in its cultivated area [155]. In conclusion, the increase in Ophiuroidea species not only disrupts the stability of the benthic community but also poses a threat to economically significant species and impacts sustainable fishery resource development.

6.3. The Biological Background of Acaudina molpadioides Semper, 1867, and the Marine Ecological Disasters It Causes in China

Acaudina molpadioides (Figure 4e), a member of the phylum Echinodermata, class Holothuroidea, order Molpadida, and family Caudinidae [176,177], is closely related to the genus *Pseudostichopus* and shows some affinity with species in the suborder Dendrochirotida but not particularly close [178]. Widely distributed in subtropical waters of Bangladesh, Australia, Japan, the Philippines, and Indonesia [179–181], it mainly inhabits coastal areas of Hainan, Guangdong, Fujian, and Zhejiang Province in China [182,183]. *Acaudina molpadioides* commonly resides in burrows from intertidal zones to a depth of 80 m [184] and typically feeds on sediments beneath clayey silt layers. It is often found in deep water areas with higher organic carbon content in sediments [183], higher seawater salinity, and a lower temperature [182]. As a typical benthic organism, *A. molpadioides* excavates and consumes organic sediments to facilitate nutrient cycling within the detrital food chain. This profoundly alters sediment distribution and composition while playing an important role in modifying sediment structure and maintaining marine ecosystem stability.

The outbreak of *A. molpadioides* primarily occurred in the vicinity of nuclear power plants in China. However, *A. molpadioides* possesses high edible and medicinal value, as its body wall contains up to 89.16% crude protein [185], with a fat content of only 0.03% and almost no cholesterol [185]. Additionally, it is rich in mineral elements such as calcium, iron, zinc, and selenium [186–188], providing essential nutrients for the human body to enhance immunity and maintain good health. The polypeptide components of *A. molpadioides* exhibit antioxidant, anti-inflammatory, and hypoglycemic effects [185,186,189], effectively inhibiting α-amylase and Dipeptidyl peptidase-4 activity while increasing insulin secretion and reducing blood pressure by inhibiting Angiotensin-Converting Enzyme [190]. Various effective components in *A. molpadioides* can be used in medical drugs: AMC-2 improves non-alcoholic fatty liver disease by inhibiting Stearoyl-CoA Desaturase activity and damaging the synthesis of monounsaturated fatty acids in the liver [191]; collagen peptide has anti-infection and anti-tumor [192], physiological regulation functions along with hypotensive physiological functions that reduce oxidative stress to prevent acute liver injury [193]; sulfated polysaccharides have anti-inflammatory properties along with liver protection and antioxidant biological activities; saponins possess pharmacological functions including anti-tumor effects as well as antifungal properties alongside liver protection [186,189]. These active components indicate great potential for the development of *A. molpadioides* in both the food industry and medical fields.

Due to the ongoing deterioration of the global marine ecological environment and the escalating eutrophication in nearshore waters, there has been a proliferation and explosive aggregation of *A. molpadioides*. This phenomenon not only disrupts the stability of the benthic marine ecosystem but also poses a significant threat to the safe operation of coastal nuclear power plants. In August 2015, the Fujian Ningde nuclear power plant in China was forced to shut down due to blockage of the cooling system's intake screen by *A. molpadioides*, causing a trip in the seawater circulation pump and triggering an emergency shutdown of the nuclear reactor. This directly impacted the safe operation of the nuclear power plant and posed a serious threat to nearby residents' lives [176,182,184, 194]. In Guangdong Province's Daya Bay Navigation Channel area and southern region, *A. molpadioides* has formed a high-density population and is dominant for most of the year, with peak biomass reaching 170 g/m^2. Assessments using the *grappler method risk index* have revealed that there is a 41.45% risk of blockage at Daya Bay Nuclear Power Plant caused by *A. molpadioides*, posing a safety threat to its stable operation [174,182]. Furthermore, investigations into large-scale benthic resources in the northern East China Sea indicate that *A. molpadioides* has consistently had the highest total biomass across all sampling areas and years. Excessive presence of *A. molpadioides* in Sanmen Bay has resulted in lower values for secondary production and production-to-biomass ratio (P/B), making it challenging to restore diversity and stability within this marine benthic ecosystem region [195].

The sudden and explosive aggregation of *A. molpadioides* in the sea area is the result of a complex interplay of factors. *Acaudina molpadioides* exhibits strong reproductive ability, with each female individual producing tens of thousands of egg cells. The environmental conditions, including the port pool and intake of the nuclear power plant, provide suitable habitat for *A. molpadioides* [184]. Under the influence of typhoons and other extreme weather conditions, as well as sea waves, seabed sediments are continuously stirred and accumulated, causing *A. molpadioides* larvae to enter the water body in a suspended state. Adult *A. molpadioides* migrate to shallow water areas to avoid clogged burrows. The weak adhesion ability of their body wall results in large quantities entering the water intake net of the nuclear power plant, forming high-density aggregation near the intake [182]. Furthermore, climate change will impact the reproductive patterns of sea cucumbers, with rising water temperatures promoting embryonic development. Additionally, eutrophication in the sea will lead to increased proliferation of planktonic plants, potentially forming harmful algal blooms. The subsequent sinking of dead organic debris from these plants will provide a substantial food source for *A. molpadioides* [183]. Moreover, sediment deposition and the erosion-reducing effects of nuclear power plant intake pipes and breakwaters may result in local sedimentation, further reducing competition for *A. molpadioides* in their habitat and benefiting their survival and reproduction [193]. It is important to note that due to their burrowing habits, limited mobility, large size, and uneven distribution, it is challenging to detect *A. molpadioides* in the ocean. This presents difficulties in monitoring and reducing the biomass of them during marine surveys [174,182].

Figure 4. In 2018, an outbreak of *Acanthaster planci* Linnaeus, 1758, occurred in the sea area of Xisha Islands [140] (**a**) (Reproduced with permission; Copyright 2021; Science China Press); in 2021, an outbreak of *Asterina pectinifera* Muller and Troschel, 1842 took place in the sea area near Laopian Island, Dalian [126] (**b**); on June 22, 2019, the Ophiuroidea species was collected in the East China Sea (**c**); in May 2017, the population of Ophiuroidea species was observed in the sea area of north Yellow Sea [173]; (**d**) (Reproduced with permission; Copyright © 2019; Chinese Society for Oceanology and Limnology; Science Press and Springer-Verlag GmbH Germany; part of Springer Natures); and *Acaudina molpadioides* Semper, 1867 was collected from the East China Sea in 2018 [196] (**e**). The white dashed arrow indicates the Ophiuroidea species collected from the seafloor sediments.

7. Discussion

7.1. Short-Term Outbreaks of Marine Biological Disasters in China Are Anticipated to Persist

The impact of climate change on marine ecosystems is profound and multifaceted, resulting in elevated seawater temperatures and CO_2 concentrations, ocean acidification, sea level rise, altered precipitation levels and oceanic structure, as well as heightened frequency of extreme weather events [197]. This impact directly influences the physiological processes and growth dynamics of marine organisms, shifts species distribution and community composition, alters interactions among species, and significantly impacts the biodiversity pattern in marine environments, posing a threat to the survival and stability of numerous species [198]. Moreover, the escalation of coastal resource development, global trade activities, and shipping operations has facilitated the introduction of non-native species through ballast water discharge [199]. This poses a risk to indigenous marine biological species while potentially reducing natural predators for explosively proliferating marine organisms. Concurrently, long-term challenges from widespread sources make marine pollution difficult to fully control [200,201]. Upon entering the oceanic environment, pollutants degrade water quality and substrate habitats, leading to diminished populations or even the extinction of sensitive species. Conversely resilient species with rapid reproduction rates can thrive in polluted settings, leading to dominance by a single species, which increases the likelihood of explosive population growth within marine biota, thereby exacerbating ecosystem instability. Consequently, it is anticipated that short-term occurrences of ecological disasters will persist within Chinese maritime territories.

Nuclear power plants have played a crucial role in meeting human energy demands, reducing reliance on fossil fuels, and mitigating greenhouse gas emissions [202,203]. However, the thermal discharges from their cooling systems have resulted in elevated water temperatures in the vicinity, creating optimal breeding conditions for certain organisms such as jellyfish and starfish. Furthermore, the disturbance of sediments around nuclear power plants has generated suitable habitats for specific benthic organisms like *A. molpadioides*, potentially leading to an upsurge in their population and disrupting ecological equilibrium. Due to the presence of protective barriers and fishing restrictions in the vicinity of nuclear power plants, certain species may experience reduced predation pressure in these areas, potentially leading to localized population surges that could impact the safe operation of the power plant. The sea areas adjacent to nuclear power plants are frequently challenging to obtain approval for conducting fishery fishing activities. Simultaneously, because of the biological enrichment effect in the vicinity of nuclear power plants, the nuclear radiation content in organisms might be relatively high, which will give rise to public apprehensions regarding the safety of related organisms in food and drug development. Consequently, the economic value derived from the outbreak organisms is considerably lower than the substantial economic losses resulting from the biological clogging accident in nuclear power plants.

China is a significant player in the development and operation of nuclear power plants [204]. Presently, there are 18 operational nuclear power plants in China, with a total of 58 generating units. According to geographical distribution, there are seven nuclear power plants located in the South China Sea, seven in the East China Sea, three in the Yellow Sea, and one in the Bohai Sea (Table 1). According to administrative divisions, Fujian Province hosts two nuclear power plants, Guangdong Province hosts four nuclear power plants, Guangxi Zhuang Autonomous Region hosts one nuclear power plant, Hainan Province hosts one nuclear power plant, Jiangsu Province hosts one nuclear power plant, Liaoning Province hosts one nuclear power plant, and Shandong Province hosts one nuclear power plant. Additionally, Taiwan Province has one nuclear power plant, and Zhejiang Province has five nuclear power plants (Table 1). Currently, at least five nuclear power plants nearby regions in China have experienced marine animal outbreaks [45,46,111,113–115,174,176,182,184,194], including Hongyanhe Nuclear Power Plant, Ling'ao Nuclear Power Plant, Daya Bay Nuclear Power Plant, Yangjiang Nuclear Power Plant, and Ningde Nuclear Power Plant, representing approximately 28% of China's nuclear power facilities (Table 1). Nuclear power plants have the potential to replace some of the current mainstream thermal power generation, which is crucial for reducing carbon emissions. However, in recent years, marine fauna disasters have frequently occurred in nuclear power plant regions. This not only causes significant damage to the ecological environment but also has adverse effects on fisheries, tourism, public health, and industrial facilities and leads to public panic [20,21,36]. Therefore, it is imperative to urgently explore effective methods and management strategies for controlling the total biomass of explosion species of marine fauna near nuclear power plants to ensure their sustainable operation and harmonious coexistence with ecological systems.

Table 1. A list of the sea areas where China's nuclear power plants are located and whether there have been outbreaks of marine faunas.

The Administrative Region Where the Nuclear Power Plant Is Located	The Name of Nuclear Power Plant	The Sea Area Where the Nuclear Power Plant Is Located	The Number of Generating Units in the Nuclear Power Plant	Whether a Biological Outbreak Disaster Exists
Fujian Province	Ningde Nuclear Power Plant	The East China Sea	4	√
Fujian Province	Fuqing Nuclear Power Plant	The East China Sea	6	-
Guangdong Province	Daya Bay Nuclear Power Plant	The South China Sea	2	√
Guangdong Province	Ling'ao Nuclear Power Plant	The South China Sea	4	√
Guangdong Province	Yangjiang Nuclear Power Plant	The South China Sea	6	√
Guangdong Province	Taishan Nuclear Power Plant	The South China Sea	2	-
Guangxi Zhuang Autonomous Region	Fangchenggang Nuclear Power Plant	The South China Sea	4	-
Hainan Province	Changjiang Nuclear Power Plant	The South China Sea	2	-
Jiangsu Province	Tianwan Nuclear Power Plant	The Yellow Sea	6	-
Liaoning Province	Hongyanhe Nuclear Power Plant	The Bohai Sea	6	√
Shandong Province	Haiyang Nuclear Power Plant	The Yellow Sea	2	-
Shandong Province	Shidao Bay Nuclear Power Plant	The Yellow Sea	1	-
Taiwan Province	The Third Nuclear Power Plant	The South China Sea	2	-
Zhejiang Province	Fangjiashan Nuclear Power Station	The East China Sea	2	-
Zhejiang Province	Sanmen Nuclear Power Plant	The East China Sea	2	-
Zhejiang Province	Qinshan Nuclear Power Plant	The East China Sea	1	-
Zhejiang Province	Qinshan Nuclear Power Plant Phase II	The East China Sea	4	-
Zhejiang Province	Qinshan Nuclear Power Plant Phase III	The East China Sea	2	-

Note: The statistical time node of the nuclear power plant that is currently operating is March 2024, and the statistical scope includes the Chinese Mainland (https://www.china-nea.cn/site/content/44813.html, access date: 15 July 2024), as well as Hong Kong Special Administrative Region, Macao Special Administrative Region, and Taiwan Province. Among them, '√' indicates the occurrence of a marine organism outbreak at the nuclear power plant, while '-' represents the current absence of an outbreak, but it does not imply whether there will be one in the future.

In addition, typhoons and other natural disasters frequently transport marine organisms from their usual offshore habitats to the shore, leading to the phenomenon of marine organisms congregating in coastal areas [182]. Strictly speaking, this may not constitute a mass proliferation-induced outbreak. It is essential to conduct on-site surveys and assessments based on actual conditions to accurately differentiate the outbreak attributes. If biological species are driven towards the shore by external forces and form an aggregation phenomenon, this process represents only a temporary alteration in their distribution and does not signify an increase in population. Classifying it as a biological distribution anomaly caused by natural disasters is more appropriate. Accurately distinguishing between natural disaster-induced biological distribution anomalies and biological mass proliferation outbreaks is crucial for marine ecological research and management. The former typically has short-term effects with relatively limited long-term impact on ecological systems, while the latter may indicate underlying issues within ecological systems that require corresponding management measures for resolution. Although natural disaster-induced biological distribution anomalies do not qualify as biological outbreaks, they still warrant serious attention. These aggregated organisms may intensify local predation competition and temporarily exert pressure on coastal ecosystems. Furthermore, they serve as a reminder to further strengthen relevant research and monitoring efforts while enhancing our understanding of these complex ecological phenomena alongside acknowledging the profound impact of extreme weather events on marine ecosystems.

7.2. The Selected and Implemented Control Measures Should Aim to Minimize Disruption to Marine Ecosystems

The complexity of marine ecosystems and the dynamic changes in the environment pose challenges for predicting marine biological outbreaks. While monitoring and early warning technologies have developed, real-time coverage still has limitations, and the implementation of preventive measures is constrained by technical, economic, and policy factors. Consequently, the primary response strategy currently involves post-disaster management, primarily encompassing physical control, chemical control, and biological control. Presently, physical salvage stands as the predominant method of control; for instance, in 2008 off the coast of Weihai, 20–50 tons of *A. aurita* were salvaged [36]; in 2019 near the intake of Daya Bay Nuclear Power Plant [113], over 20 tons of *A. chinensis* were captured; while in a single day in 2021 at Jiaozhou Bay's bottom-culture area [126], starfish weighing up to 50 tons were caught.

Fortunately, chemical control methods are seldom employed due to potential adverse effects such as water quality deterioration and non-target species damage. These methods can lead to water pollution and substrate environment destruction, which complicates their efficacy while impacting fishery resources. Similarly infrequent are regional biological control methods that require complex evaluations considering ecological relationships between target species and introduced ones along with adaptability assessments for potential risks associated with spread within local ecological systems. Furthermore, long-term environmental impact assessments are essential to prevent ecosystem imbalances resulting from introducing biological controls that could trigger outbreaks among other or introduced species. The application of biological controls necessitates stringent regulation and monitoring, ensuring sustainability while averting new ecological threats. Physical control methods generally offer greater environmental friendliness, albeit progress being slow with high labor costs, but minimizing benthic ecosystem disturbance results in relatively minor impacts on marine ecosystems overall healthiness. By reducing disturbances in benthic environments, nutrient release can be curtailed, thus preventing eutrophication occurrences alongside harmful algal bloom outbreaks, thereby safeguarding benthic organisms' well-being as well as overall ecosystem integrity. Combining long-term monitoring scientific management technological innovation like developing more efficient fishing tools and improving operational methodologies will further optimize physical controls, enhancing efficiency and sustainability.

7.3. Effective Resource Utilization of Outbreak Species Is a Long-Term Solution

When marine organisms become the single or dominant species in a particular area and significantly impact human production and living activities, they are often classified as having the potential for a biological outbreak. From an objective standpoint, these outbreak marine organisms also represent high-quality biological resources, possessing both known and yet-to-be-developed edible, medicinal, and industrial values. It is imperative to develop effective pathways for utilizing these outbreak marine organisms (Figure 5); this approach can not only prevent the loss of biological resources but also rapidly reduce the biomass of marine organism waste while effectively controlling the scale of the outbreak in affected areas, thereby fostering advancements in the fishery industry [205,206]. Although the utilization of resources from outbreak organisms represents an approach to converting 'disaster' into 'resource', this utilization method is challenging to compensate for the extensive losses they inflict on the ecosystem and socio-economic domains. Hence, it can merely function as an ancillary method and cannot substitute for more fundamental ecological management and disaster prevention and control measures.

Figure 5. Due to their culinary appeal and distinctive flavor, starfish are frequently utilized as accompaniments in Chinese hotpot cuisine (**a**); some species of jellyfish have edible oral arms, which are commonly prepared as cold shredded jellyfish in China (**b**).

In response to the outbreaks of Cnidaria (jellyfish organisms), Annelida (*U. unicinctus*), Mollusca (*P. kinglipini*), Arthropoda (*A. chinensis*), and Echinodermata (Asteroidea organisms, Ophiuroidea organisms, and *A. molpadioides*) in China, three key issues must be addressed to further advance the sustainable utilization of these organisms. Firstly, it is imperative to establish effective storage methods for the collected or harvested outbreak species, necessitating a series of experimental simulations in product development, preservation techniques, drying processes, and crude product production. Additionally, formulation of execution standards for product production (e.g., food products) is essential. Secondly, there are limitations in the existing forms of products derived from outbreak species and an inadequate research foundation for certain outbreak species, such as *P. kinglipini*; their biological value remains incompletely explored. Thirdly, products derived from outbreak species require a comprehensive food supply chain to achieve long-term stability and development. This entails integration across upstream and downstream industries; only through the formation of stable market demand can resourceful utilization of outbreak species cease to be an issue.

7.4. Enhancing and Releasing Activities Require Careful Selection of Species to Be Released

At present, the implementation of aquatic organism enhancement and release activities necessitates prudence. Uninformed release practices may detrimentally impact the ecological environment. The execution of marine enhancement and release activities demands a robust scientific foundation and fundamental research. The Regulations on Aquatic Organism Enhancement and Release issued by the Ministry of Agriculture and Rural Affairs of the People's Republic of China explicitly prohibit the utilization of foreign species, hybrid species, genetically modified species, or other species that do not align with ecological requirements for enhancement and release activities. It is imperative to conduct enhancement and release activities for endangered species, nationally protected wild animals, or significantly depleted fishery resources such as *Larimichthys crocea* Richardson, 1846, *Acipenser sinensis* Gray, 1835, *Trachidermus fasciatus* Heckel, 1837, *Hippocampus kelloggi* Jordan and Snyder, 1901, *Tridacna crocea* Lamarck, 1819, *Branchiostoma belcheri* Gray, 1847, and *Tachypleus tridentatus* Leach, 1819, in order to restore their population size and genetic diversity levels [207–209]. Nevertheless, detailed guidelines for enhancing and releasing economically important species are currently lacking.

Humans often prioritize the economic and ecological value of a select few species, neglecting the interconnectedness and holistic nature of ecological systems. For instance, the ongoing outbreak of golden tides in China caused by *Sargassum horneri* (Turner) C. Agardh, 1820, serves as a pertinent example [6,210]. In recent years, local governments have introduced a substantial number of *S. horneri* seedlings into nearshore waters in Zhejiang Province and Shandong Province to establish marine pastures. The precise origin of the current golden tide outbreak in China remains unknown, necessitating further research to ascertain whether the introduction of *S. horneri* has significantly contributed to this phenomenon. It is imperative to consider that artificially promoting the dominance of some particular species within a localized area may not represent an optimal solution for marine ecological restoration. The process of enhancing and releasing marine fauna should also carefully evaluate its potential impact on disrupting marine ecological balance. Notably, there has been a sudden proliferation of *U. unicinctus* along beaches in Yantai City, Shandong Province. This surge can be attributed to continuous releases by local government breeding programs involving numerous indigenous marine species, including *U. unicinctus* larvas; it is important to recognize that such activities may inadvertently facilitate the proliferation of ecologically adaptable species like *U. unicinctus* and disrupt the original ecological equilibrium.

In summary, the selection of species for enhancement and release activities should be conducted with careful consideration. It is imperative for scientists to prioritize understanding the holistic nature of the ocean and its inherent capacity for self-repair, particularly focusing on mitigating marine environmental pollution.

7.5. Further Enhance Fundamental Research to Ensure the Sustainable Development of Offshore Ecosystems

The frequent occurrences of marine ecological disasters present a significant threat to the sustainable development of marine ecosystems. Ensuring the sustainability of marine ecosystems is essential for maintaining ecological equilibrium, conserving biodiversity, fostering economic activities, and mitigating the impact of disasters. Various types of marine ecological disasters actually provide insights into the evolutionary processes shaping the structure and function of marine ecosystems and may even indicate changes in their structure.

To effectively address marine biological outbreaks and ensure the sustainable development of nearshore ecosystems, a comprehensive approach is essential to bolster fundamental research. This entails conducting systematic studies of marine ecosystems and their components to comprehend ecological behaviors, population dynamics, and relationships with environmental factors in order to uncover key triggers of biological outbreaks and make predictive assessments. Furthermore, fostering interdisciplinary collaboration to integrate knowledge from diverse fields such as marine biology, ecology, climatology, and environmental science will enhance comprehension of intricate ecosystems and facilitate the development of effective management strategies. Additionally, advancing new technologies and methodologies, including high-resolution remote sensing technology, genomics, and data analysis tools, will augment the capacity for ecosystem study while aiding scientists in identifying emerging ecological threats and opportunities. Building upon these studies supported by advanced technologies enables the reinforcement of comprehensive integrated research at an ecosystem level that can forecast future trends, thereby providing a scientific foundation and technical backing for the sustainable development of China's nearshore ecosystems and ultimately promoting robust functioning of marine ecosystems while effectively addressing potential future marine biological disasters.

8. Conclusions

This review focuses on the current status of marine biological disasters, the fundamental biological background, and the causes of these outbreaks in China's marine ecosystem. These outbreak species not only seriously damage fishery resources, tourism, and coastal industries but also pose a significant threat to the stability of the marine ecosystem, especially in sensitive areas such as nuclear power plants and mariculture areas. Existing studies show that marine biological outbreaks are the result of the combined effects of multiple factors. In this regard, we can further strengthen basic research, carry out resource utilization of outbreak organisms, formulate more scientific strategies for enhancement and release activities or mariculture, and adopt prevention and control measures with less damage to the marine ecology. However, the existing prevention and control strategies mostly focus on post-outbreak treatment, and future research and management should pay more attention to the early monitoring and prevention of disasters. Currently, the marine ecological disaster monitoring system in China only covers green tides and red tides with no inclusion of outbreaks of marine organisms such as Cnidaria (jellyfish organisms), Annelida (*U. unicinctus*), Mollusca (*P. kinglipini*), Arthropoda (*A. chinensis*), and Echinodermata (Asteroidea organisms, Ophiuroidea organisms, and *A. molpadioides*) within the standardized regulatory scope. It is recommended that more severe marine fauna disasters be gradually incorporated into China's regulatory system for marine disasters based on observed trends in marine ecosystem changes and potential threats posed by outbreaks of marine fauna to the environment. This approach will bolster ecological protection efforts, mitigate economic losses, safeguard public health, and advance scientific research.

Author Contributions: Conceptualization, J.L. (Jinlin Liu); Methodology, J.L. (Jinlin Liu); Software, J.L. (Jinlin Liu); Validation, P.H., L.Y., Z.X. and J.L. (Jinlin Liu); Formal Analysis, J.L. (Jinlin Liu) and J.L. (Jiye Li); Investigation, J.L. (Jinlin Liu) and J.L. (Jiye Li); Resources, P.H., L.Y., Z.X., J.L. (Jinlin Liu) and J.L. (Jiye Li); Data Curation, P.H. and J.L. (Jinlin Liu); Writing—Original Draft Preparation, P.H., L.Y., Z.X., J.L. (Jinlin Liu), and J.L. (Jiye Li); Writing—Review and Editing, P.H., L.Y., Z.X., J.L. (Jinlin Liu), and J.L. (Jiye Li); Visualization, P.H., L.Y., Z.X., J.L. (Jinlin Liu), and J.L. (Jiye Li); Supervision, P.H. and J.L. (Jinlin Liu); Project Administration, P.H. and J.L. (Jinlin Liu); Funding Acquisition, P.H. and J.L. (Jinlin Liu). All authors have read and agreed to the published version of the manuscript.

Funding: This work was encouraged by the National Key Research & Development Program of China (Grant Nos. 2022YFC3106004 and 2022YFC3106001).

Acknowledgments: Jinlin Liu expresses his gratitude to Qunhui Yang (Project Management Office of China National Scientific Seafloor Observatory, Tongji University) for providing ample time for him to organize his previous research data and write this manuscript and offers his highest respect to Yang. Also, Jinlin Liu thanks the North China Sea Environmental Monitoring Center of the State Oceanic Administration for their long-term cooperation and support. Without their support, Liu believes that he would not have achieved his current academic and research advancements.

Conflicts of Interest: The authors declare no conflicts of interest.

References

1. Xu, C.; Yang, Y.; Zhang, F.; Li, R.; Li, Z.; Wang, Y.P.; Jia, J. Spatial-temporal distribution of tropical cyclone activity on the eastern sea area of China since the late 1940s. *Estuar. Coast. Shelf Sci.* **2022**, *277*, 108067. [CrossRef]
2. Rabinovich, A.B.; Titov, V.V.; Moore, C.W.; Eblé, M.C. The 2004 Sumatra tsunami in the Southeastern Pacific Ocean: New Global Insight from Observations and Modeling. *J. Geophys. Res. Ocean.* **2017**, *122*, 7992–8019. [CrossRef]
3. Lahcene, E.; Ioannou, I.; Suppasri, A.; Pakoksung, K.; Paulik, R.; Syamsidik, S.; Bouchette, F.; Imamura, F. Characteristics of building fragility curves for seismic and non-seismic tsunamis: Case studies of the 2018 Sunda Strait, 2018 Sulawesi-Palu, and 2004 Indian Ocean tsunamis. *Nat. Hazards Earth Syst. Sci.* **2021**, *21*, 2313–2344. [CrossRef]

4. Manda, A.; Nakamura, H.; Asano, N.; Iizuka, S.; Miyama, T.; Moteki, Q.; Yoshioka, M.K.; Nishii, K.; Miyasaka, T. Impacts of a warming marginal sea on torrential rainfall organized under the Asian summer monsoon. *Sci. Rep.* **2014**, *4*, 5741. [CrossRef] [PubMed]
5. Blunden, J. State of the climate in 2021. *Bull. Am. Meteorol. Soc.* **2022**, *103*, S1–S465. [CrossRef]
6. Liu, J.; Xia, J.; Zhuang, M.; Zhang, J.; Sun, Y.; Tong, Y.; Zhao, S.; He, P. Golden seaweed tides accumulated in *Pyropia* aquaculture areas are becoming a normal phenomenon in the Yellow Sea of China. *Sci. Total Environ.* **2021**, *774*, 145726. [CrossRef]
7. Xia, Z.; Liu, J.; Zhao, S.; Sun, Y.; Cui, Q.; Wu, L.; Gao, S.; Zhang, J.; He, P. Review of the development of the green tide and the process of control in the southern Yellow Sea in 2022. *Estuar. Coast. Shelf Sci.* **2024**, *302*, 108772. [CrossRef]
8. Chen, W.; Liu, W.; Liang, H.; Jiang, M.; Dai, Z. Response of storm surge and M2 tide to typhoon speeds along coastal Zhejiang Province. *Ocean Eng.* **2023**, *270*, 113646.
9. Mori, N.; Takahashi, T.; Yasuda, T.; Yanagisawa, H. Survey of 2011 Tohoku earthquake tsunami inundation and run-up. *Geophys. Res. Lett.* **2011**, *38*, L00G14. [CrossRef]
10. Dade, M.; Downing, A.; Benessaiah, K.; Falardeau, M.; Lin, M.; Rieb, J.; Rocha, J. Inequalities in the adaptive cycle: Reorganizing after disasters in an unequal world. *Ecol. Soc.* **2022**, *27*, 10. [CrossRef]
11. Yuan, S.; Li, Y.; Bao, F.; Xu, H.; Yang, Y.; Yan, Q.; Zhong, S.; Yin, H.; Xu, J.; Huang, Z.; et al. Marine environmental monitoring with unmanned vehicle platforms: Present applications and future prospects. *Sci. Total Environ.* **2023**, *858*, 159741. [CrossRef] [PubMed]
12. Zheng, J.; Xu, W.; Tao, A.; Fan, J.; Xing, J.; Wang, G. Synergy between coastal ecology and disaster mitigation in China: Policies, practices, and prospects. *Ocean Coast. Manag.* **2023**, *245*, 106866. [CrossRef]
13. Ministry of Natural Resources. *China China Maritime Disaster Bulletin 2023*; Ministry of Natural Resources of the People's Republic of China: Beijing, China, 2024. (In Chinese)
14. Sun, Y.; Yao, L.; Liu, J.; Tong, Y.; Xia, J.; Zhao, X.; Zhao, S.; Fu, M.; Zhuang, M.; He, P.; et al. Prevention strategies for green tides at source in the Southern Yellow Sea. *Mar. Pollut. Bull.* **2022**, *178*, 113646. [CrossRef]
15. Yu, Z.; Song, X.; Cao, X.; Liu, Y. Mitigation of harmful algal blooms using modified clays: Theory, mechanisms, and applications. *Harmful Algae* **2017**, *69*, 48–64. [CrossRef]
16. Xia, Z.; Liu, J.; Zhao, S.; Cui, Q.; Bi, F.; Zhang, J.; He, P. Attached *Ulva meridionalis* on nearshore dikes may pose a new ecological risk in the Yellow Sea. *Environ. Pollut.* **2023**, *332*, 121969. [CrossRef]
17. Zhou, M.J.; Shen, Z.L.; Yu, R.C. Responses of a coastal phytoplankton community to increased nutrient input from the Changjiang (Yangtze) River. *Cont. Shelf Res.* **2008**, *28*, 1483–1489. [CrossRef]
18. Liu, J.; Xia, J.; Zhuang, M.; Zhang, J.; Yu, K.; Zhao, S.; Sun, Y.; Tong, Y.; Xia, L.; Qin, Y.; et al. Controlling the source of green tides in the Yellow Sea: NaClO treatment of *Ulva* attached on *Pyropia* aquaculture rafts. *Aquaculture* **2021**, *535*, 736378. [CrossRef]
19. Gao, Y.; Hao, P.; Wei, Z.; Li, S.; Song, J.; Yu, C. Dynamic causes contribute to the increasing trend of red tides in the east China sea during 2020–2022. *Mar. Environ. Res.* **2024**, *198*, 106521. [CrossRef]
20. Wei, Z.; Zhang, X.; Chen, Y.; Liu, H.; Wang, S.; Zhang, M.; Ma, H.; Yu, K.; Wang, L. A new strategy based on a cascade amplification strategy biosensor for on-site eDNA detection and outbreak warning of crown-of-thorns starfish. *Sci. Total Environ.* **2024**, *927*, 172258. [CrossRef]
21. Feng, S.; Sun, S.; Li, C.; Zhang, F. Controls of *Aurelia coerulea* and *Nemopilema nomurai* (Cnidaria: Scyphozoa) blooms in the coastal sea of China: Strategies and measures. *Front. Mar. Sci.* **2022**, *9*, 946830. [CrossRef]
22. An, L.N.; Wang, L.; Ou, D.Y.; Jia, C.; Li, W.W.; Ding, Y.; You, C.M.; Liao, J.J.; Huang, H. The ecological mechanisms of *Acetes* blooms as a threat to the security of cooling systems in coastal nuclear power plants. *J. Coast. Conserv.* **2021**, *25*, 55. [CrossRef]
23. Duan, Z.L.; Yu, Y.N.; Wang, J.; Zhang, J.L.; Chen, N.S. Morphological and molecular analyses of a *Philine kinglipini* outbreak in summer of 2022 in Jiaozhou Bay, China. *J. Oceanol. Limnol.* **2024**, *42*, 831–848. [CrossRef]
24. Ding, D.W.; He, P.M.; Ye, S.F.; Chen, B.; Huang, L.F.; Wang, H.L.; Yang, J.L.; He, W.H.; Sun, S.; Yang, H.S.; et al. *Coastal Marine Ecology in China: Research and Management*; Science Press: Beijing, China, 2022. (In Chinese)
25. Wang, P.P. *The Diet and Trophic Relationship of Three Disaster Causing Jellyfish in Chinese Coastal Waters*; Institute of Oceanology of the Chinese Academy of Sciences: Qingdao, China, 2020. (In Chinese)
26. Sun, S. New perception of jellyfish bloom in the East China Sea and Yellow Sea. *Oceanol. Et Limnol. Sin.* **2012**, *43*, 406–410. (In Chinese)
27. Sun, S.; Sun, X.X.; Jenkinson, I.R. Preface: Giant jellyfish blooms in Chinese waters. *Hydrobiologia* **2015**, *754*, 1–11. [CrossRef]
28. Du, C. *Molecular Identification and Collagen Activity of Twojellyfish in the Northern South China Sea*; Yantai University: Yantai, China, 2023. (In Chinese)
29. Deng, P.P.; Shi, Y.H.; Xu, J.B.; Liu, Y.S.; Jiang, F. Influence factors and harmfulness of water jellyfish outbreak for larval rearing ponds in Hangzhou Bay. *Fish. Sci. Technol. Inf.* **2018**, *45*, 280–284. (In Chinese)
30. Tang, C.S. *Study on Biological Control of Larval Stage of Giant Jellyfish*; Institute of Oceanology of the Chinese Academy of Sciences: Qingdao, China, 2020. (In Chinese)
31. Qu, C.F.; Song, J.M.; Li, N. The effects of jellyfish decomposition on marine ecological environments. *Acta Ecol. Sin.* **2015**, *35*, 6224–6232. (In Chinese)
32. Chiarelli, P.G.; Pegg, R.B.; Kumar, G.D.; Solval, K.M. Exploring the feasibility of developing novel gelatin powders from salted, dried cannonball jellyfish (*Stomolophus meleagris*). *Food Biosci.* **2021**, *44*, 101397. [CrossRef]

33. Masuda, A.; Baba, T.; Dohmae, N.; Yamamura, M.; Wada, H.; Ushida, K. Mucin (Qniumucin), a glycoprotein from jellyfish, and determination of its main chain structure. *J. Nat. Prod.* **2007**, *70*, 1089–1092. [CrossRef]
34. Sugahara, T.; Ueno, M.; Goto, Y.; Shiraishi, R.; Doi, M.; Akiyama, K.; Yamauchi, S. Immunostimulation Effect of the Jellyfish Collagen. *Biosci. Biotechnol. Biochem.* **2006**, *70*, 2131–2137. [CrossRef]
35. Putra, A.B.N.; Nishi, K.; Shiraishi, R.; Doi, M.; Sugahara, T. Jellyfish collagen stimulates maturation of mouse bone marrow-derived dendritic cells. *J. Funct. Foods* **2015**, *14*, 308–317. [CrossRef]
36. Dong, Z.J.; Liu, D.Y.; Keesing, J.K. Jellyfish blooms in China: Dominant species, causes and consequences. *Mar. Pollut. Bull.* **2010**, *60*, 954–963. [CrossRef] [PubMed]
37. Wang, X.C.; Jin, Q.Q.; Yang, L.; Jia, C.; Guan, C.J.; Wang, H.N.; Guo, H. Aggregation process of two disaster-causing jellyfish species, *Nemopilema nomurai* and *Aurelia coerulea*, at the intake area of a nuclear power cooling-water system in Eastern Liaodong Bay, China. *Front. Mar. Sci.* **2023**, *9*, 1098232. [CrossRef]
38. Feng, S.; Wang, S.W.; Zhang, G.T.; Sun, S.; Zhang, F. Selective suppression of in situ proliferation of scyphozoan polyps by biofouling. *Mar. Pollut. Bull.* **2017**, *114*, 1046–1056. [CrossRef] [PubMed]
39. Palmieri, M.G.; Barausse, A.; Luisetti, T.; Turner, K. Jellyfish blooms in the Northern Adriatic Sea: Fishermen's perceptions and economic impacts on fisheries. *Fish. Res.* **2014**, *155*, 51–58. [CrossRef]
40. Fuentes, V.L.; Angel, D.L.; Bayha, K.M.; Atienza, D.; Edelist, D.; Bordehore, C.; Gili, J.M.; Purcell, J.E. Blooms of the invasive ctenophore, span the Mediterranean Sea in 2009. *Hydrobiologia* **2010**, *645*, 23–37. [CrossRef]
41. Cruz-Rivera, E.; Abu El-Regal, M. A bloom of an edible scyphozoan jellyfish in the Red Sea. *Mar. Biodivers.* **2016**, *46*, 515–519. [CrossRef]
42. Keesing, J.K.; Gershwin, L.A.; Trew, T.; Strzelecki, J.; Bearham, D.; Liu, D.Y.; Wang, Y.Q.; Zeidler, W.; Onton, K.; Slawinski, D. Role of winds and tides in timing of beach strandings, occurrence, and significance of swarms of the jellyfish *Crambione mastigophora* Mass 1903 (Scyphozoa: Rhizostomeae: Catostylidae) in north-western Australia. *Hydrobiologia* **2016**, *768*, 19–36. [CrossRef]
43. Jiang, H.; Cheng, H.Q.; Xu, H.G.; Francisco, A.S.; Manuel, J.Z.R.; Pablo, D.M.L.; William, J.F.L.Q. Trophic controls of jellyfish blooms and links with fisheries in the East China Sea. *Ecol. Model.* **2008**, *212*, 492–503. [CrossRef]
44. Yuan, X.B.; Liu, Z.L.; Xue, L.Y.; Chen, X.; An, Y. Review on the occurrence regularity and current situation of jellyfish disaster in Qinhuangdao coastal area. *Hebei Fish.* **2021**, *6*, 12–17. (In Chinese)
45. Li, C. Review on the current situation and perspective of Chinese jellyfish disasters' research. *J. Fish. Res.* **2018**, *40*, 156–162. (In Chinese)
46. Zhang, C.W.; Guan, C.J.; Xu, P.; Liu, G.Z.; Xu, Q.M.; Ye, J.Q.; Wang, X.; Zhao, B.Q. Analysis on risk organisms for the cold source water of nuclear power plantin the eastern waters of Liaodong bay. *Mar. Environ. Sci.* **2019**, *38*, 41–45. (In Chinese)
47. Du, C.; He, J.; Sun, T.T.; Wang, L.; Wang, F.H.; Dong, Z.J. Molecular identification on the causative species jellyfish blooms in the northern South China Sea in 2019. *J. Trop. Oceanogr.* **2022**, *41*, 142–148. (In Chinese)
48. Liu, J.M.; Rao, Q.H. Analysis of the causes of jellyfish outbreaks in the nearshore waters of Qinhuangdao and countermeasures against them. *Hebei Fish.* **2016**, *1*, 55–57. (In Chinese)
49. Wei, Z.Z.; Ma, S.L.; Jin, B.W. Causes and effects of jellyfish outbreaks in the ocean. *Hebei Fish.* **2016**, *7*, 71–72. (In Chinese)
50. Tseng, L.C.; Chou, C.; Chen, Q.C.; Hwang, J.S. Jellyfish assemblages are related to interplay waters in the southern east China Sea. *Cont. Shelf Res.* **2015**, *103*, 33–44. [CrossRef]
51. Sun, M.; Dong, J.; Purcell, J.E.; Li, Y.L.; Duan, Y.; Wang, A.Y.; Wang, B. Testing the influence of previous-year temperature and food supply on development of *Nemopilema nomurai* blooms. *Hydrobiologia* **2015**, *754*, 85–96. [CrossRef]
52. Ye, L.J.; Peng, S.J.; Ma, Y.Q.; Zhang, W.J.; Wang, L.; Sun, X.Y.; Zhang, C.; Yeasmin, M.; Zhao, J.M.; Dong, Z.J. Biodiversity and distribution patterns of blooming jellyfish in the Bohai Sea revealed by eDNA metabarcoding. *BMC Ecol. Evol.* **2024**, *24*, 37. [CrossRef]
53. Dong, J.; Sun, M.; Purcell, J.E.; Chai, Y.; Zhao, Y.; Wang, A.Y. Effect of salinity and light intensity on somatic growth and podocyst production in polyps of the giant jellyfish *Nemopilema nomurai* (Scyphozoa: Rhizostomeae). *Hydrobiologia* **2015**, *754*, 75–83. [CrossRef]
54. Xu, Y.J.; Ishizaka, J.; Yamaguchi, H.; Siswanto, E.; Wang, S.Q. Relationships of interannual variability in SST and phytoplankton blooms with giant jellyfish (*Nemopilema nomurai*) outbreaks in the Yellow Sea and East China Sea. *J. Oceanogr.* **2013**, *69*, 511–526. [CrossRef]
55. Wu, J.W.; Kong, J.; Laws, E.A.; Liu, X.; Wang, C.Y.; Chen, J.X.; Chen, M.R.; Yao, Q.Z.; Wang, Y.T.; Zhen, Y.; et al. The Link between Marine Thermal Discharges and *Nemopilema Nomurai* Blooms Around Nuclear Power Plants. *Ecosyst. Health Sustain.* **2023**, *9*, 9. [CrossRef]
56. Zang, W.X.; Zhang, F.; Sun, Y.; Xu, Z.Q.; Sun, S. Benthic ecosystem determines jellyfish blooms by controlling the polyp colony development. *Mar. Pollut. Bull.* **2023**, *193*, 115232. [CrossRef] [PubMed]
57. Cheng, J.H.; Li, S.F.; Ding, F.Y.; Yan, L.P. Primary analysis on the jellyfish blooms and its cause in the East China Sea and the Yellow Sea. *Fish. Inf. Strategy* **2004**, *5*, 10–12. (In Chinese)
58. Wu, Z.G. *Phylogenetic Analyses of Complete Mitochondrial Genome of Urechis unicinctus (Echiura) Support That Eehiurans Are Derived Annelids*; Institute of Oceanology of the Chinese Academy of Sciences: Qingdao, China, 2009. (In Chinese)
59. Lv, H.C.; Li, B.J.; Jiao, X.D. Research progress in molecular biology of *Urechis unicinctus*. *Bull. Biol.* **2020**, *55*, 1–4. (In Chinese)

60. Zhang, S.S.; Li, M.; Sun, Y.; Shang, H.X.; Wang, L.; Yang, T.T.; Ma, L.; Chen, Y.; Zhang, B.; Liu, T.; et al. Analysis of the relationship between geography and body color with the genetic diversity in the Echiura worm *Urechis unicinctus* based on the mitochondrial COI and D-loop sequences. *Mitochondrial DNA Part B-Resour.* **2021**, *6*, 1380–1386. [CrossRef] [PubMed]
61. Chen, W.B.; Zhang, S.S.; Sun, Y.; Tian, B.; Song, L.J.; Xu, Y.; Liu, T. Effects of substrate on the physiological characteristics and intestinal microbiota of Echiura worm (*Urechis unicinctus*) juveniles. *Aquaculture* **2021**, *530*, 735710. [CrossRef]
62. Abe, H.; Sato-Okoshi, W.; Tanaka, M.; Okoshi, K.; Teramoto, W.; Kondoh, T.; Nishitani, G.; Endo, Y. Swimming behavior of the spoon worm *Urechis unicinctus* (Annelida, Echiura). *Zoology* **2014**, *117*, 216–223. [CrossRef]
63. Cao, M.; Xin, Y.; Chen, J.S.; Wang, X.Q.; Liang, D.; Wei, S.Y.; Sun, M.C.; Xu, M.; Yu, R.C. Characteristics of *Urechis unicinctus* and breeding Techniques. *Mod. Agric. Sci. Technol.* **2021**, *20*, 178–180. (In Chinese)
64. Li, H.H.; Liu, X.; Sun, N.; Zuo, B.N.; Li, H.Y. Ecological Artificial Breeding Techniques for *Urechis unicinctus*. *Sci. Fish Farming* **2021**, *1*, 66–68. (In Chinese)
65. Jia, Y.; Du, M.R.; Li, W.H.; Jiang, W.W.; Lin, F.; Yao, L.; Wu, Y.P.; Jiang, Z.J. Effect of bioturbation of *Urechis unicinctus* on the diffusion flux of nitrogen and phosphorus at the sediment-water interface. *J. Fish. China* **2023**, *47*, 102–112. (In Chinese)
66. Jia, Y. *Effects of Bioturbation by Urechis Unicinctus on Key Process of Sediment Nitrogen Cycle in Coastal Aquaculture Area*; Shanghai Ocean University: Shanghai, China, 2023. (In Chinese)
67. Wang, Y. *Effect of Digestion Process of Filter Feeders on Silicon Cycle in the Ocean*; Qingdao University of Science and Technology: Qingdao, China, 2022. (In Chinese)
68. Song, X.Y.; Hou, M.Y.; Luo, J.; Liu, R.Y.; Chen, W.B. Experiment on the purification effect of the *Urechis unicinctus* on Aquaculture Tail Water. *China Fish.* **2024**, *2*, 97–99. (In Chinese)
69. Li, N.; Song, S.L.; Tang, Y.Z.; Meng, R.P.; Wang, B.G. Analysis of Amino Acid Composition and Content in the Body Wall of the *Urechis unicinctus*. *Shandong Fish.* **2000**, *5*, 26–27+49. (In Chinese)
70. Fu, X.Y.; Sun, C.H.; Wang, C.W.; Du, F.; Li, B.F. Technology research of preparation of seafood flavor condiment base from *Urechis unicinctus*. *J. Anhui Agric. Sci.* **2021**, *49*, 168–171+176. (In Chinese)
71. Cheng, Z.Y.; Zhao, J.R.; Xiao, H.; Gu, M.H.; Ji, A.G.; Song, S.L. Study on the bioactive components of *Urechis Unicinctus* and the development of condiments. *Food Sci. Technol.* **2020**, *45*, 253–258. (In Chinese)
72. Zhang, Y.X.; Song, J.; Chen, Y.N.; Liu, F.; Liu, C.E. Preparation of flavor seasoning by maillard reaction of enzymatic hydrolysate of *Urechis unicinctus*. *Food Sci. Technol.* **2023**, *48*, 243–250. (In Chinese)
73. Gong, J.; Zhao, R.P.; Deng, J.H.; Zhao, Y.C.; Zuo, J.C.; Huang, L.; Jing, M.D. Genetic diversity and population structure of penis fish (*Urechis unicinctus*) based on mitochondrial and nuclear gene markers. *Mitochondrial DNA Part A* **2018**, *29*, 1261–1268. [CrossRef] [PubMed]
74. Wang, L.; Liu, H.M.; Peng, G.; Che, C.Y.; Gong, H.S.; Zhang, Z.M. Research advance on the nutrient component and bioactive substance in *Urechis unicinctus*. *J. Ludong Univ. (Nat. Sci. Ed.)* **2011**, *27*, 342–345. (In Chinese)
75. Bi, Q.Q.; Han, B.Q.; Feng, Y.L.; Jiang, Z.Q.; Yang, Y.; Liu, W.S. Antithrombotic effects of a newly purified fibrinolytic protease from *Urechis unicinctus*. *Thromb. Res.* **2013**, *132*, E135–E144. [CrossRef]
76. Li, J.J.; Lu, J.J.; Asakiya, C.; Huang, K.L.; Zhou, X.Z.; Liu, Q.L.; He, X.Y. Extraction and Identification of Three New *Urechis unicinctus* Visceral Peptides and Their Antioxidant Activity. *Mar. Drugs* **2022**, *20*, 293. [CrossRef]
77. Ryu, B.; Kim, M.; Himaya, S.W.A.; Kang, K.H.; Kim, S.K. Statistical optimization of high temperature/pressure and ultra-wave assisted lysis of *Urechis unicinctus* for the isolation of active peptide which enhance the erectile function in vitro. *Process Biochem.* **2014**, *49*, 148–153. [CrossRef]
78. Yuan, C.Y.; Liu, P.; Han, X.; Cui, Q.M. Hypoglycemic Effects of Glycosaminoglycan from *Urechis unicinctus* in Diabetic Mice. *J. Med. Food* **2015**, *18*, 190–194. [CrossRef]
79. Wang, S.F. *The Oxidative Detoxification and Metabolic Adaptation of Urechis Unicinctus to Sulfide*; Ocean University of China: Qingdao, China, 2007. (In Chinese)
80. Xv, X.H.; Liu, T.H.; Zhu, X.Y.; Ding, Z.Y.; Liu, S.; Wang, S.L.; Xu, G.C. Acute Toxicity of Pb, Cd, Cr, and Zn heavy metal salts to echiuran worm *Urechis unicinctus*. *Fish. Sci.* **2022**, *41*, 1029–1035. (In Chinese)
81. Zhu, X.Y.; Gan, H.T.; Meng, X.; Yao, H.Y.; Xu, G.C.; Xu, J.T.; Liu, T.H.; Cao, L.X.; Xu, X.H. Effects of Cadmium on Non-specific Immunity and Bioaccumulation of *Urechis unicinctus*. *Asian J. Ecotoxicol.* **2019**, *14*, 106–115. (In Chinese)
82. Cai, L.Z.; Rao, Y.Y.; Zhao, X.Y.; Yang, D.Y.; Lin, J.X.; Fu, S.J.; Zhou, X.P. Two risk indices for benthic macrofauna entrapment evaluation on the water intake systems in coastal nuclear power plants. *J. Appl. Oceanogr.* **2022**, *41*, 655–662. (In Chinese)
83. Zhang, W.B.; Sun, W.; Xu, X.; Wang, H.B.; Zheng, B.Q.; Lu, C.; Li, J.; Mu, J.L. Ecological environment and the potential hazard-causing organisms in the sea area near the nuclear power plant. *Mar. Sci.* **2022**, *46*, 32–43. (In Chinese)
84. Cao, J.Y. *Studies of Artificial Breeding and Early Cultivation of Urechis Unicinctus*; Dalian Ocean University: Dalian, China, 2022. (In Chinese)
85. Zhang, N.X.; Gao, S.; Wang, L.; Zhang, S.P.; Zhang, H.Z.; Ji, Y.L.; Pu, S.C. A Fishing Device for *Philine kinglipini*. Patent for Invention CN202310311601.3, 6 June 2023. (In Chinese)
86. Liu, X.S.; Ni, D.P.; Zhong, X.; Zhang, Z.N. Structure of benthic food web and trophic relationships of the macrofauna in the Yellow Sea. *Period. Ocean Univ. China* **2020**, *50*, 20–33. (In Chinese)

87. Zhang, C.X.; Chen, J.; Ji, Y.L.; Chen, L.L.; Li, H.H.; Wang, Q.C.; Li, B.Q.; Xing, R.L. Benthic food web structure of Xiaoqing River Estuary adjacent sea area near revealed by carbon and nitrogen stable isotope analysis. *Haiyang Xuebao* **2022**, *44*, 89–100. (In Chinese)
88. Ma, Q.Y. *Trophic Structure of Major Species in Jiaozhou Bay Considered through Stable-Isotope Analysis*; Ocean University of China: Qingdao, China, 2015. (In Chinese)
89. Zhang, T.; Ma, P.Z.; Wang, H.Y.; Lin, Z.S.; Wang, J.H.; Song, H.; Song, A.H.; Shi, Y. *Identification and Classification of Philine Species in Jiaozhou Bay and Analysis of the Causes of Population Outbreaks*; China Fisheries: Beijing, China, 2022. (In Chinese)
90. Yu, H.Y.; Li, X.Z.; Li, B.Q.; Wang, J.B.; Wang, H.F. The species diversity of microbenthic fauna in Jiaozhou Bay. *Acta Ecol. Sin.* **2006**, *2*, 416–422. (In Chinese) [CrossRef]
91. Tu, L.L.; Liu, W.X.; Sui, J.X.; Qu, F.Y.; Zhao, F.Q.; Zhong, H.X.; Zhang, M.S.; Yu, Z.S. The community structure of macrobenthos in the southern coastal waters nearby the Shandong peninsula in summer. *Prog. Fish. Sci.* **2018**, *39*, 27–36. (In Chinese)
92. Wang, Z.; Qu, F.Y.; Sui, J.X.; Wang, Z.Z.; Ji, X.X.; Zhao, N.; Yu, Z.S. Community structure and diversity of macrobenthos in the western waters of Liaodong Bay during summer. *Mar. Sci.* **2016**, *40*, 40–47. (In Chinese)
93. Gao, A.G.; Dong, Y.T.; Wang, H.Z.; Wang, Y.H. Preliminary study on the distribution of mollusca ecology in sublittoral of Nanji Islands. *J. Mar. Sci.* **1998**, *2*, 50–55. (In Chinese)
94. Zhang, P.C. *Long-Term Changes of the Macrobenthic Community Structure in the Cold Water Mass of the Southern Yellow Sea*; Shanxi Normal University: Taiyuan, China, 2018. (In Chinese)
95. Ren, B.B. *Assessing the Restoration Effect of Artificial Reefs by Macrobenthos in Shuangdao Bay*; Shanghai Ocean University: Shanghai, China, 2016. (In Chinese)
96. Ministry of Ecology and Environment of the People's Republic of China. *China Ecological Environment Status Bulletin 2022*; Ministry of Ecology and Environment of the People's Republic of China: Beijing, China, 2023. (In Chinese)
97. Zhang, Y.Q. *Community Structure, Feeding Ecology, and Movement Behavior of Demersal Fish around Qiansan Islets*; University of Chinese Academy of Sciences (Institute of Oceanology): Qingdao, China, 2016. (In Chinese)
98. Chen, L.J.; Yang, F.; Zhong, X.M.; Song, D.D.; Li, G.D.; Kang, Z.J.; Xiong, Y. Review of the life history of *Acetes chinensis*. *J. Shanghai Ocean Univ.* **2022**, *31*, 1032–1040. (In Chinese)
99. Kim, S.; Kim, J.; Choi, H.G.; Park, J.K.; Min, G.S. Complete mitochondrial genome of the northern mauxia shrimp *Acetes chinensis* (Decapoda, Dendrobranchiata, Sergestoidae). *Mitochondrial DNA* **2012**, *23*, 28–30. [CrossRef]
100. Sun, Y.Y.; Zhang, S.M.; Shi, Y.C.; Tang, F.H.; Chen, J.L.; Xiong, Y.; Dai, Y.; Li, L. YOLOv7-DCN-SORT: An algorithm for detecting and counting targets on *Acetes* fishing vessel operation. *Fish. Res.* **2024**, *274*, 106983. [CrossRef]
101. Li, G.D.; Li, D.J.; Xiong, Y.; Zhong, X.M.; Tang, J.H.; Song, D.D.; Shi, J.J.; Yang, F.; Kang, Z.J.; Yan, X.; et al. Changes in the resource distribution of *Acetes chinensis* and patterns of species replacement in Haizhou Bay in summer based on BeiDou VMS data. *Reg. Stud. Mar. Sci.* **2022**, *56*, 102655. [CrossRef]
102. Wu, X.R.; Song, D.D.; Xiong, Y.; Zhong, X.M.; Li, G.; Yang, F.; Kang, Z.J.; Li, G.D.; Li, D.J.; Shi, J.J.; et al. Population biological characteristics and exploitation status of *Acetes chinensis* in Haizhou Bay. *Oceanol. Et Limnol. Sin.* **2023**, *54*, 573–582. (In Chinese)
103. Zan, H. *Feasibility Study on Development of Acetes Chinensis Seafood Sauce*; Zhejiang Ocean University: Zhoushan, China, 2011. (In Chinese)
104. Cao, W.H.; Zhang, C.H.; Chen, S.H.; Hong, P.Z. Analysis and Evaluation of Nutrients of *Acetes Chinensis*. *J. Fish. Res.* **2001**, *1*, 8–14. (In Chinese)
105. Chen, X.F. *Key Technology in Preparation of Acetes Chinensis Peptide Chelating Calcium and Development of the Paste Enriched with Peptide Chelating Calcium*; Zhejiang University: Hangzhou, China, 2017. (In Chinese)
106. Yang, Z.Y.; Hui, T.T.; Zhu, B.H.; Xu, C.C.; Li, Y.; Li, X.H. Effects of Peptides from *Acetes chinensis* on Immunoregulation in Immunocompromised Mice. *Sci. Technol. Food Ind.* **2023**, *44*, 380–386. (In Chinese)
107. Wang, H.T. *The Study of Active Peptide from Acetes Chinensis with Inhibitory Activity on Neuraminidase of Influenza Virus*; Ocean University of China: Qingdao, China, 2014. (In Chinese)
108. Cao, W.H.; Zhang, C.H.; Hong, P.Z.; Ji, H.W.; Hao, J.M. Purification and identification of an ACE inhibitory peptide from the peptic hydrolysate of *Acetes chinensis* and its antihypertensive effects in spontaneously hypertensive rats. *Int. J. Food Sci. Technol.* **2010**, *45*, 959–965. [CrossRef]
109. He, H.L.; Chen, X.L.; Sun, C.Y.; Mang, Y.Z.; Zhou, B.C. Analysis of novel angiotensin-I-converting enzyme inhibitory peptides from protease-hydrolyzed marine shrimp *Acetes chinensis*. *J. Pept. Sci.* **2006**, *12*, 726–733.
110. Cao, W.H.; He, X.Q.; Zhang, C.H.; Hao, J.M.; Zhao, Z.K. Inhibition type of two angiontensin I-converting enzyme inhibitory peptides from *Acetes chinensis*. *Food Mach.* **2013**, *29*, 4–7+34. (In Chinese)
111. Shen, P.P.; Yu, X.B.; Meng, B.F.; Wu, C.C. Classification and characteristics of the risk organisms in cold source water intake area of coastal nuclear power plants. *Mar. Sci.* **2023**, *47*, 131–140. (In Chinese)
112. Xu, C.; Hu, Z.C. Construction and Research of Cooling-water Interception System for Coastal Nuclear Power Plants. *Electr. Saf. Technol.* **2019**, *21*, 45–48. (In Chinese)
113. Xu, G.S. *Real-Time Detection System for Underwater Disaster-Causing Organisms in Coastal Nuclear Power Plant*; Xiamen University: Xiamen, China, 2020. (In Chinese)
114. Zeng, L.; Chen, G.B.; Wang, T.; Yang, B.Z.; Yu, J.; Liao, X.L.; Huang, H.H. Acoustic detection and analysis of *Acetes chinensis* in the adjacent waters of the Daya Bay Nuclear Power Plant. *J. Fish. Sci. China* **2019**, *26*, 1029–1039. (In Chinese)

115. Sha, C.; Yang, J.; Zhang, W.J.; Zhang, R.Y.; Bai, W. Investigation and analysis for marine biological monitoring technologies of nuclear power plants' water intake. *Water Wastewater Eng.* **2020**, *56*, 13–16. (In Chinese)
116. Dai, L.; Zhang, Z.; Zhi, Y.F.; Zhang, R.Y. Analysis on the origin of *Acetes chinensis* outbreak in the water intake area of nuclear power plant. *Reg. Gov.* **2022**, *25*, 141–144. (In Chinese)
117. An, L.N.; Wang, L.; Huang, H.; Ou, D.Y.; Li, W.W. Population dynamics of *Acetes chinensis* and its response to environmental factors in western Daya Bay. *J. Appl. Oceanogr.* **2021**, *40*, 403–412. (In Chinese)
118. Xing, X.F.; Zhang, Z.L.; Tang, J.M.; Guo, T.X.; Wang, S.; Tao, Z. Study on the lssue about Water Intake Blockage by Marine Organism in Nuclear Power Plant. *Nucl. Saf.* **2021**, *20*, 103–109. (In Chinese)
119. Guo, W.C.; Ding, X.Q.; Liu, J.H.; Zhao, B.B. Species, Distribution, and Comprehensive Utilization of Starfish Resources in China. *Spec. Econ. Anim. Plants* **2013**, *16*, 9–13. (In Chinese)
120. Li, S.Y. *The Breliminary Studies on Behavior and Control Strategies of Asterias amurensis*; Ocean University of China: Qingdao, China, 2015. (In Chinese)
121. Liu, J. *Preliminary Study of Foraging Behavior and Reproductive Biology of the Sea Star Asterias amurensis*; Ocean University of China: Qingdao, China, 2013. (In Chinese)
122. Xia, J.M.; Miao, Z.; Xie, C.L.; Zhang, J.W.; Yang, X.W. Chemical Constituents and Bioactivities of Starfishes: An Update. *Chem. Biodivers.* **2020**, *17*, e1900638. [CrossRef]
123. Gale, A.S.; Jagt, J.W.M. The fossil record of the family Benthopectinidae (Echinodermata, Asteroidea), a reappraisal. *Eur. J. Taxon.* **2021**, *755*, 149–190. [CrossRef]
124. Xiao, N.; Liao, Y.L.; Liu, R.Y.; Liu, J.Y. Records of the genus *Henricia* Gray, 1840 (Echinodermata: Asteroidea: Echinasteridae) from Chinese waters. *Zootaxa* **2011**, *3115*, 1–20. [CrossRef]
125. Xiao, N.; Liao, Y.L. Three new records of deep-water goniasterids (Echinodermata: Asteroidea: Goniasteridae) from China seas. *J. Oceanol. Limnol.* **2013**, *31*, 1036–1044. [CrossRef]
126. Wang, Y.; Gu, Y.B.; Guo, H.; Cao, L.Q.; Jin, Y. Advances and Perspectives on the Research of Starfish Outbreak in Northern China. *Chin. J. Appl. Ecol.* **2023**, *34*, 1146–1152. (In Chinese)
127. Zhou, D.Y.; Zhou, X.; Liu, X.Y.; Shen, Y.; Yu, Z.L. Characterization of Lipid Profiles in Different Body Parts of *Asterias amurensis*. *J. Dalian Polytech. Univ.* **2018**, *37*, 313–319. (In Chinese)
128. Li, H.J.; Li, S.Q.; Wang, F.; Lan, W.J. Chemical Constituents and Its Biological Analyses of the Fatty Acids from Whole Body of Starfish *Acanthaster planci*. *Acta Sci. Nat. Univ. Sunyatseni* **2009**, *48*, 55–60. (In Chinese)
129. Zhao, J.; Zhao, J.; Fan, T.J.; Yuan, W.P.; Zhang, Z.; Cong, R.S. Purification and Anti-tumor Activity Examination of Water-Soluble asterosaponin from *Asterias rollestoni bell*. *J. Shandong Univ. (Nat. Sci.)* **2013**, *48*, 30–35+42. (In Chinese)
130. Haywood, M.D.E.; Thomson, D.P.; Babcock, R.C.; Pillans, R.D.; Keesing, J.K.; Miller, M.; Rochester, W.A.; Donovan, A.; Evans, R.D.; Shedrawi, G.; et al. Crown-of-thorns starfish impede the recovery potential of coral reefs following bleaching. *Mar. Biol.* **2019**, *166*, 99. [CrossRef]
131. Keesing, J.K.; Bradbury, R.H.; Devantier, L.M.; Riddle, M.J.; Death, G. Geological evidence for recurring outbreaks of the crown-of-thorns starfish: A reassessment from an ecological perspective. *Coral Reefs* **1992**, *11*, 79–85. [CrossRef]
132. Uthicke, S.; Schaffelke, B.; Byrne, M. A boom-bust phylum? Ecological and evolutionary consequences of density variations in echinoderms. *Ecol. Monogr.* **2009**, *79*, 3–24. [CrossRef]
133. Babcock, R.C.; Milton, D.A.; Pratchett, M.S. Relationships between size and reproductive output in the crown-of-thorns starfish. *Mar. Biol.* **2016**, *163*, 234. [CrossRef]
134. Yao, Q.C.; Yu, K.F.; Liao, Z.H.; Chen, B.; Yu, X.P.; Wei, F.; Hu, B.Q. A Review of Crown-of-Thorns Starfish and Their Ecological Effects on Coral Reefs. *Acta Ecol. Sin.* **2022**, *42*, 7517–7528. (In Chinese)
135. Wang, Y.L.; Wang, Y.X.; Yang, Y.J.; Ni, G.; Li, Y.L.; Chen, M.Y. Chromosome-level genome assembly of the northern Pacific seastar *Asterias amurensis*. *Sci. Data* **2023**, *10*, 767. [CrossRef] [PubMed]
136. Zhao, S.J. *Heavy Metal Content and Bioaccumulation Characteristics of Acanthaster Planci in Coral Reef of South China Sea*; Guangxi University: Nanning, China, 2023. (In Chinese)
137. Qi, Z.H.; Wang, J.; Mao, Y.Z.; Zhang, J.H.; Fang, J.G. Prey Selection and Feeding Rate of Sea Star *Asterias amurensis* and *Asterina pectinifera* on Three Bivalves. *Acta Ecol. Sin.* **2013**, *33*, 4878–4884. (In Chinese)
138. Wang, J.; Duan, Z.L.; He, Z.Y.; Chen, N.S. Molecular Analysis of the *Asterias amurensis* from Starfish Outbreak in Jiaozhou Bay in Summer 2022. *Oceanol. Et Limnol. Sin.* **2023**, *54*, 1656–1671. (In Chinese)
139. Reimer, J.D.; Kise, H.; Wee, H.B.; Lee, C.L.; Soong, K. Crown-of-thorns starfish outbreak at oceanic Dongsha Atoll in the northern South China Sea. *Mar. Biodivers.* **2019**, *49*, 2495–2497. [CrossRef]
140. Li, Y.C.; Wu, Z.J.; Liang, J.L.; Chen, S.Q.; Zhao, J.M. Analysis on the Outbreak Period and Cause of *Acanthaster planci* in Xisha Islands in Recent 15 Years. *Chin. Sci. Bull. -Chin.* **2019**, *64*, 3478–3484.
141. Dai, K.T.; Li, J.; Guan, C.T.; Zhang, Y.; Zhao, X.W. Feeding Selectivity Feeding on Five Species Bivalve and Feeding Rhythm of *Asterias amurensis*. *Prog. Fish. Sci.* **2015**, *36*, 97–102. (In Chinese)
142. Li, Y.; Hu, Z.B.; Xie, S.J.; Liu, S.G.; Miao, X.; Wang, W.; Xiao, J.G.; Wang, R.; Lin, L.S. Discussion on Species Validity of Crown-of-thorns Starfish in the South China Sea. *J. Appl. Oceanogr.* **2024**, *43*, 201–207. (In Chinese)
143. Liu, J.H.; Wang, P.C.; Wang, Z.H.; Xia, W.T.; Xie, S.G.; Song, Y.Q. Preliminary Study on the Reproductive Biology of Crown-of-Thorns Starfish (*Acanthaster solaris*) in the Xisha Islands. *J. Trop. Oceanogr.* **2024**, 1–13. (In Chinese)

144. Heng, W.K.; Ho, M.J.; Kuo, C.Y.; Huang, Y.Y.; Ko, C.Y.; Jeng, M.S.; Chen, C.A. Crown-of-thorns starfish outbreak at Taiping Island (Itu Aba), Spratlys, South China Sea. *Bull. Mar. Sci.* **2022**, *98*, 101–102. [CrossRef]
145. Su, N.; Zhang, Y.; Yang, L.T.; Chen, H.Z.; Gao, Q.; Chen, C.; Yang, Y.F. Effects of Four Microalgae on the Survival and Growth of *Acanthaster* spp. Larvae. *J. Appl. Oceanogr.* **2023**, *42*, 633–642. (In Chinese)
146. Zheng, J.H.; Ren, G.B.; Hu, Y.B.; Zhang, F.F.; Ma, Y.; Li, M.J.; Wang, R.F. High Resolution Remote Sensing Monitoring and Analysis of Coral Reef Degradation Caused by Outbreaks of Biological Natural Enemies: A Case Study of the Taiping Island in the South China Sea. *Trop. Geogr.* **2023**, *43*, 1856–1873. (In Chinese)
147. Yan, Z.C.; Xing, J.J.; Cai, W.Q.; Zhang, K.D.; Wu, Z.J.; Li, Y.C.; Tang, J.; Zhou, Z. Study on the population distribution of *Acanthaster planci* in the reef area of the Xisha Islands based on environmental DNA technology. *Haiyang Xuebao* **2023**, *45*, 76–83. (In Chinese)
148. Xia, R.L.; Ning, Z.M.; Yu, K.F.; Fang, C.; Huang, X.Y.; Wei, F. Study on the impacts of crown-of-thorns starfish on nutrient dynamics in the coral reef sediments. *Haiyang Xuebao* **2022**, *44*, 23–30. (In Chinese)
149. Matthews, S.A.; Mellin, C.; Pratchett, M.S. Larval connectivity and water quality explain spatial distribution of crown-of-thorns starfish outbreaks across the Great Barrier Reef. *Adv. Mar. Biol.* **2020**, *87*, 223–258.
150. Bos, A.R.; Gumanao, G.S.; Mueller, B.; Saceda-Cardoza, M.M.E. Management of crown-of-thorns sea star (*Acanthaster planci* L.) outbreaks: Removal success depends on reef topography and timing within the reproduction cycle. *Ocean Coast. Manag.* **2013**, *71*, 116–122. [CrossRef]
151. Clements, C.S.; Hay, M.E. Size matters: Predator outbreaks threaten foundation species in small Marine Protected Areas. *PLoS ONE* **2017**, *12*, e0171569. [CrossRef]
152. Eichsteller, A.; Taylor, J.; Stöhr, S.; Brix, S.; Martìnez Arbizu, P. DNA Barcoding of Cold-Water Coral-Associated Ophiuroid Fauna from the North Atlantic. *Diversity* **2022**, *14*, 358. [CrossRef]
153. Chen, Z.Q.; McNamara, K. End-Permian extinction and subsequent recovery of the Ophiuroidea (Echinodermata). *Palaeogeogr. Palaeoclimatol. Palaeoecol.* **2006**, *236*, 321–344. [CrossRef]
154. Shi, B.Y.; Lu, Y.L.; Pan, Y.J.; Guo, C.H.; Qu, J.Y. Progress on the Biological Research of Ophiuroidea. *J. Hainan Trop. Ocean Univ.* **2016**, *23*, 92–98. (In Chinese)
155. Yu, N. *Study on the Mechanism of Quantity Variations of Brittle Stars in the Zhangzi Island Area*; University of Chinese Academy of Sciences: Qingdao, China, 2022. (In Chinese)
156. Thuy, B. Temporary expansion to shelf depths rather than an onshore-offshore trend: The shallow-water rise and demise of the modern deep-sea brittle star family Ophiacanthidae (Echinodermata: Ophiuroidea). *Eur. J. Taxon.* **2013**, *48*, 1–242. [CrossRef]
157. Liao, Y.L.; Xiao, N. Species composition and faunal characteristics of echinoderms in China seas. *Biodivers. Sci.* **2011**, *19*, 729–736. (In Chinese)
158. Li, Y.X. *Population Structure and Genetic Divergence of Ophiura Sarsii Vadicola from Yellow Sea*; First Institute of Oceanography: Qingdao, China, 2021. (In Chinese)
159. Yang, Z.Y. On the discovery of a scythic ophiuroid from Kueichou, China. *Acta Palaeontol. Sin.* **1960**, *2*, 97–102. (In Chinese)
160. Feng, R.L. New discovery of fossil ophiuroids from Guizhou and Southern Sichuan, China. *Acta Palaeontol. Sin.* **1985**, *3*, 337–343+374–376. (In Chinese)
161. Liao, Z.T.; Wang, X.J. Ophiuroid fossils from the Middle Permian Tongzi Formation, Longyan, Fujian Province. *Acta Palaeontol. Sin.* **2002**, *3*, 396–398. (In Chinese)
162. Chen, W.Y.; Zhang, D.S.; Wang, C.S. Two new records of genus Ophiurothamnus (Ophiuroidea, Ophiacanthidae) from a deep-sea seamount of the South China Sea. *Oceanol. Et Limnol. Sin.* **2020**, *51*, 649–655. (In Chinese)
163. Li, Q.H.; Li, Y.X.; Na, J.Y.; Han, X.Q.; Paterson, G.L.J.; Liu, K.; Zhang, D.S.; Qiu, J.W. Description of a new species of *Histampica* (Ophiuroidea: Ophiothamnidae) from cold seeps in the South China Sea and analysis of its mitochondrial genome. *Deep Sea Res. Part I Oceanogr. Res. Pap.* **2021**, *178*, 103658. [CrossRef]
164. Nethupul, H.; Stöhr, S.; Zhang, H.B. Order Euryalida (Echinodermata, Ophiuroidea), new species and new records from the South China Sea and the Northwest Pacific seamounts. *ZooKeys* **2022**, *1090*, 161–216. [CrossRef]
165. Nethupul, H.; Stöhr, S.; Zhang, H.B. New species, redescriptions and new records of deep-sea brittle stars (Echinodermata: Ophiuroidea) from the South China Sea, an integrated morphological and molecular approach. *Eur. J. Taxon.* **2022**, *810*, 1–95. [CrossRef]
166. Nethupul, H.; Stöhr, S.; Zhang, H.B. Review of *Ophioplinthaca* Verrill, 1899 (Echinodermata, Ophiuroidea, Ophiacanthidae), description of new species in *Ophioplinthaca* and *Ophiophthalmus*, and new records from the Northwest Pacific and the South China Sea. *ZooKeys* **2022**, *1099*, 155–202. [CrossRef]
167. Yang, H.S.; Xiao, N.; Zhang, T. Present Status and Prospect of the Study of Echinoderms. *Stud. Mar. Sin.* **2016**, *51*, 125–131. (In Chinese)
168. Fang, J.H.; Zhang, J.H.; Jiang, Z.J.; Mao, Y.Z.; Zang, Y.Q.; Fang, J.G. The Analysis of Nutrient Components of Three Brittle Star Species in the North Yellow Sea. *Prog. Fish. Sci.* **2015**, *36*, 17–21. (In Chinese)
169. Cong, J.S.; Zhen, J.R.; Qu, J.Y.; Guo, C.H. Analysis of Nutrient Constituents in *Stegophiura sladeni*. *Acta Nutr. Sin.* **2016**, *38*, 308–310. (In Chinese)
170. Yu, N.; Sun, S.; Zhang, G.T.; Zhang, F. Reproductive cycle of *Ophiopholis mirabilis* (Echinodermata: Ophiuroidea) in Zhangzi Island area, northern Yellow Sea. *J. Oceanol. Limnol.* **2021**, *39*, 173–184. [CrossRef]

171. Xu, Y.; Sui, J.X.; Li, X.Z.; Wang, H.F.; Zhang, B.L.; Shuai, L.M. Variations of Macrofaunal Community Classification in the South Yellow Sea. *Guangxi Sci.* **2016**, *23*, 339–345. (In Chinese)
172. Zhang, P.C.; Xu, Y.; Li, X.Z.; Wang, H.F.; Zhang, B.L.; Shuai, L.M.; An, J.M. Comparison in macrobenthic community composition inside and outside the Cold Water Mass in the Southern Yellow Sea in summer. *Oceanol. Et Limnol. Sin.* **2017**, *48*, 312–326. (In Chinese)
173. Yu, N.; Sun, S.; Wang, S.W.; Liu, Q.; Zhang, G.T.; Zhang, F.; Sun, X.X. An enhanced underwater camera apparatus for seabed observation of megabenthic epifauna in the northern Yellow Sea. *J. Oceanol. Limnol.* **2020**, *38*, 1799–1810. [CrossRef]
174. Lin, X.C.; Cai, L.Z.; Yang, D.Y.; Zhao, X.Y.; Lin, J.X.; Zhou, X.P. Potential risk assessment of macrozoobenthos blocking nuclear power cold source system in three subtidal zone of Fujian, Guangdong and Hainan coasts. *Mar. Environ. Sci.* **2023**, *42*, 286–293. (In Chinese)
175. Li, X.Z.; Wang, J.B.; Wang, H.H. Community Structure of Macrobenthos and Seasonal Variations in the Zhangzi Island Waters of the Northern Yellow Sea. *Stud. Mar. Sin.* **2016**, *51*, 245–257. (In Chinese)
176. Zheng, B.Q.; Ji, X.; Xia, Y.J.; Lin, J.; Huang, W.; Cheng, H.; Liu, S.H.; Qin, Y.T.; Liu, W. Morphological Characteristics and Phylogenetic Analysis of the Sea Cucumber *Acaudina molpadioides* in the Fujian Coastal Waters. *Mar. Sci. Bull.* **2022**, *41*, 427–435. (In Chinese)
177. Wang, G.L.; Li, X.; Wang, J.H.; Zhang, J.; Liu, W.; Lu, C.; Guo, Y.C.; Dong, B. The complete mitochondrial genome and phylogenetic analysis of *Acaudina molpdioides*. *Mitochondrial DNA Part B-Resour.* **2019**, *4*, 668–669. [CrossRef]
178. Li, H.; Li, T.W.; Su, X.R.; Qin, Y.M.; Wang, M.Q.; He, J.J.; Hu, G.Z.; Huang, S.F.; Zhou, J. The Phylogenetic Relation of Sea Cucumber *Acaudina molpadioides*. *Fish. Sci.* **2009**, *28*, 733–736. (In Chinese)
179. Tang, Y.F. *Study on Typical Disaster-causing Organisms of Coastal Nuclear Power Operation Safety: Taking Ningde Nuclear Power as an Example*; Shanghai Ocean University: Shanghai, China, 2019. (In Chinese)
180. Zhao, Y.H.; Li, B.F.; Ma, J.J.; Dong, S.Y.; Liu, Z.Y.; Zeng, M.Y. Purification and Synthesis of ACE Inhibitory Peptide from *Acaudina molpadioidea* Protein Hydrolysate. *Chem. J. Chin. Univ.* **2012**, *33*, 308–312.
181. Hu, S.W.; Chang, Y.G.; Wang, J.F.; Xue, C.H.; Shi, D.; Xu, H.; Wang, Y.M. Fucosylated chondroitin sulfate from *Acaudina molpadioides* improves hyperglycemia via activation of PKB/GLUT4 signaling in skeletal muscle of insulin resistant mice. *Food Funct.* **2013**, *4*, 1639–1646. [CrossRef]
182. Li, J.W.; Rao, Y.Y.; Liao, X.L.; Dai, M.; Huang, H.H. Spatial Distribution and Habitat Condition of *Acaudina molpadioides* in Western Daya Bay. *J. Appl. Oceanogr.* **2022**, *41*, 286–293. (In Chinese)
183. He, Y.L.; Liu, S.H.; Yuan, Y.M.; Jiang, K.J.; Wang, T.; Qin, Y.T.; Ren, Y.H. The Potential Suitability Habitat Prediction of *Acaudina molpadioides* Based on Maxent Model. *J. Oceanogr.* **2021**, *43*, 65–74. (In Chinese)
184. Cheng, H. *Research on the Causes, Prevention and Control Measures of Bio-Blogging on Nuclear Power Cold Source by Acaudina molpadioides*; Shanghai Ocean University: Shanghai, China, 2019. (In Chinese)
185. Wang, Q.; Yi, X.T.; Wang, N.L.; Pei, D.; Wei, J.T.; Di, D.L.; Wang, L.T. Evaluation of In Vitro Hypoglycemic Activity in Vitro for Polypeptides from *Acaudina molpadioides*. *Food Sci. Technol.* **2018**, *43*, 245–248. (In Chinese)
186. Zhang, X.Y. *Isolation, Purification, Identification, and Activity Evaluation of Hypoglycemic Peptides from Acaudina molpadioides*; Lanzhou University of Technology: Lanzhou, China, 2023. (In Chinese)
187. Hu, S.W.; Wang, J.F.; Li, S.J.; Jiang, W.; Song, W.D. Anti-Inflammatory Effects of Fucoidan from *Acaudina molpadioides* in the Liver of Insulin Resistant Mice. *Acta Nutr. Sin.* **2016**, *38*, 24–29+35. (In Chinese)
188. Gu, Z.C.; Wu, Y.; Chen, Z.W.; Wang, P.Y. Study on the Distribution of Heavy Metals and Nutrients in *Acaudina molpadioides*. *Chin. Prev. Med.* **2016**, *17*, 55–58. (In Chinese)
189. Wang, N.L.; Yu, P.L.; Zhao, L.C.; Zhang, J.Y.; Wang, Q.Q.; Fan, K.Q.; Tian, K.; Pei, D. Determination of Active Substances and Evaluation of Hypoglycemic Activity in *Acaudina molpadioides* from Different Producing Areas. *Chin. J. Mar. Drugs* **2019**, *38*, 11–16. (In Chinese)
190. Wang, N.L.; Wei, J.T.; Wang, Y.; Zhu, X.; Zhang, J.F.; Pei, D.; Di, D.L. In Vitro Evaluation of ACE Inhibitory Peptides from *Acaudina molpadioides*. *Food Sci. Technol.* **2017**, *42*, 224–226+232. (In Chinese)
191. Xu, J.; Wang, Y.M.; Feng, T.Y.; Zhang, B.; Sugawara, T.; Xue, C.H. Isolation and Anti-Fatty Liver Activity of a Novel Cerebroside from the Sea Cucumber *Acaudina molpadioides*. *Biosci. Biotechnol. Biochem.* **2011**, *75*, 1466–1471. [CrossRef]
192. Jin, H.X.; Xu, H.P.; Li, Y.; Zhang, Q.W.; Xie, H. Preparation and Evaluation of Peptides with Potential Antioxidant Activity by Microwave Assisted Enzymatic Hydrolysis of Collagen from Sea Cucumber *Acaudina Molpadioides* Obtained from Zhejiang Province in China. *Mar. Drugs* **2019**, *17*, 169. [CrossRef]
193. Li, J.; Li, Y.; Lin, S.J.; Zhao, W.; Chen, Y.; Jin, H.X. Collagen peptides from *Acaudina molpadioides* prevent CCl4-induced liver injury via Keap1/Nrf2-ARE, PI3K/AKT, and MAPKs pathways. *J. Food Sci.* **2022**, *87*, 2185–2196. [CrossRef] [PubMed]
194. Lu, C.; Guo, Y.C.; Liu, W.; Zheng, B.Q.; Huang, Q.Z.; Wang, H.; Cheng, H.; Huan, F. Risk Analysis Aggregation Behavior of *Acaudina molpadioides* on Nuclear Power Plant Cooling Sources. *Nucl. Saf.* **2024**, *23*, 37–44. (In Chinese)
195. Han, Q.X.; Wang, X.B. Macrobenthic Biomass and Secondary Production in the Northern East China Sea and the Relative Importance of Environmental Variables. *Pac. Sci.* **2019**, *73*, 249–259. [CrossRef]
196. Zheng, B.Q.; Lu, C.; Liu, W.; Guo, Y.C.; Huang, W.; Cheng, H.; Lin, J. Gonadal development of the sea cucumber *Acaudina molpadioides* in north coast of Fujian. *Haiyang Xuebao* **2021**, *43*, 133–140. (In Chinese)

197. Li, J.S.; Zhao, C.Y.; Wu, X.P.; Qu, R. The impact of climate change on species distribution patterns. *7th Int. Forum Biodivers. Conserv. China* **2010**, *7*, 269–280. (In Chinese)
198. Jing, X.; Jiang, S.J.; Liu, H.Y.; Li, Y.; He, J.S. Complex relationships and feedback mechanisms between climate change and biodiversity. *Biodivers. Sci.* **2022**, *30*, 293–311. (In Chinese) [CrossRef]
199. Gollasch, S. The importance of ship hull fouling as a vector of species introductions into the North Sea. *Biofouling* **2002**, *18*, 105–121. [CrossRef]
200. Huang, J.; Xu, C.C.; Ridoutt, B.G.; Wang, X.C.; Ren, P.A. Nitrogen and phosphorus losses and eutrophication potential associated with fertilizer application to cropland in China. *J. Clean. Prod.* **2017**, *159*, 171–179. [CrossRef]
201. Li, K.; Jiang, R.T.; Qiu, J.Q.; Liu, J.L.; Shao, L.; Zhang, J.H.; Liu, Q.G.; Jiang, Z.J.; Wang, H.; He, W.H.; et al. How to control pollution from tailwater in large scale aquaculture in China: A review. *Aquaculture* **2024**, *590*, 741085. [CrossRef]
202. Rahmanta, M.A.; Al Hasibi, R.A.; Tambunan, H.B.; Ruly; Syamsuddin, A.; Aditya, I.A.; Susanto, B. Towards a Net Zero-Emission Electricity Generation System by Optimizing Renewable Energy Sources and Nuclear Power Plant. *Energies* **2024**, *17*, 1958. [CrossRef]
203. Liu, L.M.; Guo, H.; Dai, L.H.; Liu, M.L.; Xiao, Y.; Cong, T.L.; Gu, H.Y. The role of nuclear energy in the carbon neutrality goal. *Prog. Nucl. Energy* **2023**, *162*, 104772. [CrossRef]
204. Li, A.T.; Liu, Y.H.; Yu, Z.Y. China's nuclear exports: Understanding the dynamics between domestic governance reforms and international market competition. *Energy Res. Soc. Sci.* **2023**, *103*, 103230. [CrossRef]
205. Sun, Y.Q.; Liu, J.L.; Xia, J.; Tong, Y.C.; Li, C.X.; Zhao, S.; Zhuang, M.M.; Zhao, X.H.; Zhang, J.H.; He, P.M. Research development on resource utilization of green tide algae from the Southern Yellow Sea. *Energy Rep.* **2022**, *8*, 295–303. [CrossRef]
206. Sierra, A.R.G.; Amador-Castro, L.F.; Ramírez-Partida, A.E.; García-Cayuela, T.; Carrillo-Nieves, D.; Alper, H.S. Valorization of Caribbean Sargassum biomass as a source of alginate and sugars for de novo biodiesel production. *J. Environ. Manag.* **2022**, *324*, 116364. [CrossRef]
207. Wu, L.N.; Zhang, N.Y.; Sun, S.; Yuan, J.G.; Chen, J.; Li, M.M.; Lin, N.; You, Y.; Wang, W.J.; Ding, S.X. Application of microsatellite markers for evaluating the effect of restocking enhancement in *Larimichthys crocea*. *J. Fish. Sci. China* **2021**, *28*, 1100–1108. (In Chinese)
208. Xu, L.X.; Zhou, L.; Wei, Q.W. Stock status and conservation dilemma of species of Acipenseriformes in the Yangtze River and relevant suggestions. *J. Fish. China* **2023**, *47*, 62–72. (In Chinese)
209. Zhu, J.H.; Wu, Z.; Feng, B.B.; Deng, S.S.; Zhen, W.Q.; Liao, Y.Y.; Jie, X.Y.; Kwan, K.Y. Global conservation of *Tachypleus tridentatus*: Present status and recommendations. *Biodivers. Sci.* **2020**, *28*, 621–629. (In Chinese) [CrossRef]
210. Yu, Z.; Tang, Y.; Gobler, C.J. Harmful algal blooms in China: History, recent expansion, current status, and future prospects. *Harmful Algae* **2023**, *129*, 102499. [CrossRef] [PubMed]

MDPI AG
Grosspeteranlage 5
4052 Basel
Switzerland
Tel.: +41 61 683 77 34

Biology Editorial Office
E-mail: biology@mdpi.com
www.mdpi.com/journal/biology

www.ingramcontent.com/pod-product-compliance
Lightning Source LLC
Chambersburg PA
CBHW061547210926
51850CB00009B/2022
9783725859771